W0090564

International Political Economy Series

Series Editor: **Timothy M. Shaw**, Visiting Professor, University of Massachusetts Boston, USA and Emeritus Professor, University of London, UK

The global political economy is in flux as a series of cumulative crises impacts its organization and governance. The IPE series has tracked its development in both analysis and structure over the last three decades. It has always had a concentration on the global South. Now the South increasingly challenges the North as the centre of development, also reflected in a growing number of submissions and publications on indebted Eurozone economies in Southern Europe.

An indispensable resource for scholars and researchers, the series examines a variety of capitalisms and connections by focusing on emerging economies, companies and sectors, debates and policies. It informs diverse policy communities as the established trans-Atlantic North declines and 'the rest', especially the BRICS, rise.

Md Saidul Islam and Md Ismail Hossain
SOCIAL JUSTICE IN THE GLOBALIZATION OF PRODUCTION
Labor, Gender, and the Environment Nexus

Geoffrey Allen Pigman
TRADE DIPLOMACY TRANSFORMED
Why Trade Matters for Global Prosperity

Kristian Coates Ulrichsen
THE GULF STATES IN INTERNATIONAL POLITICAL ECONOMY

Eleonora Poli
ANTITRUST INSTITUTIONS AND POLICIES IN THE GLOBALISING ECONOMY

Andrea C. Simonelli
GOVERNING CLIMATE INDUCED MIGRATION AND DISPLACEMENT
IGO Expansions and Global Policy Implications

Victoria Higgins
ALLIANCE CAPITALISM, INNOVATION AND THE CHINESE STATE
The Global Wireless Sector

Andrei V. Belyi
TRANSNATIONAL GAS MARKETS AND EURO-RUSSIAN ENERGY RELATIONS

Silvia Pepino
SOVEREIGN RISK AND FINANCIAL CRISIS
The International Political Economy of the Eurozone

Ryan David Kiggins (*editor*)
THE POLITICAL ECONOMY OF RARE EARTH ELEMENTS. RISING POWERS
AND TECHNOLOGICAL CHANGE

Seán Ó Riain, Felix Behling, Rossella Ciccia and Eoin Flaherty (*editors*)
THE CHANGING WORLDS AND WORKPLACES OF CAPITALISM

Alexander Korolev and Jing Huang
INTERNATIONAL COOPERATION IN THE DEVELOPMENT OF RUSSIA'S
FAR EAST AND SIBERIA

Roman Goldbach
GLOBAL GOVERNANCE AND REGULATORY FAILURE
The Political Economy of Banking

Kate Ervine and Gavin Fridell (*editors*)
BEYOND FREE TRADE
Alternative Approaches to Trade, Politics and Power

Ray Kiely
THE BRICS, US 'DECLINE' AND GLOBAL TRANSFORMATIONS

Philip Fountain, Robin Bush and R. Michael Feener (*editors*)
RELIGION AND THE POLITICS OF DEVELOPMENT
Critical Perspectives on Asia

Markus Fraundorfer
BRAZIL'S EMERGING ROLE IN GLOBAL SECTORAL GOVERNANCE
Health, Food Security and Bioenergy

Katherine Hirschfeld
GANGSTER STATES
Organized Crime, Kleptocracy and Political Collapse

Matthew Webb and Albert Wijeweera (*editors*)
THE POLITICAL ECONOMY OF CONFLICT IN SOUTH ASIA

Matthias Ebenau, Ian Bruff and Christian May (*editors*)
NEW DIRECTIONS IN COMPARATIVE CAPITALISMS RESEARCH
Critical and Global Perspectives

Jeffrey Dayton-Johnson
LATIN AMERICA'S EMERGING MIDDLE CLASSES
Economic Perspectives

Andrei Belyi and Kim Talus
STATES AND MARKETS IN HYDROCARBON SECTORS

Dries Lesage and Thijs Van de Graaf
RISING POWERS AND MULTILATERAL INSTITUTIONS

Jerry Haar and Ricardo Ernst (*editors*)
INNOVATION IN EMERGING MARKETS

International Political Economy Series
Series Standing Order ISBN 978–0–333–71708–0 hardcover
Series Standing Order ISBN 978–0–333–71110–1 paperback

You can receive future titles in this series as they are published by placing a standing order. Please contact your bookseller or, in case of difficulty, write to us at the address below with your name and address, the title of the series and one of the ISBNs quoted above.

Customer Services Department, Macmillan Distribution Ltd, Houndmills, Basingstoke, Hampshire RG21 6XS, England

---

Innovation in Emerging Markets

Edited by

Jerry Haar
Professor of Management and International Business, Florida International University, senior research fellow, McDonough School of Business, Georgetown University, and Global Fellow, Woodrow Wilson International Center for Scholars, USA

and

Ricardo Ernst
Professor of Operations and Global Logistics, Managing Director of the Global Business Initiative, Managing Director of the Latin American Board, and former Deputy Dean, all at the McDonough School of Business, Georgetown University, USA

INNOVATION IN EMERGING MARKETS

Introduction, conclusion, editorial matter and selection © Jerry Haar and Ricardo Ernst, 2016
Individual chapters © Respective authors, 2016
Foreword © Ana Patricia Botín, 2016
Foreword © Samuel Lewis Navarro, 2016

Softcover reprint of the hardcover 1st edition 2016 978-1-137-48028-6

All rights reserved. No reproduction, copy or transmission of this publication may be made without written permission. No portion of this publication may be reproduced, copied or transmitted save with written permission. In accordance with the provisions of the Copyright, Designs and Patents Act 1988, or under the terms of any licence permitting limited copying issued by the Copyright Licensing Agency, Saffron House, 6–10 Kirby Street, London EC1N 8TS.

Any person who does any unauthorized act in relation to this publication may be liable to criminal prosecution and civil claims for damages.

First published 2016 by
PALGRAVE MACMILLAN

The authors have asserted their rights to be identified as the authors of this work in accordance with the Copyright, Designs and Patents Act 1988.

Palgrave Macmillan in the UK is an imprint of Macmillan Publishers Limited, registered in England, company number 785998, of Houndmills, Basingstoke, Hampshire RG21 6XS.

Palgrave Macmillan in the US is a division of Nature America, Inc., One New York Plaza, Suite 4500 New York, NY 10004-1562.

Palgrave Macmillan is the global academic imprint of the above companies and has companies and representatives throughout the world.

ISBN 978-1-349-69390-0   ISBN 978-1-137-48029-3 (eBook)
DOI: 10.1057/9781137480293

Distribution in the UK, Europe and the rest of the world is by Palgrave Macmillan®, a division of Macmillan Publishers Limited, registered in England, company number 785998, of Houndmills, Basingstoke, Hampshire RG21 6XS.

Library of Congress Cataloging-in-Publication Data
Names: Haar, Jerry, editor. | Ernst, Ricardo, editor.
Title: Innovation in emerging markets / Jerry Haar, Ricardo Ernst, [editors].
Description: Houndmills, Basingstoke, Hampshire ; New York, NY : Palgrave Macmillan, 2016. | Includes index.
Identifiers: LCCN 2015038913 | ISBN 9781137480286 (hardback)
Subjects: LCSH: Technological innovations—Economic aspects—Developing countries. | Diffusion of innovations—Developing countries. | Economic development—Developing countries.
Classification: LCC HC59.72.T4 I5256 2016 | DDC 338/.064091724—dc23
LC record available at http://lccn.loc.gov/2015038913

A catalog record for this book is available from the Library of Congress.

A catalogue record for the book is available from the British Library

Typeset by MPS Limited, Chennai, India.

To all the innovators in emerging markets and the public and private institutions and enterprises that strive to create, sustain, and expand an ecosystem of innovation

Contents

ix

List of Figures and Tables

Figures

Tables

Foreword

Innovation has always powered progress. Today, thanks to digital technology in particular, that power is greater than ever. Never has it been more important to foster innovation, as a nation's prosperity relies on it.

Every company has a role to play, but the role of small and medium-sized enterprises (SMEs) is critical. At Santander we believe our purpose is to help people and businesses prosper, so we take a very close interest in what we can do in working with SMEs. Each day, we see the contribution they make to the world economy – especially in emerging economies. Representing over 85 percent of private enterprise, SMEs are the powerhouse of economic growth, generating jobs and the tax revenue that pays for public services.

If these companies are to thrive, economies need to be built on a culture of enterprise, innovation, and ambition. To help turn these words into reality, Santander is increasing its support to entrepreneurs – including disruptive ones. Meanwhile, we are helping to encourage entrepreneurs through 1,100 partnership agreements with universities in over 20 countries. This funding allows students, professors, and researchers to pursue their entrepreneurial ideas and promote an entrepreneurial culture and technology transfer from universities to business.

We can always do more to encourage innovation, but what more should be done? Professors Ernst and Haar have made an important contribution to this debate by analyzing and discussing trends of innovation in emerging markets. Each chapter, authored by an expert in his or her field, delves broadly and deeply into innovation by region, industry, or function. Business people, policy makers, academics and researchers, and graduate students should keep this timely, relevant study close by as they, like us, seek to nurture innovation.

Ana Patricia Botín
Executive Chairman, Santander Group

Foreword

Innovation is an increasingly important driver of competitiveness for companies, industries, sectors, and entire nations. While we commonly think of innovation as focusing on products such as smartphones, folding bikes, and biotech and pharmaceuticals, or services such as e-banking, loyalty cards, and low-cost airline carriers, in fact the reach of innovation transcends the private sector – industrial and commercial – to encompass the public and nonprofit sectors as well.

To illustrate, when I had the privilege of serving as Vice-president and Foreign Minister of the Republic of Panama in the government of President Martín Torrijos, we introduced a number of sweeping innovative initiatives. These included expansion of the Panama Canal, reforms of social security and pensions, and a push for Panama to build its own science with emphasis on the niches of biotechnology, infectious diseases, and bioprospecting. Through grant programs, scholarships, and other forms of human capital investment, the government of former president Torrijos hugely supported innovation for a knowledge economy. Government policy, in consort with the private sector, nonprofit organizations, and multilateral agencies, was also instrumental in creating Panama's Ciudad de Saber (City of Knowledge), the former Clayton military base, now home to academic organizations, technology companies, and nongovernmental organizations.

This comprehensive, highly readable book is a most welcome addition to the literature on innovation, especially with respect to emerging markets. Ricardo Ernst and Jerry Haar, along with the experts contributing to more than a dozen chapters, are to be commended for their important contribution. One hopes that they and other scholars in the field continue to examine the dynamic relationship between innovation and economic growth and development and that policy makers and business leaders heed the advice and recommendations that emanate from their work.

Samuel Lewis Navarro
Vice-president and Foreign Minister of the
Republic of Panama, 2004–2009

Preface and Acknowledgements

Whether a consumer product (iPhone) or service (Uber), a firm (3M, Embraer), sector (biopharmaceutical), industry (mobile payments), or country (Taiwan, Israel), *innovation* is a driving, catalytic force in both domestic and global commerce today.

Research confirms that innovative enterprises generally achieve stronger growth or are more successful than those that do not innovate. There is a strong correlation between market performance and new products; and while new products can capture and boost market share and profitability, nonprice factors such as design and quality can increase competitive sales growth.[1]

But the performance encompasses not just products but processes as well, such as Toyota's production system, Benetton's IT-led production network, online trading and shopping – along with positioning (low-cost airlines, fully online higher education) and platform innovation (Boeing's 737 airliner, car makers).[2]

Innovation occurs in many different places – government and corporate research labs, universities, incubators, basements, and garages. In fact, the *intrapreneurial* spirit is alive and well in the "new corporate garage" where today's most innovative and world-changing thinking is taking place, in firms such as Medtronic, IBM, Unilever, and Syngenta.[3]

The "domains" of innovation – where it takes place – include products, processes, services, and business models. As for the forms of innovations, approaches and strategies, there are many. Three in particular stand out:

1. *Disruptive innovation.* Disruptive innovation, a term of art coined by Clayton Christensen, describes a process by which a product or service takes root initially in simple applications at the bottom of a market and then relentlessly moves upmarket, eventually displacing established competitors. Disruptors such as mini-mills, disrupting integrated steel mills; cell phones versus fixed line telephony; retail medical clinics overtaking many traditional doctors' offices; Uber and Lyft disrupting taxi services; and firms like Tesla and Airbnb.
2. *Lean manufacturing.* For many, lean is the set of "tools" that assist in the identification and steady elimination of waste. As waste is eliminated quality improves, while production time and cost are reduced. Lean thinking changes the focus of management from optimizing separate technologies, assets, and vertical departments to optimizing the flow of products and services through entire value streams that flow horizontally

across technologies, assets, and departments to customers. Although commonly associated with Japanese automotive vehicle manufacturing systems, businesses in all industries and services, including health care and governments, are using lean principles. Illustrative is Aldi, a German-based leading global discount supermarket chain with over 9,000 stores in over 18 countries. In its supply chain management, up to 60 percent of Aldi's fruit and vegetables are sourced locally where possible, reducing the need for long and costly delivery journeys. This demonstrates a time-based management approach. The characteristics of a lean organization and supply chain are described in the seminal volume *Lean Thinking* by Womack and Jones.[4]

3. *Blue ocean strategy.* Based on a study of 150 strategic moves (spanning more than 100 years across 30 industries), the authors W. Chan Kim and Renée Mauborgne argue that lasting success comes not from battling competitors but from creating "blue oceans" – untapped new market spaces ripe for growth. Southwest Airlines, a US company, made history as the world's first low-cost carrier, positioning itself as a competitor to the car – not to other airlines.[5]

In all cases of innovation, human capital is the most critical factor – it is the fountain of innovative thought and action. In research personnel rankings, a nation's size is inconsequential as the leaders are Finland, Iceland, Denmark, Israel, and Singapore according to Bloomberg's *Global Innovation Index*. In terms of R&D – with human capital at its core – the champions are South Korea, Israel, Finland, Sweden, and Japan.

The aim of the volume is to advance the knowledge and understanding of innovation in emerging markets as an important force in the global economy. These innovations include: wind-up radios in Africa; microcredit lending in South Asia and Latin America; the Jaipur foot, an artificial limb designed in India and simple to assemble, making use of low-tech materials; and CT scanners in China made by GE not only for that market but for rural locales in industrialized nations, innovation for and from emerging nations.

With this focus on the emerging markets of Asia, Africa, Latin America, the Middle East, and Central Europe, the contributors to this volume address such critically important themes and issues as releasing trapped value, catch-up innovation, reverse innovation, and social inclusion innovation as well as innovation in financial and nonfinancial services, health care, education, media, and social enterprises.

This information should be of great interest to policy makers; academics in business, engineering and information technology, public policy, and economic development; national, regional, and local economic development agencies; business associations; and innovators and entrepreneurs. Innovation, as a force of economic and social change, is reshaping the world. We analyze the trends, parameters, conditions, and outcomes of this transformative force.

The editors wish to extend their deepest gratitude to the Santander Group, its CEO Ana Patricia Botín, and to entrepreneur and former Panamanian Vice-president Samuel Lewis Navarro for their generous support of the publication of this book. We also wish to acknowledge the indispensable contributions of Surabhi Agrawal, Shanthala Gorur Ashwath, Johanna Hedman, Charles Rice, Leonor Dominguez, the Center for International Affairs at Harvard University, the McDonough School of Business, Georgetown University, and Professor Timothy Shaw and the International Political Economy Series team at Palgrave Macmillan.

Notes

1. Tidd, J. (2006). *From Knowledge Management to Strategic Competence: Measuring Technological, Market and Organizational Innovation*. London: Imperial College Press.
2. Bessant, J. and Tidd, J. (2011). *Innovation and Entrepreneurship*. New York: John Wiley & Sons.
3. Anthony, S. D. (2012). "The New Corporate Garage," *Harvard Business Review*, September: 3–11.
4. Womack, J. P. and Jones, D. T. (2003). *Lean Thinking*. New York: Productivity Press.
5. Kim, W. C. and Mauborgne, R. (2015). *Blue Ocean Strategy*. Boston, MA: Harvard Business Review Press.

List of Contributors

Emira Bečić is a senior advisor for the Directorate for Science and Technology of the Ministry of Science, Education and Sports, Croatia.

Lourdes Casanova is a senior lecturer and Academic Director at the Emerging Markets Institute, Johnson School of Business, Cornell University, USA.

Bhaskar Chakravorti is Senior Associate Dean of International Business and Finance and Professor of the Practice of International Business at The Fletcher School of Law and Diplomacy, Tufts University, USA.

Tory Colvin is Account Director at Forum One, USA, a digital agency that crafts solutions for foundations, think tanks, nonprofits, and government agencies.

Leslie R. Crutchfield is senior advisor to FSG, USA, a consulting firm that helps foundations, corporations, governments, and nonprofits worldwide, and a senior research fellow at McDonough School of Business, Georgetown University

Marina Dabić is Professor of Entrepreneurship and International Business at the University of Zagreb, Croatia, and Nottingham Trent University

Carl Dahlman is Head of both the Thematic Division and the Global Development Research at the OECD Development Centre, France.

Jeff Dayton-Johnson is Professor of Development Practice and Policy and Dean, Middlebury Institute of International Studies, USA.

Mark A. Dutz is a lead economist at the Trade and Competitiveness Global Practice, World Bank Group, USA.

Ricardo Ernst is Professor of Operations and Global Logistics, Managing Director of the Global Business Initiative, Managing Director of the Latin American Board, Co-Director of the Global Logistics Research Program and former Deputy Dean, all at the McDonough School of Business, Georgetown University, USA.

Rebecca A. Fannin is Founder of Silicon Dragon Ventures, USA, and contributes to Forbes.com.

Nils Olaya Fonstad is Research Scientist for Europe and LATAM at the MIT Center for Information Systems Research, USA.

Vijay Govindarajan is Coxe Distinguished Professor at the Tuck School of Business, Dartmouth College, USA, and Marvin Bower Fellow at the Harvard Business School.

Jerry Haar, professor of management and international business, Florida International University, senior research fellow, McDonough School of Business, Georgetown University, research affiliate of the David Rockefeller Center of Latin American Studies, Harvard University, and Global Fellow, Woodrow Wilson International Center for Scholars, USA.

Sukriti Jain is a research assistant at the Emerging Markets Institute, Johnson School of Business, Cornell University, USA.

Kurt Larsen is a senior Education Specialist for the Global Education Practice at the World Bank, USA.

Esperanza Lasagabaster is Practice Manager of Trade and Competitiveness for the South Asia Region, World Bank Group, USA.

Graham Macmillan is Director of Partnerships for Citi Corporate Citizenship, Citi, USA.

Kyle Peterson is Managing Director at FSG, USA.

Ravi Ramamurti is Distinguished Professor of International Business and Strategy and Director at the Center for Emerging Markets, Northeastern University, USA.

Krzysztof Rybinski is Rector at the New Economic University, Almaty, Kazakhstan.

Gabriel Sanchez Zinny is Founder and President of KUEPA and Managing Director of Blue Star Strategies, USA.

Christopher M. Schroeder is a venture investor based in the USA. He is the author of *Startup Rising: The Entrepreneurial Revolution Remaking the Middle East.*

Norean R. Sharpe is Senior Associate Dean and Distinguished Teaching Professor of Statistics and Operations at McDonough School of Business, Georgetown University, USA.

Michael Shoag is Managing Director, Forum One, USA, a digital agency that crafts solutions for foundations, think tanks, nonprofits, and government agencies.

Tony Siesfeld is Director of Monitor Deloitte, USA, and a senior leader of Monitor Institute.

Françoise Simon is a professor emerita and senior lecturer at the Mailman School of Public Health, Columbia University, and a senior faculty at the Icahn School of Medicine at Mount Sinai

Jadranka Švarc is a scientific advisor at the Institute of Social Sciences Ivo Pilar, Zagreb, Croatia.

David Wernick is a senior lecturer at the College of Business, Florida International University, USA.

List of Abbreviations

CITEC	Commission for Incorporation of Technologies
CMMi	Capability Maturity Model Integration
CNDRL	National Drug Reimbursement List
CORFO	Production Development Corporation
CRIATEC	Brazilian seed capital fund
CSR	Corporate Social Responsibility
DBT	Department of Biotechnology
DFID	Department for International Development (UK)
DMNEs	developed country multinational enterprises
ECG	electrocardiogram
ECU	engine control unit
EIT	European Institute for Innovation and Technology
EMBRAPA	Brazilian Agricultural Research Corporation
EMNEs	emerging market multinational enterprises
ESFRI	European Strategic Forum on Research Infrastructures
ESI	European Structural and Investment Funds
ETI	Global Enabling Trade Index
FCPA	Foreign Corrupt Practices Act
FedC	Community Pharmacies
FINEP	Funding Authority for Studies and Projects
FONTAR	Argentine Technological Fund
GATT	General Agreement on Tariffs and Trade
GCC	Gulf Cooperation Council
GELP	Global Educational Leaders' Program
GEM	Global Entrepreneurship Monitor
GII	Global Innovation Index
GNI	Gross National Income
HPS	Huff Power Systems
HRST	human resources sciences and technology
HVC	hybrid value chain

ICT	Information and Communications Technology
IFC	International Finance Corporate
INOVAR	Innovation Program (Brazil)
INPI	National Industrial Property Institute
IPR	intellectual property rights
ITIF	Information Technology Innovation Foundation
JIC	South Moravian Innovation Centre
LAVCA	Latin American Venture Capital Association
LGTs	local growth teams
MENA	Middle East and North Africa
MIF	Multilateral Investment Fund
MIM	Monitor Inclusive Markets
MIT	Massachusetts Institute of Technology
MOH	Ministry of Health
MSTQ	metrology, standards, testing and quality
MVNO	Mobile Virtual Network Operator
NAFTA	North American Free Trade Agreement
NBA	Network Behavior Analysis
NBDS	National Biotechnology Development Strategy
NICE	British National Institute of Health and Clinical Excellence
NRF	National Research Foundation
NRI	Networked Readiness Index
NSVF	New School Venture Fund
ODA	Office of Development Assitance
OECD	Organization for Economic Cooperation and Development
OEI	Organization of Ibero-American States
PATH	Program for Appropriate Technology in Health
PISA	Program for International Student Achievement
PSL	priority sector lending
RBI	Reserve Bank of India
SBA	Small Business Act
SEBRAE	Brazilian Service of Support for Micro and Small Companies
SIBA	Sustainable and Inclusive Business Activities
SIPO	State Intellectual Property Office

SITEAL	Latin American Educational Tendencies Information System
SPH	Shanghai Pharmaceuticals
STIC	Shanghai Technology Innovation Center
SUS	United Health System
TFP	total factor productivity
TRIPS	Trade-Related Aspects of Intellectual Property Rights
UAE	United Arab Emirates
UNIFESP	Federal University of São Paulo
USAID	US Agency for International Development
USPTO	US Patent and Trademark Office
WIPO	World International Property Organization

1
Introduction

Jerry Haar and Ricardo Ernst

Consider for a moment two anecdotes on groundbreaking innovation:

> Grace Choi – a 30-year-old Harvard Business School graduate, born to Korean immigrant parents and raised in Brooklyn – is an activist and businesswoman. Her company, Mink, hopes to use an innovative application of 3D printing technology to decentralize the cosmetic industry, and place the power to define beauty back in the hands of the consumer. On the way, she'll be disrupting a $55 billion market.[1]

> The Chinese e-commerce giant Alibaba raised $25 billion with its initial public offering in the fall of 2014, the largest recorded IPO [initial public offering] in history. Some critics argue that Alibaba is only a derivative of American market leaders like Amazon and eBay. While Alibaba did not invent the e-commerce platform, the innovative application of existing technologies and business models has allowed it to dominate the world's largest consumer market.[2]

As a driver of economic productivity, social progress, and ultimately human achievement, innovation is impacting every region and functional dimension of the global economy. Whether it is corporate giant Samsung spending $13.4 billion on research and development (R&D) in 2014 (ranking second globally) in an effort to dominate the hypercompetitive tech market,[3] or agrichemical firm Syngenta introducing innovative crop chemical packaging and training for smallholder farmers in Kenya,[4] innovation is inextricably tied to success and profit. Boston Consulting Group's annual innovation survey found that 75 percent of CEOs saw innovation as a top-three priority, but a falling number of executives were confident in their company's ability to innovate.[5] Innovation is receiving full recognition as both an opportunity and a challenge.[6]

1

While both developed and developing markets are capable of producing market-disrupting innovation, each tends to produce innovation with distinct characteristics and aspects. With existing stocks of infrastructure and ready access to finance, the developed world continues to lead the way in the most capital-intensive R&D, as well as high-tech innovation. Emerging market actors are particularly well suited to providing pragmatic and innovative solutions, for example mobile magnetic resonance imaging (MRI) in India, that address specific market gaps and failures. There is also clearly room for synthesis – innovating methods or ideas based upon opportunities for seizing upon (and combining) comparative advantages of developed and developing actors. For example, many developing countries are working towards establishing national health care systems, and at the same time, advanced Western nations are struggling to maintain their own systems in the face of shrinking workforces and rising costs – there is clearly room for reverse and mutual learning in this arena.[7]

Understanding the trends and frontiers of innovation is no longer an abstract or novel branch of knowledge. There is an imperative to harness innovation in the face of global competition, and governments, research institutions, corporations, and individuals are pursuing this goal in every corner of the globe. While innovation is often a globally integrated process, this book will specifically assess innovation as a driver of productivity, profit, and economic transformation in emerging markets around the world.

Emergence of an innovation ecosystem

It is first and foremost necessary to identify broad global trends that have helped prompt the current wave of innovation sweeping the global economy. While any number of specific factors might be referenced, the key advancements that have facilitated innovative economic progress, particularly in emerging markets, are fluid financial capital, technological advancement, an integrated and open global market, and the economic emergence of the developing world.

- *Expanding access to finance*: Whether it is venture capital in California, or development finance loans for small and medium enterprises (SMEs) in sub-Saharan Africa, innovators around the world have greater access to the financial resources needed to spur growth than in previous eras. In established markets, major banks, venture capital funds, and even novel investment platforms like Kickstarter help identify and fund innovative ideas. Emerging economies, which attracted more Foreign Direct Investment (FDI) than developed counterparts for the first time ever in 2012,[8] also have growing access to financial markets and resources, which has fueled the innovation explosion in these regions.

- *Technology explosion*: Modern technological breakthroughs have fueled the ability to travel, connect, and share information in ways unimaginable to previous generations. Online platforms like Kayak, Travelocity, and Expedia have made air travel increasingly affordable and accessible, and social media allow us to globally broadcast images and news from our lives in an instant. Technology can itself be an innovation, but it can also facilitate innovative activity and models. Global business structures with complex value chains are only possible because of the speed of communication and travel facilitated by current technology. While global GDP rose at about 3 percent annually from 2000 to 2008, world trade expanded at a rate of 5.4 percent per year over the same period.[9] Computing and information technologies, in particular, have helped spread connectivity and access to previously isolated regions and markets, and created business competition that spans the globe.

- *Global markets*: Often outpaced by this technological evolution, institutional changes and neoliberal policy reform, both in individual countries and in the global economic structure, have served to open markets and connect individual countries to the global economic system. Global institutions like the General Agreement on Tariffs and Trade (GATT) and its successor the World Trade Organization (WTO), as well as significant bilateral and regional trade deals, have significantly reduced protectionism and increased economic interconnectedness around the world. The WTO records that the average import tariff had fallen from 14 percent in 1952 to 3.9 percent by 2005.[10]

- *Emerging economies*: Absolute poverty has fallen by nearly half since 1990,[11] and these rising incomes paired with greater access to information and capital have made emerging markets hotbeds for innovation and business creation. With increasingly skilled workforces and rising disposable income, emerging economies are fertile grounds for both innovation and end market success. Today the global middle class includes about 2 billion people, but it is projected to more than double to 4.9 billion by 2030. The vast majority of this expansion will occur in emerging economies, and billions of new BOP (bottom of the pyramid) consumers will fundamentally alter national and global markets.

While innovation has been a force of evolution and advancement throughout human existence, the unique characteristics of our current global system have placed innovation at the heart of progress for all units of economic society. Moving back into human history, innovative activity was once the province of singularly creative individuals. "Gutenberg's press" and "Edison's lightbulb" are products inextricably linked to the inventors that birthed them. Today innovation determines the wealth and success of nations. Stephen Ezell of the Information Technology Innovation Foundation (ITIF) asserts that more than 90 percent of income growth per worker across countries is attributable to innovation.[12]

In our modern global economy, innovation is occurring in the garages and dorm rooms of individual entrepreneurs, but also in innovation units at the largest global corporations, government-funded research labs, and incubators and accelerators around the world. Miami's Venture Hive accelerator program offers grants, office space, and training programs for tech start-ups to establish an entrepreneurial ecosystem in downtown Miami.[13] In the emerging market context, China's Ministry of Science and Technology's Torch Program for High Technology Industrial Development has helped drive China's economic growth over the last 25 years, and relies on innovation clusters as a core principle.[14] Economic activity ties together a series of actors in a complicated series of interactions, and increasingly, innovation can occur at any juncture in this chain.

Individuals, companies, and countries feel that they live and die based on how effectively they innovate. By some measures corporate life spans have decreased by 50 percent over the last half-century;[15] in the words of management guru Gary Hamel, "out there in some garage is an entrepreneur who's forging a bullet with your company's name on it."[16] As the process of creative destruction has accelerated, all economic actors face the pressure to innovate, or be destroyed. This threat is equally real for a small business in Kenya, a start-up in Silicon Valley, and a corporate titan like Exxon. The recognition of innovation as a transformative economic force – even institutions like the World Economic Forum have identified innovation as a driver of global competitiveness[17] – has significantly altered the competitive landscape in the global economy. For a range of actors, innovation has become an economic and commercial priority.

Today there is a conscious and structured pursuit of innovation as an economic activity, a shift that has prompted our current "age of innovation." Governments are looking to national innovation policies and institutions to help establish environments conducive to innovation. Corporations funnel huge resources to largely autonomous innovation units, reflecting a tolerance for higher-risk investment without clearly defined deliverables or timelines. For individuals in innovation hotbeds, entire ecosystems of venture capital, incubators, and informal networks help identify and funnel resources to the most promising entrepreneurs.

These clusters are simultaneously a key driver and an outcome of innovation: as an example, the University of Cambridge established a strategic partnership with the surrounding community under the name "Silicon Fen." The cluster included 1500 tech-based firms working with various research centers and faculty. In the past 20 years, this has led to the creation of more than 300 high-tech ventures, 14 of which can claim revenues of over \$1 billion. This clustering of innovative activity and firms has been both observed and replicated elsewhere around the world – the cooperation, coordination, and group learning in these clusters help foster competitiveness for smaller firms, which is particularly relevant for operators in the developing market context. Some notable clusters are included in Table 1.1.

This book will provide a comprehensive manual of the dimensions, impacts, and forms of innovation in the context of emerging markets around the world. We will do so by examining efforts to promote innovation at three distinct operational levels: national policy, facilitating institutions, and firm-level behavior. In this framing, innovation is at once a familiar force, but also unique in its specific manifestation. Innovation is already of critical importance, yet we have only scratched its potential as a force of economic transformation in developing markets. In order to fully unleash transformative innovation, policy makers, researchers, and private sector actors all need to understand their respective roles, and the best practices for achieving their goals.

In our estimation, innovation is the single most powerful force for driving economic evolution and progress globally, but particularly in the emerging market context where it is critical in driving job creation and income growth. Innovation holds this elevated economic importance for a number of reasons.

First, innovation is key to sustainability and progress. There is no doubt that innovation is an economic priority; questions remain on "how" we innovate, and "how much." Sustainable energy sources and power infrastructure have been one of the top goals for innovators around the world. Functionally, this type of innovation can occur in a number of ways – solar breakthroughs, hydraulic fracturing, and more efficient power and

Table 1.1 Innovative clusters worldwide

UK (Silicon Fen, Cambridge)
Taiwan (Hsinchu Park)
Netherlands (Technopolis Innovation Park, Delft)
Israel (Silicon Wadi, Tel Aviv)
France (Aerospace Valley, Toulouse)
Canada (ICT, Waterloo)
Chile (Santiago, technology and "Start-up Chile")
India (Bangalore, IT)
Singapore (IT)
Brazil (Campinas, IT)
U.S. (Albany Technology Valley, NY, nanotechnology)
Colombia (Medellín, multi-sector clusters)

---

utilities grids are all distinct innovations that address the same essential market need.

Second, this impact is magnified because innovation is occurring in every region around the world, at every link in the economic value chain, and at every organizational level. A 2010 report in *The Economist* titled "The World Turned Upside Down" noted that the developing world now rivals rich countries in business innovation.[18] Additionally, innovation has become a priority for all businesses at each point in their value proposition and functional structure, driving innovation in every domain from product creation to packaging and delivery.

While it is easy to view innovation as a global trend, it is critical to note that innovation is unique in the developing world as compared to its developed counterparts, and warrants its own line of study. Innovation in emerging markets has been driven by suppliers seeking to capture the growing demand of low-end consumers, who will soon constitute a majority of world demand. Affordability, flexibility, and functionality are key in meeting challenges and providing innovative solutions in these markets.

Finally, innovation provides the opportunity for market disruption and leapfrogging progress, which could help frontier markets "equalize" or even pass developed markets on a number of fronts. Developing economies commonly suffer from a lack of established infrastructure and industry; however, this can be viewed as an opportunity. Rather than follow the incremental process of advancement and industrialization utilized by the OECD countries of the West, developing countries can leapfrog forward by effectively utilizing new technology. Innovative application of mobile ICT and sustainable energy sources are particularly relevant to this form of progress.

Quantifying innovation

Both intuition and anecdotal evidence suggest clearly the power of innovation as a force for economic creation and progress, but it is somewhat more complicated to abstract its full range of impacts and benefits. While there is data available that points towards effects of innovation, a simple statistical review does not capture the full consequences that innovation, and the pursuit of innovation, have on individual firms, markets, and the global economy. That said, it does provide some context for its impact.

R&D spending can serve as a proxy indicator for commitment to innovation, and is generally correlated with innovative outcomes in both firms and markets. Switzerland, which tops the 2014 Global Innovation Index,[19] is one of the top per capita R&D spenders in the world at about 3 percent of GDP in 2009, or about 4.4 percent of total global R&D spending.[20] Private firms are also contributing huge sums to R&D activity; the top five private sector spenders allocated nearly $52 billion to R&D, and had revenues in excess of $600 billion in 2013. The top 40 economies are projected to have spent $1.478 trillion on R&D in 2014, with the US ($447 billion) and China ($232 billion) leading the way.[21] While not all innovation is the product of R&D, the latter no doubt helps facilitate groundbreaking technological innovation. PricewaterhouseCoopers constructs a list of the top innovating firms each year, and no member of the top ten spent less than $200 million on R&D in 2014 (see Table 1.2).[22]

Intellectual property and patent applications provide another quantitative indicator by which we may map the growth and trajectory of global innovation – if R&D is the simplest statistic for input into innovation, patent data is the most primary measure of innovation output. The World Intellectual Property Organization (WIPO) indicates that in 2011, 2.35 million patent applications were filed worldwide; this represents a growth rate of 9.2 percent from the previous period, the highest growth rate in 18 years.[23] While the majority of these applications come from the US, China, and other established economies, impressive patent growth is occurring in emerging markets. Mexico (9 percent), Turkey (13.4 percent), and Bangladesh (15.7 percent) all exhibited a significant increase in patent application filings,[24] indicating a potential uptick in innovative activity.

Determining the employment impact of innovation at the firm or national level can be a challenge; however, there is clear reason to believe that innovation has an unambiguously positive effect of job quality and creation. On the front end, R&D has definitively positive employment impacts – in the US $465 billion in R&D spending will directly create

Table 1.2 Top innovating firms, 2014

Rank	Company	Location	Industry	R&D spent ($bn)
1	Apple	United States	Computing and electronics	4.5
2	Google	United States	Software and Internet	8
3	Amazon	United States	Software and Internet	6.6
4	Samsung	South Korea	Computing and electronics	13.4
5	Tesla Motors	United States	Automotive	0.2
6	3M	United States	Industrials	1.7
7	General Electric	United States	Industrials	4.8
8	Microsoft	United States	Software and Internet	10.4
9	IBM	United States	Computing and electronics	6.2
10	Procter & Gamble	United States	Consumer	2

2.7 million jobs, and will support an additional 6 million employment opportunities. There is also good evidence that suggests a similar relationship in less developed countries. A 2012 paper from Gustavo Crespi and Ezequiel Tacsir at the Inter-American Development Bank assessed the impact of innovation in select Latin American countries, and found that employment growth in innovative firms is higher than in noninnovative ones.[25] Broadly, the private sector figures to be the prime driver of employment in the developing world. The International Finance Corporation (IFC), the World Bank's private sector investment arm, estimates that the private sector provides nine out of ten jobs in emerging economies.[26] Given innovation's role in instigating progress and profit in the private sector, there are clear employment implications.

Innovation and its related activities provide economic benefits on a range of fronts, making it difficult to provide a discrete and quantifiable impact. Particularly in emerging markets, where there are often opportunities for revolutionary restructuring of the economic order, the scale of potential benefit is massive. Sudanese entrepreneur Mo Ibrahim founded African telecom giant Celtel at a time when the continent had essentially zero cell users[27] – today, he is a billionaire and the African continent is projected to have hit 1 billion cell users by 2015.[28] M-PESA is a mobile-based money transfer and microfinancing system which has taken off in a number of developing markets, primarily in Africa. M-PESA has subsequently expanded to over 17 million users in Kenya alone, about two thirds of the adult population, and nearly 25 percent of Kenya's GDP flows through the platform.[29] Celtel helped establish a mobile market in Africa, which M-PESA utilized to deliver a decentralized financial system to an underserved market – both innovations have had booming reverberations for economic progress and structure on the African continent.

Forms of innovation

The global business environment has undergone massive change in recent decades, and so too has our understanding of what constitutes innovation. New technology and connectivity between the developed and developing world have driven much of this change – as information, people, ideas, and capital move around the world more freely, there are growing opportunities for pairing a market need with a solution. Innovation is in this regard more than creativity – it is, in the words of Vijay Govindarajan, the commercialization of creative thought.[30]

Part of the challenge in assigning a clear demarcation for "innovation" arises because innovation can occur in a number of meaningful ways. An iPad is clearly an innovation, but so is the WaterCredit Initiative, which provides microfinance loans for clean water and sanitation in countries like Uganda, Kenya, and Bangladesh.[31] For the purposes of this analysis, innovation will be defined as the creation of better or more effective products, processes, services, technologies, or ideas that are readily available to markets. These are as follows:

- *Product innovation*: Broadly, product innovation is the development of new products for sale in the market, or the incremental alteration or improvement of an existing product to reduce cost or better satisfy market demand. Innovative products provide firms with a way to differentiate from their competition, and can help drive growth, expansion, and competitive advantage in the market.
- *Process innovation*: The OECD's Oslo Manual, which provides guidelines for collecting and interpreting innovation data, defines process innovation as the implementation of a new or significantly improved production or delivery method, including significant changes to technique, equipment, or software.[32] Process innovation aims to decrease unit costs of production, increase quality, or significantly improve the methods for the creation and provision of services. Henry Ford's assembly method is perhaps the most famous example of process innovation, but increasingly the integration of big data and ICT in production and delivery processes is driving this form of innovation. Today, companies like Siemens are turning to microgrids integrated with IT systems to deliver more secure and efficient power.[33] As global systems and value chains become more complex, this process innovation will only be more critical.
- *Service innovation*: Service innovation was first proposed as a topic of study in the last two decades, and is in some cases difficult to differentiate from the process of product innovation. Bart van Ark provides a useful working definition for service innovation: "a new or considerably changed service concept, client interaction channel, service delivery system, or technological concept that individually, but most likely in combination, leads to one or more (re)new(ed) service functions that are new to the firm and do change the service/good offered on the market and do require structurally new technological, human or organizational capabilities of the service organization."[34] The meteoric rise of companies like Uber and Lyft demonstrates how an innovative form of service delivery can disrupt an entire market, and replicate across the globe.[35] While service innovation is often tied to the integration of new technology into the service, it is not necessarily so.
- *Business model innovation*: Business model innovation is unique from the other forms listed above in that it relies on existing products, technologies, and markets – the innovation occurs in the key decisions that collectively determine business operations. Boston Consulting Group sees a business model as being jointly comprised of a value proposition and an

organizational model.[36] Companies must determine what they offer, and to whom they offer it. Subsequently, companies must determine how they can profitably deliver based on customer demand. Unique configurations of target consumers, product and service offerings, and compensation drive a variety of distinct value propositions, alongside a wide array of value chains and organizational models that propel profitable delivery. Innovation at any of these points constitutes a business model innovation.[37] Chinese e-commerce giant Alibaba offers an excellent example of effectively adapting a business model proven elsewhere for a new market. Alibaba's record-breaking $25 billion IPO in 2014 reflects the company's success in dominating a domestic market of 1.3 billion people.[38]

As our understanding of innovation has broadened to include a growing number of functional forms, innovation is at once elevated as a business priority, and further complicated as a practical goal. Firms need to understand what type of innovation they are best suited to producing, and then pursue that specific subset of activities. This manual will offer a broad review of each form of innovation, provide specific examples of leading success stories, and analyze the practices and methods that lead to this success.

Innovation in emerging markets

Today, the recognized global leaders in innovation are still established industrial economies – quality infrastructure, finance, and human capital are readily available for sophisticated businesses pursuing innovation agendas. The annual Global Innovation Index (GII) ranks national economies based upon innovativeness, and in 2014 the top 25 performers are all high-income countries.[39] Despite the ongoing dominance of developed countries in the innovation sphere, emerging market economies are making significant strides towards meaningful innovation – the GII's largest regional improvement on innovation ranking occurred in sub-Saharan Africa, where 17 of 33 countries included in the index improved their ranking.[40]

Indexes like the GII may not be capturing the full impact and scale of innovation in emerging markets, because the process of innovation is often distinct from what is observed in more established markets. Rather than duplicating the processes and systems employed in developed economies, emerging markets increasingly rely on original innovation ecosystems that leverage specific advantages to address local market needs. In some ways, emerging economies are leading the charge on global innovation – underdeveloped markets present huge opportunity, and the transformative impact of innovation can be particularly meaningful in these settings.

Urbanization, rising stocks of skilled human capital, global connectivity, growing incomes, and a number of other broad global trends are driving innovation in emerging markets around the globe. The difference in population between the developed and developing countries is significant – 1.2 billion against 5.9 billion – presenting enormous potential for growth and opportunity in these markets.[41] The growth rate in these countries has also been a full percentage point higher than in developed economies over recent decades.[42] Emerging markets, and China in particular, are producing a steadily increasing supply of college graduates and highly skilled workforces, and are closing the gap with their high-income counterparts (see Figure 1.1).[43]

This new generation of emerging market innovators are taking advantage of the huge opportunity before them, and are changing their countries and the world through innovation that is uniquely suited to their context.

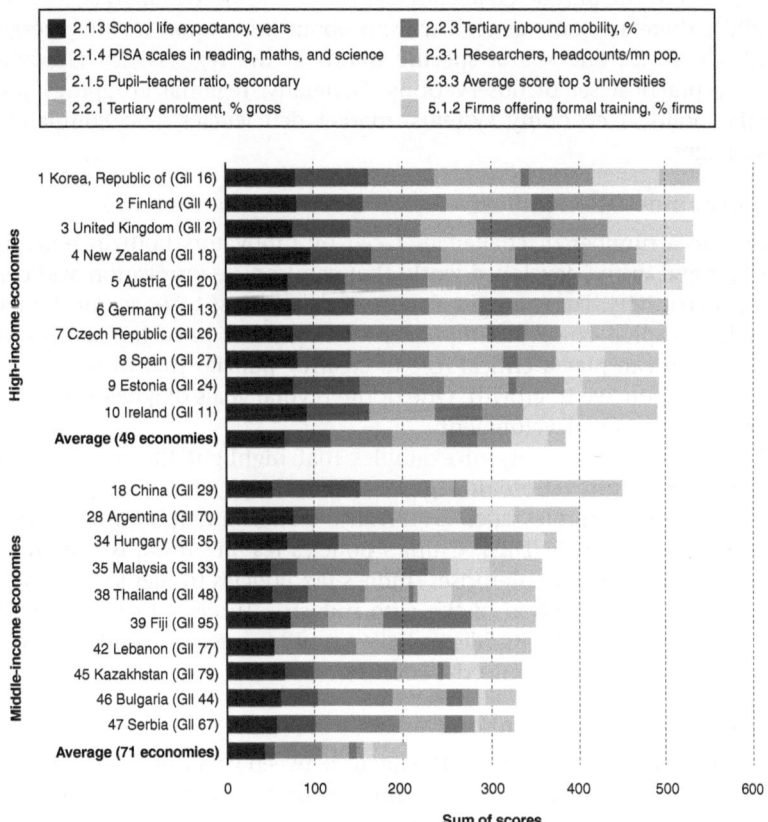

Figure 1.1 Education as a human aspect of innovation: top ten high- and top ten middle-income economies

This is to say, innovation in emerging markets is fundamentally distinct from the innovation activities occurring in the US, UK, or other developed markets.

For example, Hindustan Unilever realized that Indians want to use shampoo but that the price point of a whole bottle was too high. They created shampoo sachets that are sold for 1 rupee, thus making it available even to the poorest segments of the population.[44] In Turkey, personal signatures are much more widely required than in other countries. Turkish law mandates that a signature is collected for processes such as receiving paychecks, holiday requests, and corporate travel reservations. In 2004, Turkey adopted a law giving electronic signatures the same legal status as "wet signatures." In response, Turkcell innovated a secure mobile signature solution service called "MobilImza."[45] The program was so successful that it ultimately spread to other countries. Both of these innovations were context specific, and suited to the unique characteristics of their respective markets.

While there are some general truisms about innovation in the emerging market context, analysis of specific trends at the regional level provides a more practical set of observations. Generally, regional groupings share similar political economy systems, market deficiencies, and comparative advantages.

National innovation policy

There are a number of challenges faced by innovators both in emerging markets and in the developed world that can impede innovation and commercial creativity. For example, factors such as access to financial, human, or natural capital can act as impediments to growth. In such instances, the government can play a critical role in creating policies that foster an environment conducive to growth. One of the pivotal goals of government is to promote economic development.

There are a wide variety of examples that highlight the differing policies regarding innovation that governments have adopted. This section will highlight examples such as Brazil's establishment of the Brazilian Development Bank (BNDES), China's policies leading towards the growth of municipally owned enterprises, India's tax policies to fuel local business growth, Japan's protection of the auto industry, Turkey's Technopolis law, Russia's development of its Silicon Valley, and Morocco's incentives to redirect diaspora funds back into the country.

Role of facilitating institutions

Facilitating institutions serve as the nexus between government and the private sector. For national systems of innovation to operate effectively, there needs to be some level of harmony and coordination between government and firm-level activities. Facilitating institutions occupy this middle ground, and can thus serve a crucial role to meld the interests of both governments and businesses in order to foster innovation. Through case examples, this section will analyze the role of both the government and the private sector in sponsoring innovation.

The government is prominent in the development of university research centers, R&D initiatives, technology centers, science parks, industrial clusters, and export zones to facilitate an engaging environment for innovation. This section will provide two specific case studies: the United Arab Emirates' (UAE) use of knowledge-based institutions to facilitate entrepreneurship, and the growth of SMEs in Brazil through industry-specific clusters. Similarly, the private sector also provides a venue for entrepreneurial growth through partnerships with these research centers, universities, and laboratories. The private sector is able to facilitate growth through incubators, accelerators, microfinance institutions, venture capital and private equity firms, private R&D labs, and business associations or industry groups that help their members innovate. Innovators use angel and venture capitalists to gain training, mentorship, and access to financing, in order to accelerate their ideas. Examples of private sector promotion of innovation via facilitating institutions may be seen across East Asia, Latin America, and Africa. Both government and private sector have been able to develop a wide variety of rich facilitating institutions that enable innovation.

Furthermore, the UAE has more than ten incubators and accelerators in operation – an increase from the three active in 2008 – in order to fuel SME business development through mentorship and support services.

In Argentina, Parque Austral is the business park created in 2008 by Universidad Austral offering intellectual capital, a collaborative workspace, access to a global network, and tax exemption to the 90 companies that it has the capacity to house.

In Rwanda, similar facilitating institutions such as kLab (knowledge Lab) have also provided an open space for IT entrepreneurs to collaborate and innovate in Kigali. kLab hosts events, workshops, boot camps, hackathons, and networking sessions to promote partnership, and provides investments, financing, and mentorship (kLab, 2015).

Innovation in the firm

In today's world, innovation is much more than new products. It is about reinventing business processes and building entirely new markets that meet untapped customer needs. Most importantly, as the Internet and globalization widen the pool of new ideas, it is about selecting and executing the right ideas and bringing them to market in record time.

While a firm can innovate its products, services, processes, and/or business models, all firms operate under a given set of environmental conditions and external factors over which the firm has no control. Given all the external dynamics, firms attempt to innovate to survive and outrival in the modern world. A number of external factors could trigger innovation within an organization. Emerging trends in demography, technology, and consumer and social behavior can be tapped to create valuable innovation that can be profitable for the firm. For instance, emerging trends in cloud computing technology are driving rapid innovation in many industries and firms. El Corte Inglés, the largest department store in Europe, is using the cloud to quickly expand online and provide customized offers, promotions, and pricing in real time (LeBlanc, 2014). Other factors like government regulations, environmental conditions, industry trends, and competition can also help lead the way for many firms on the possible directions they can take in being innovative. Social networks have taken the business process of innovation and entirely changed the dynamic and underlying rules of how work gets done. In today's world, several companies use social networks to communicate and collaborate with customers and business partners to cocreate and generate ideas for new products and services. "My Starbucks Idea" is such an initiative by Starbucks to collect new ideas from customers and partners for improving their products and services.

Large firms

Large firms have considerable amounts of human and financial capital, which makes it easier for established firms to venture into new investments in innovation. These firms can also afford to invest hugely in market research, infrastructure development, distribution channel development, and brand building which go hand in hand with the innovation the firms carry on. For example, Samsung's heavy marketing efforts in popularizing its brand for smartphones was a key factor that helped it gain a large global market share and overpower its competitors. Larger firms also have an inherent advantage when it comes to cushioning up market shocks in unstable markets. They are usually more vertically integrated to internalize value chains and reduce external risks. Small failures would not stop them from reinvesting in research activities. Large firms can also mobilize resources on a large scale to conduct heavy R&D, create dedicated teams for driving innovation within the firm, and allocate large budgets to future innovations. PARC, a Xerox company founded in 1970, was such an effort by Xerox to establish a strong R&D capability internal to the firm. This initiative has resulted in many successful innovations for Xerox in information technology and hardware fields. This is an advantage that only established firms can make use of in echoing their innovative positions.

The culture of innovation

All firms, whether they are successful with innovation or not, have one thing in common: they have their own "personalities," their own particular organizational culture – the shared experiences, values, norms, assumptions, and beliefs that shape individual and group behavior, and that ultimately impact their success.

One of the main factors that affect culture within a firm is leadership. How the organization's leaders influence strategic direction and day-to-day operations, both through their explicit decisions and their more subtle behavior, makes a great impact on the culture of the company. The behavior of leadership is arguably the single most important factor in driving culture change.

Amazon is known to be a very successful firm in the modern world that has instilled a gene of innovation in its culture. Amazon focuses on a "spectrum" of innovation. While the big breakthroughs get all the attention – Kindle, cloud computing, Prime – the firm also strives for small innovations and daily improvements that reduce costs, save time, or improve quality.

Hotbeds of innovation

Innovation has long been thriving in the developed world in various well-known innovation hotbeds, such as Silicon Valley, the Greater Boston area, North Carolina Research Triangle, Tech City London, and the multi-sector innovation hub in Paris-Saclay. Over the past decade, hotbeds of innovation have surfaced all over emerging markets, particularly in Asia and Africa. Hotbeds of innovation such as Nairobi, Bangalore, and Beijing have become major breeding grounds for innovation, helping emerging markets to quickly catch up with innovation in the developed world (*MIT Technology Review*, 2013).

Innovation is a natural response, in part, to scarcity and a dense and rapidly growing population. Poverty, lack of resources, legal challenges, and poor infrastructure force people to develop creative alternative solutions using existing resources. At the same time, a majority of the world's population lives in the developing world and the wealth and level of education there are increasing at a much faster pace than in the developed world. Consumption and demand are continuously rising as the middle class in the developing world expands exponentially. The rural to urban exodus occurring in emerging markets is historically unprecedented, creating highly dense urban clusters with enormous potential along with challenges which will spur innovation (Pilling, 2010). Innovation hotbeds will continue to flourish given the unceasing rapid growth of the global population, especially in developing markets, along with diminishing natural resources. This represents great potential profits for innovators who can provide the market with affordable products and services that increase survival (such as potable water, food, and medical services) and quality of life (such as banking, communications, waste management, transportation, and utilities).

Innovation in emerging markets occurs on many different levels. We will review a few of the most important ones.

Types of innovation

Reverse innovation

Up until the last decade, "innovation" was associated nearly exclusively with the transfer of knowledge, products, services, and business forms from the industrialized nations of the world to developing countries. However, the groundbreaking book *Reverse Innovation* by Vijay Govindarajan and Chris Trimble of Dartmouth's Tuck School of Business, ignited major interest in this form of innovation – one consisting of scaling growth in emerging markets, and importing low-cost and high-impact innovations to mature ones (see Table 1.3).

With the rise of innovation hotbeds in emerging markets, innovation activities have gradually moved from the West to the developing world. Historically, products were innovated and produced for the developed world for eventual expansion to the developing world as production volumes increased and prices decreased. However, over time a greater percentage of innovations has come to be adopted first in emerging markets to eventually spread to the developed world.

GE Healthcare developed MAC400, a low-cost, portable, battery-operated electrocardiogram machine in India which has since spread to numerous developed markets due to its functionality and attractive price (Govindarajan, 2012). The machine can be easily transported to rural areas in the developing world and provides a crucial medical service at a fraction of the traditional cost. Mobility-impaired patients can undergo the procedure directly in their bed, preventing the pain and trouble caused by moving.

Table 1.3   Reverse innovations

	Innovation	Description	Country
1	GE MAC 400 electrocardiogram machine	Portable, low-cost electrocardiogram machine	India
2	Lyft	Ride sharing/peer-to-peer taxi service	Zimbabwe
3	John Deere 5003 series tractor	Basic tractor model	India
4	The Odon device	Low-cost labor device	Argentina
5	Tata Nano car	Basic, low-cost car model	India
6	Allview tablets	Affordable tablet	Romania
7	Walmart Bodegas Aurrera	Small format stores	Mexico
8	Microsoft starter	Simple operating system	India/Thailand
9	Synoptix diagnostic device	Breast cancer early detection device	Saudi Arabia
10	Minute Maid Pulpy	Orange juice with pulp	China

In Zimbabwe, a lack of infrastructure and financial means has made ride sharing a common means of transportation. While visiting the country, American entrepreneur Logan Green got the idea that eventually led to founding Lyft, a car-sharing/semi-pro taxi service. In 2013, Lyft was able to raise $60 million from venture capital firms and service has spread to cities all over the US (Carlson, 2013). This is a great example of reverse innovation of a business model.

Suppliers focus on catering to the growing demand of low-end consumers in emerging markets. There is an enormous potential in serving the bottom of the wealth pyramid. Basic "no frills" products are innovated to the price point. Volume is increased, allowing for attractive profits despite tight margins. Because piracy is rampant in emerging markets, companies have to continuously upgrade products. These upgrades can also help companies eventually expand to the developed world, which demands more sophisticated products.

Deere & Company has a history of producing high-end tractors for affluent developed markets. The tractors typically feature a multitude of options including air conditioning, GPS, and powerful engines. However, Deere & Company developed a very simple and basic tractor in its India-based R&D center. It was initially manufactured only for sale in India. However, when Indian competitor Mahindra & Mahindra began selling their basic tractor to small-scale farmers and amateurs in the US, Deere & Company swiftly responded by beginning to sell their 5003 series no-frills tractor in the US, with great success (Madhavan, 2010).

In emerging markets there is a high demand for products and services that are affordable, flexible, and functional. Affordability is key to reach less affluent mass consumers. Innovation to the price point is essential – creating a good enough product/service at an accessible price point (reducing the product down to serve just the most basic need).

The Odon device is an inexpensive polyethylene sheath that is used to pull a baby out during prolonged labor. The innovation was created by a car mechanic in Argentina, who got the idea while trying to get a cork out of a wine bottle with a plastic bag. It is easy to use and a safer alternative to vacuum extractors or forceps in cases where there is a risk of vaginally transmitted infections such as HIV. It is currently undergoing trials in Argentina and South Africa. If approved, it will likely be used in both the developed and the developing world (McNeil Jr., 2013).

Products and services need to be flexible to overcome infrastructure challenges. For example, Mi-Fone, a mobile phone manufacturer headquartered in Mauritius and focused on the African market, innovated a mobile smart feature phone (model Mi3000) with a 60-day battery standby time and edge connectivity. This mobile phone allows people in rural areas with limited access to power to have steady access to a phone and to the Internet (Gicheru, 2012).

Functionality is key and products that can serve a secondary purpose or are easy to transport and assemble are in high demand. "The Hippo Water Roller" is a water drum that can be rolled on the ground and was innovated in Africa to facilitate the transportation of water over extended distances. It reduces the time spent on transporting water (typically done by women and children), leading to more time being available for educational and economic opportunities while also preventing injuries caused by carrying heavy loads of water (Walker, 2009).

In India there is a well-known Hindi term used to describe innovation driven by frugal improvised solutions to everyday problems. Innovation and leadership strategist Navi Radjou and his colleagues Jaideep Prabhu and Simone Ahuja wrote the book *Jugaad Innovation* (2012) exploring this phenomenon in detail. The term *jugaad* has a similar meaning to the English term "hack," albeit with a stronger focus on finding survival solutions than on being mere intellectual work-arounds to simplify processes (Radjou et al., 2010). *Jugaad* was originally used to describe makeshift vehicles built from water pump engines mounted on carts, a great example of a clever improvised solution using few resources to serve a pressing transportation need. *Jugaad* is a concept deeply embedded in Indian culture and *jugaad* innovating takes place on a daily basis throughout India.

In another example, Renault-Nissan gave three teams in France, Japan, and India the same engineering problem to solve. All three teams produced solutions that were comparable in quality. However, the Indian team, being used to the frugal engineering approach, produced a solution that cost only a fifth of the French and Japanese teams' solutions (Madhavan, 2010). Recognizably, emerging markets have become world leaders in innovation in several areas, such as microfinance, mobile banking, and desalination, and there is ample indication that this trend will continue.

For example, Kenya has emerged as a world leader in mobile banking. A very large percentage of the Kenyan population has access to cell phones but does not have bank accounts. Kenyan mobile network operator Safaricom developed M-PESA, a mobile banking service. Mobile banking allows people without a bank account to safely save and send money using their cell phones. Users buy digital funds at an agent and send a text message to the recipient, who claims the physical funds at an agent near them. M-PESA also functions as an electronic account, allowing users to securely save money by depositing cash in return for digital funds. Mobile banking is now rapidly spreading to the developed world, as complex banking systems and rigid credit requirements have caused many people to be unbanked (Fengler, 2012).

Another example of a frugal innovation is a solar-powered, cashless Water ATM created by Indian company Sarvajal, providing potable water in poor, underserved areas at all times. The Water ATM is remotely managed and monitored thanks to cloud technology.

Seven other arenas of innovation are shaping the landscape for the development of processes, products, and services. We examine them below.

Packaging/price innovation

Hindustan Unilever, a multinational consumer goods company, realized that there were many Indians who wanted to use shampoo but found the price point of a whole bottle too high. They created shampoo sachets that are sold for 1 rupee, thus making shampoo available even to the poorest segments of the population. This caused shampoo to become very popular and widely used in India. With the global recession, both Unilever and Nestlé have successfully applied the same packaging innovation to staples such as coffee and hygiene products in particularly troubled areas of Europe and the US (O'Connor, 2012).

Social media innovation

According to the World Bank, China has the world's largest Internet user base, consisting of more than 620 million users, which is more than twice that of the US with a user base of 266 million. Chinese social media company Tencent created "WeChat," a mobile multimedia communication service providing voice messaging, broadcast (one-to-many) messaging, and photo/video sharing. It has become China's largest social media service. The Internet is one of the ways the Chinese attempt to get around censorship, since it is much more difficult for the government to monitor than traditional communication methods. Even so, the Chinese government is doing what it can to limit access to Western search engines. In preparation for the twenty-fifth anniversary of the Tiananmen Square protests, the Chinese censorship authorities blocked access to most Google services (Mozur, 2014).

Distribution innovation

Home-cooked lunches are very popular and highly in demand in India. However, because of commuting it is logistically impossible for most Indians to have lunch at home. Dabbawalla, an Indian company, innovated a distribution network which picks up freshly cooked lunch boxes at their customers' homes and delivers it to their workplaces. They later pick up the empty lunch boxes and return them to the customer's home. Dabbawalla relies on various modes of transportation, most commonly bicycles. Dabbawalla also innovated a unique marking system to identify each lunch box, as most of their transportation employees are illiterate (Rai, 2007).

Advertising innovation

To promote their Clinique brand in China, Estée Lauder created a web-based drama series called "Sufei's Diary" on a dedicated website as well as on TV screens in public spaces and buses. The series is perceived as entertainment for the public, while promoting Clinique's brand awareness by frequently displaying skin care products in the plot. It reaches a wide market and is much cheaper than conventional advertising. It has also proved to be much more efficient in creating brand awareness and demand, as it connects with the viewer on a more profound and emotional level than regular ads (Fong, 2009).

Sustainable innovation

The developing world is a leader in creating innovative solutions to recycle waste, thus gaining access to recycled materials and reducing contamination and ever-expanding landfills. "Bottle brick" technology was developed in Latin America and India as a method whereby plastic bottles are packaged with mud or sand to be used as a replacement for actual bricks in construction. It is much cheaper and more durable than bricks and provides more heat shelter due to its thickness of its perimeter. Guatemalan schoolchildren are instructed in building "bottle schools" by filling plastic bottles with plastic wrappers. The bottles are stacked and covered in cement and serve primarily as insulation, rather than structural support, and provide the students with classrooms (Hopkins, 2014). See Table 1.4.

In Lagos, Nigeria, human waste is converted into biogas through the relatively simple addition of inexpensive retrofit entry waste pipes to existing underground septic tanks. The waste pipes remove oxygen from the decaying waste and produce a combustible gas. It solves the problem of airborne disease and provides self-produced energy to fuel Nigerian households (Webster, 2011).

Brazilian companies SOL Embalagens and Braskem innovated a new concept of disposable boxes for food transport (particularly for exports), since bacteria proliferation and breakage were a common issue with existing boxes and compliance with sanitary standards was at stake. The new

Table 1.4 Sustainable innovations

Innovation	Description	Country
Hug It Forward	Bricks of garbage-filled bottles	Guatemala
Polypropylene Box	Disposable refrigeration box	Brazil
Human Biogas	Energy from human sewage	Nigeria
Hippo Water Roller	Efficient transport of water	South Africa
Advantix AC	AC using salt and water to cool air	Israel
Easy Latrine	Simple, low-cost toilet	Cambodia
Bottle Light	Lightbulb created by bottle filled with water and bleach	Brazil
Mitticool Fridge	Nonelectric clay refrigerator	India
Soletek	Multisource renewable heating	Estonia
WiseSoil	Accelerator transforming organic waste into energy	Russia

boxes are easy to assemble and made of 100 percent polypropylene. They are light, extremely durable, and bacteria resistant. They have an on-screen injection structure using heat exchange for cooling, leading to lower power consumption than traditional cooling boxes. They are easy to clean and can be reused over and over.

Digital innovation

In Turkey, personal signatures are much more widely required than in other countries. Turkish law mandates that a signature is collected for processes such as receiving paychecks, holiday requests, and corporate travel reservations. In 2004, Turkey adopted a law giving electronic signatures the same legal status as "wet signatures." In response, Turkcell innovated a secure mobile signature solution service called "MobilImza." It has become very popular and has spread to other countries such as Moldova and Finland (MobilImza et al., 2012).

Disruptive innovation

A disruptive innovation is one that spurs new market creation and value networks, and eventually disrupts an existing market – although this may take several years or even decades. It leads to displacement of an earlier technology. Disruptive innovations are those that improve a product or service in ways that the market does not anticipate, usually first by designing for a different set of consumers in a new market and later by lowering prices in the existing market. Harvard Business School professor Clayton Christensen is the architect of and the world's foremost authority on disruptive innovation, a concept that has had a monumental influence on business executives and academic researchers. Although focusing on developed nation markets, his book *The Innovator's Dilemma* applies equally to emerging markets (Christensen, 2011).

Undisrupted industries are characterized by products and services that are so costly and complex that only educated people of financial means have access to them. On the other hand, disruptive innovation is the process by which complicated and costly products and services are transformed into simple, affordable ones, gaining a competitive advantage. Quality is gradually increased while maintaining a price advantage. These products may disrupt and compete with the market incumbents in the developed world depending on how the products develop over time and how the incumbents respond (Markides, 2012).

The health care industry is a prime example of an industry that is continuing to be disrupted by innovations which increase access and affordability. Portable medical devices similar to GE's mobile electrocardiogram machine are disrupting the health care industry. For example, Dr Hayat Sindi, a medical researcher from Saudi Arabia, founded the nonprofit institution Diagnostics for All, with the goal to create affordable and user-friendly diagnostic tools, primarily for the developing world. Dr Sindi was involved in the development of a device for early detection of breast cancer and is working on developing a portable MRI scanner (Atwood, 2013).

Another example is that of Toyota who started out creating inferior, low-frills cars such as the Corona and the Tercel models. US automobile manufacturers did not pay a lot of attention to the new Japanese competitor, which focused on a segment from which American manufacturers did not reap a lot of profit. However, slowly Toyota began improving their products, manufacturing higher-quality cars while maintaining a price advantage, eventually disrupting the American automobile industry. Over time, Toyota launched the Lexus range and began competing even with German luxury automobile brands such as Mercedes, BMW, and Audi (Togo and Wartman, 1993).

Typically, the traditional leaders in an industry become victims of disruptive innovation, rather than being initiators of it. Business leaders often view disruption and change as a threat which they need to respond to, taking a reactive rather than a proactive approach. Existing players that want to remain competitive need to be open to innovation and initiating experimentation with new concepts that may be disruptive but lead to long-term gain.

The rapidly growing Hungarian presentation software company Prezi was founded in 2009 and is redefining presentations. It introduced users to a dynamic, nonlinear approach, said to be conducive to memorization. It remains to be seen if Prezi will effectively disrupt the presentation market leader PowerPoint (Butcher, 2014).

Polycentric innovation

Multinational corporations (MNCs) are increasingly outsourcing R&D to emerging markets as globalization and Internet network opportunities continue to grow. This outsourcing trend is known as polycentric innovation and the benefits are manifold. Emerging markets like India and China have enormous and continuously growing workforces, which are highly educated and available at a much lower cost than their Western counterparts. By basing R&D locally in emerging markets, MNCs are able to innovate according to the demands of nearby developing countries. Local innovators are familiar with the needs and wants of these economies and have an enhanced understanding to aid in creating innovations that are relevant to their market.

Microsoft's second-largest R&D facility is located in Beijing, which is their main research hub in the Asia-Pacific region. According to the Microsoft website, the Beijing lab is home to more than 230 researchers and developers and the host for more than 250 guest scientists and students. Microsoft also has smaller labs in other emerging markets such as India and Egypt.

Close ties are created between the MNCs and the emerging markets where the R&D is distributed. Trust, psychological ownership, and political connections are generated, as vital R&D functions are entrusted to emerging markets and job opportunities are created. Lastly, frugal innovations created in emerging markets can be imported back to the West, enabling basic needs to be met on a larger scale, while resulting in attractive profits for the MNCs.

For instance, Xerox is one major MNC which has realized the vast potential of opening an R&D center in Chennai and strategically building innovation networks with local incubators and technology and science institutes (Radjou et al., 2010). Cisco has also invested heavily in creating a global development center outside of the US in India. The development center in Bangalore employs more than half of all Cisco employees and spans over 1 million square feet. The R&D center is responsible for creating disruptive business models to create new channels, processes, and technologies for emerging markets. Cisco is vying to become the foremost IT services company in India by 2016, led by their Bangalore development center (Menezes, 2013).

According to Navi Radjou, MNCs generally implement the polycentric innovation model in four stages. Initially, the driver is low-cost talent, to be subsequently replaced as a driver by the opportunities in the emerging market. At stage 0, most R&D activities are located in the West. The MNC's primary operations in emerging markets are sales and marketing. At stage 1, the MNC begins to move some of its R&D to emerging markets like India, which offer high-quality professionals at low cost. At this stage, R&D is mostly focused on innovating for the West. A typical example of this is the outsourcing of Western IT services to India. At stage 2, the MNC realizes the vast opportunities available in emerging markets and creates separate R&D projects focused on the needs of these markets. However, profit and loss (P&L) functions remain in the West. At stage 3, the MNC begins to capitalize on the synergy potential between R&D in the West and in emerging markets. As a result, new products and ideas are born. P&L activities are progressively transferring to emerging markets. At the final stage, P&L responsibilities are entirely transferred to and owned by the R&D units located in emerging markets (Radjou, 2009).

Innovation hubs in emerging markets

Innovation in emerging markets is specifically concentrated in "clusters." Three good examples are Bangalore in India, Shanghai in China, and Skolkovo in Russia. These centers of innovation arise when synergy is developed between different sectors such as the public, private, civil, and academic sectors. In a 2012 survey conducted by KPMG, more than 40 percent of technology executives responded "it is likely for the world's technology innovation center to shift from Silicon Valley to another country in the next four years" (KPMG Technology Innovation Survey, 2012). Respondents to KPMG's Technology Innovation Survey 2012 identified

China as possessing the highest potential to be the next world innovation center. Around the world, innovation hubs are popping up, many within emerging markets, that prove attractive due to cheap qualified labor and huge growth opportunities.

During the late 1980s, Shanghai started to transform into the rapidly growing modern financial and commercial center of China, as the Chinese government sought to promote the city and encourage investment. It became the natural epicenter of innovation in China. At the core of innovation is the Shanghai Technology Innovation Center (STIC), a nonprofit institution founded in 1988 and led by the Shanghai municipality. Its tenets are to "transfer technology achievements, incubate technology enterprises, and cultivate technology entrepreneurs."

Shanghai Technology Innovation Center has also established a dozen technology business incubators. Numerous MNCs, such as Hershey, PepsiCo, and Medtronic, have set up their own innovation centers in Shanghai. The Chinese government is proactively working on a profound economic overhaul that is being tested out in Shanghai. A free trade zone is being created in Shanghai, promoting full currency convertibility, interest rate liberalization, and innovations within the financial sector (Barboza, 2013). China's strength is innovation by commercialization. Products are tested in the markets very early on and imperfections are accepted. This is quite different from the West, where innovations are tested and perfected over many rounds before making it to the market.

In 2013, the Hershey Company announced the opening of a 22,000-square foot R&D center in Shanghai. China is Hershey's fastest-growing market. Their goal is to develop, test, and launch new products that appeal to the Chinese and Asian markets. Hershey's plan is to mainly hire locally, tapping into the regional tastes and forging close relationships with local universities and dealers (*Business Wire*, 2013).

Bangalore is home to India's IT sector and has been popularly dubbed the "Silicon Valley of India." The vision of an innovation hub was created in the 1970s when real estate outside of Bangalore was allocated for the construction of an "electronic city." It took off when Texas Instruments opened an office in the mid-1980s. During the 1990s, India went through an economic liberalization as Bangalore grew rapidly into one of the largest global innovation and technology hubs. The main advantages of Bangalore are its inexpensive and highly educated English-speaking workforce, making it ideal for outsourcing and back-office services (Chengappa, 2011).

In Russia, the Skolkovo Innovation Center was established outside of Moscow in 2010 to serve as a scientific and technological innovation center by concentrating global intellectual capital. It is mainly funded through the Skolkovo Foundation, a nonprofit organization supported by the Russian government. The idea is to create a culture of entrepreneurship and reduce Russia's dependency on natural resources. Skolkovo has been granted special laws and tax benefits to make it attractive for foreign companies, and investments have been made in residential and social infrastructure. Skolkovo even has its own border controls. It is home to the Skolkovo Institute of Science and Technology (SkTech) and numerous incubators, as well as Skolkovo Technopark offering R&D centers for established companies and technology services to start-up companies. Skolkovo is organized into five different clusters, each focusing on innovation within a certain arena. These clusters are IT, space, biomed, energy efficiency, and nuclear technology. Skolkovo is still in the early stages, and major challenges include corruption, bureaucracy, and uncertain political support. In 2011, IBM agreed to establish a science and technology center within Skolkovo. The partnership with the Skolkovo Foundation is aimed at benefiting both parties by creating solutions for the oil and gas industry, mobile payments, and road transportation safety (Luhn, 2013).

Central Mexico is home to a thriving automobile innovation cluster spanning multiple major cities such as Aguascalientes, Cuernavaca, and San Luis Potosí. Nissan has been assembling vehicles in Mexico for more than 50 years and within the near future new Mexican plants will manufacture luxury vehicles such as Audi, Mercedes-Benz, and BMW. As of 2014, Mexico produces more cars than Brazil (Johnson, 2014). Mexico offers a competitive combination of infrastructure, inexpensive quality labor, skilled engineers, and numerous free trade agreements. In addition, the geographical proximity to the US market offsets lower Chinese wages by drastically reducing transportation time and cost. Mexico is also conveniently located close to the growing South American market and relatively close to the European market compared to Asia. According to the World Bank, FDI in Mexico more than doubled between 2008 and 2013 (World Bank, 2014).

Start-up proliferation

In addition to major global innovation hubs, there is a global trend of start-up proliferation in the form of incubators providing ecosystems and support for promising new entrepreneurs. Many governments in emerging markets recognize the benefit of attracting and encouraging start-ups to further their national progress and development. Joint ventures between governments and private companies are created, providing a framework of financial, legal, and technological support for start-ups, and acting as accelerators.

The Chilean capital Santiago has a dynamic and thriving start-up scene supported by the government through initiatives such as "Start-Up Chile" which launched in 2010. This incubator program provides grants and temporary work permits for handpicked foreign start-up companies with potential. The open data platform Junar was founded by two Argentine graduates of the program and successfully secured $1.2 million in financing as well as reaching the semifinals in an MIT entrepreneurship competition (Geromel, 2012). Other innovation initiatives in Santiago include incubators such as

Innova as well as hosting global innovation conferences and hackathons. The city was dubbed "Chilecon Valley" by *The Economist* in 2012 and programs have been created for entrepreneurs to give back to the community.

Start-Up Brazil is a government-supported program similar to that of Start-Up Chile. It matches domestic and foreign start-ups with mostly private accelerators. Brazil is the home of several start-up communities in cities such as Belo Horizonte, Campinas, and Curitiba, besides the large cities São Paulo and Rio de Janeiro. Belo Horizonte's innovation cluster, dubbed "San Pedro Valley," provides start-ups with an infrastructure of support networks and financing (*The Economic Times*, 2013). It is also the home of BH-TEC, a technological park resulting from a joint venture between players such as the government, local universities, and industrial federations. Campinas, outside Sao Paulo, is known as one of Brazil's main cradles for tech start-ups, largely due to the city having several prominent universities heavily focused on entrepreneurship. E-commerce and f-commerce (Facebook commerce) are hugely popular in Brazil. They are led by start-ups such as LikeStore that provide vendors a merchant platform to sell their goods on Facebook through an app (Heim, 2011).

Singapore is the home of one of Asia's most prominent incubator scenes. The English-speaking city is business friendly and attractive as a gateway to the Asian market. Incubator fund Golden Gate Ventures is investing heavily in facilitating partnerships between Asia and Silicon Valley. One of its general partners Jeffrey Paine has created an intermediary service to help people connect with start-ups and investors in Asia's many start-up communities. It currently covers communities in Malaysia, Vietnam, Thailand, the Philippines, Indonesia, Taiwan, and Sri Lanka (Wee, 2012). Paine also founded the successful start-up Pyrks, offering small businesses corporate savings and perks by pooling them and creating a network.

Geoarbitrage/tropicalization

As the venture capital industry invests more and more in emerging markets, "geo-arbitrage" or "tropicalization" is taking off. It is the practice of financially supporting start-up companies that are taking an established and proven business model and adapting it for success in an emerging market. Emerging markets offer higher potential returns and growth prospects. It is also more likely that an established business model will succeed again in another setting such as an emerging market. There is also less competition in emerging markets than in the developed world. Concept business ideas like the discount sale websites Groupon and Gilt have been cloned in countries like China (Lashou and Letao) and Turkey (Trendyol). Baidu is a successful Chinese copy of Google. Ozon is the Russian clone of Amazon. However, cloned businesses are vulnerable and run the risk of losing market share if the original company enters the emerging market. This happened to Facebook replica Sonico, which was very big in Latin America before Facebook entered the market (*The Economist*, 2012).

Summary

Innovation in emerging markets takes many different forms and institutionally is driven, initially, by necessities and frugal solutions that are flexible and functional. However, increasingly more and more Western corporations are opening R&D centers in emerging markets, tapping into low-cost talent and the growing market. Innovation is situated most often within clusters – economic agglomeration areas – where business, government, academe, and supporting institutions fuel the dynamos of new business development and knowledge creation.

This volume addresses innovation in the global environment, with a focus on emerging markets – where tremendous growth and opportunity reside for the present and future. An introduction to the various facets and dimensions of innovation, along with the forces and drivers that will continue to shape innovation in emerging markets, will be followed by case studies of innovation in emerging market regions and we conclude with a synthesis and summation of the key features and outlook for innovation over the next decade. The framework that will be presented, and within which each regional chapter will be posited, is threefold: (1) innovation as national policy; (2) facilitating institutions – universities, research labs, accelerators and incubators, and business associations; and (3) firm-level innovation – products, services, processes, and business models.

The cases and frameworks presented in this book introduce the state-of-the-art thinking surrounding innovation by calling upon the experiences of experts in relevant fields. We aim to provide a deeper understanding of the core issues for *any* person involved in the private sector, academia, or public policy and intrigued by the challenge and opportunity of innovating in this dynamic global world.

Notes

1. Restauri, Denise. "A Harvard Woman Is Blowing Up the $55 Billion Beauty Industry with 3D Printed Makeup." *Forbes.* 30 June 2014. Web. 5 Jan. 2015.
2. Browne, Andrew. "Alibaba IPO: Innovation Chinese Style." *The Wall Street Journal.* Dow Jones & Company, 16 Sept. 2014. Web. 5 Jan. 2015.
3. "Top 20 R&D Spenders 2005–2014." *Top 20 R&D Spenders 2005–2014.* PricewaterhouseCoopers, n.d. Web. 5 Jan. 2015.
4. Kanani, Rahim. "Business Model Innovation Is the Fastest Path to Greatness." *Forbes.* 2 Oct. 2012. Web. 5 Jan. 2015.
5. "Bcg.perspectives – Most Innovative Companies." *Bcg.perspectives – Most Innovative Companies.* Boston Consulting Group, n.d. Web. 5 Jan. 2015.
6. Ibid.
7. Crisp, Nigel. "Mutual Learning and Reverse Innovation – Where Next?" *Globalization and Health* 10 (2014): 14. *PMC.* Web. 5 Jan. 2015.
8. UNCTAD, *Global Value Chains: Investment and Trade for Development.* Geneva: United Nations Conference on Trade and Development. 2013. Print.

9. maribus. "World Ocean Review," Chapter 8 "Maritime Highways of Global Trade," mareverlag, Hamburg, 2010.
10. *World Trade Report 2008: Trade in a Globalizing World.* Rep. World Trade Organization, 9 July 2008. Web, p. 82.
11. "Poverty Overview." *Poverty.* The World Bank Group, n.d. Web. 4 Jan. 2015.
12. Ezell, Stephen. *Mexico Innovation Policymakers' Forum,* presentation at the Woodrow Wilson International Center for Scholars, Washington, DC, 20 Nov. 2014.
13. Brinkmann, Paul. "Meet Venture Hive's New Spring 2014 Class of Tech Startups." *South Florida Business Journal.* American City Business Journals, 15 Jan. 2014. Web.
14. <http://www.chinatorch.gov.cn/english/xhtml/index.html>
15. Anthony, Scott. "The New Corporate Garage." *Harvard Business Review* (2012): n.p. Harvard Business Publishing, 1 Sept. 2012. Web. 5 Jan. 2015.
16. "Innovation: An A–Z of Business Quotes." *The Economist.* 17 Aug. 2012. Web. 5 Jan. 2015.
17. Schwab, Klaus, Xavier Sala-i-Martin, and Borge Brende. "The Global Competitiveness Report, 2013–2014." *Global Competitiveness and Benchmarking Network* (2013): n.p. World Economic Forum. Web.
18. "The World Turned Upside Down." *The Economist.* 17 Apr. 2010. Web. 5 Jan. 2015.
19. Cornell University, INSEAD, and WIPO (2014): *The Global Innovation Index 2014: The Human Factor in Innovation,* second printing. Fontainebleau, Ithaca, and Geneva, p. xxiv.
20. "R&D Policy: Switzerland." The World Bank Group, n.d. Web. <http%3A%2F%2Fsiteresources.worldbank.org%2FECAEXT%2FResources%2F258598-1284061150155%2F7383639-1323888814015%2F8319788-1324485944855%2F07_switzerland.pdf%2C>, p. 68.
21. *2014 Global R&D Funding Forecast.* Rep. Battelle Memorial Institute, Dec. 2013. Web. <http://www.battelle.org/docs/tpp/2014_global_rd_funding_forecast.pdf>, p. 7.
22. "The Top Innovators and Spenders." *Innovation 1000 Study.* Pricewaterhouse Coopers, 2014. Web. 5 Jan. 2015. <http://www.strategyand.pwc.com/global/home/what-we-think/global-innovation-1000/top-innovators-spenders>
23. Cornell University, INSEAD, and WIPO, *The Global Innovation Index 2014,* p. 4.
24. Ibid., p. 12.
25. Crespi, Gustavo, and Ezequiel Tacsir. "The Effects of Innovation on Employment in Latin America." *Inter-American Development Bank* (2012): n.p. Dec. 2012. Web. <http://idbdocs.iadb.org/wsdocs/getdocument.aspx?docnum=37375940>, p. 18.
26. *IFC Jobs Report: Assessing Private Sector Contributions to Job Creation and Poverty Reduction.* Rep. International Finance Corporation, Jan. 2013. Web. <http://www.ifc.org/wps/wcm/connect/0fe6e2804e2c0a8f8d3bad7a9dd66321/IFC_FULL+JOB+STUDY+REPORT_JAN2013_FINAL.pdf?MOD=AJPERES>, p. 1.
27. Ibrahim, Mo. "Celtel's Founder on Building a Business on the World's Poorest Continent." *Harvard Business Review* (2012): n.p. Harvard Business Publishing, 1 Oct. 2012. Web. 5 Jan. 2015.
28. Jidenma, Nmachi. "How Africa's Mobile Revolution Is Disrupting the Continent – CNN.com." *CNN.* Cable News Network, 24 Jan. 2014. Web. 5 Jan. 2015. <http://edition.cnn.com/2014/01/24/business/davos-africa-mobile-explosion/>
29. "Why Does Kenya Lead the World in Mobile Money?" *The Economist.* 27 May 2013. Web. 5 Jan. 2015. <http://www.economist.com/blogs/economist-explains/2013/05/economist-explains-18>

30. Govindarajan, Vijay. "Innovation Is Not Creativity." *Harvard Business Review* (2010). Harvard Business Publishing, 3 Aug. 2010. Web. 5 Jan. 2015.
31. "WaterCredit: Bringing Microfinance to Water and Sanitation Sector." *WaterCredit.* Foundation Center, n.d. Web. 5 Jan. 2015. <http://washfunders.org/Finding-Solutions/Case-Studies/WaterCredit>
32. *Oslo Manual: Guidelines for Collecting and Interpreting Innovation Data.* Rep. Organization for Economic Co-operation and Development, 2005. Web, p. 49.
33. Dohn, Robert L. *The Business Case for Microgrids.* Rep. Siemens AG, 2011. Web. <http://w3.usa.siemens.com/smartgrid/us/en/microgrid/Documents/The%20business%20case%20for%20microgrids_Siemens%20white%20paper.pdf>
34. van Ark, B., L. Broersma, and P. den Hertog. 2003. *Services Innovation, Performance and Policy: A Review,* Synthesis Report in the Framework of the Project SIID (Structural Information Provision on Innovation in Services), The Hague.
35. Huet, Ellen. "Why Uber and Lyft Should Be Focusing Overseas." *Forbes.* 11 Sept. 2014. Web. 5 Jan. 2015. <http://www.forbes.com/sites/ellenhuet/2014/09/11/uber-lyft-slowing-growth-rate/>
36. Lindgart, Zhenya, Martin Reeves, George Stalk, and Michael S. Deimler. *Business Model Innovation: When the Game Gets Tough, Change the Game.* Rep. The Boston Consulting Group, Inc., Dec. 2009. Web. <https://www.bcg.com/documents/file36456.pdf>, p. 1.
37. The four forms or arenas of innovation cited here are not meant to be exclusive. There are many types of innovation which may also encompass networks, organization, channel, brand, and customer engagement. See: Larry Keeley et al., *Ten Types of Innovation: The Discipline of Building Breakthroughs.* New York: Wiley, 2013.
38. Browne, "Alibaba IPO."
39. Cornell University, INSEAD, and WIPO, *The Global Innovation Index 2014*, p. xvii.
40. Ibid.
41. "World Population Data Sheet 2013." *2013 World Population Data Sheet.* Population Reference Bureau, 2013. Web. 5 Jan. 2015. <http://www.prb.org/Publications/Datasheets/2013/2013-world-population-data-sheet.aspx>
42. "Beyond Economic Growth," Chapter 4. *Economic Growth Rates.* The World Bank Group, n.d. Web. 5 Jan. 2015. <http://www.worldbank.org/depweb/english/beyond/global/chapter4.html>
43. Cornell University, INSEAD, and WIPO, *The Global Innovation Index 2014.*
44. "Affordable Products." *Affordable Products.* Unilever, n.d. Web. 5 Jan. 2015. <http://www.unilever.com/sustainable-living-2014/enhancing-livelihoods/understanding-our-economic-impacts/affordable-products/>
45. "Turkcell Selects Gemalto for World's Largest Mobile Signature Rollout." Gemalto, 2 Apr. 2007. Web. 5 Jan. 2015. <http://www.gemalto.com/press/Pages/news_164.aspx>

References

Attwood, Ed. (2013). "Saudi Arabia's Dr Hayat Sindi Appointed to Senior UN Role." *Arabian Business.* Retrieved from http://www.arabianbusiness.com/saudi-arabia-s-dr-hayat-sindi-appointed-senior-un-role-524006.html

Barboza, David. (2013). "Experimental Free-Trade Zone Opens in Shanghai." *NY Times.* Retrieved from http://www.nytimes.com/2013/09/30/business/international/experimental-free-trade-zone-opened-in-shanghai.html

Business Wire. (2013). "Hershey Opens Asia Innovation Center in Shanghai." Retrieved from http://www.businesswire.com/news/home/20130522005033/en/Hershey-Opens-Asia-Innovation-Center-Shanghai#.VCWI7vldV8E

Butcher, M. (2014). "Prezi Puts Out 10 Mn Users inside 5 Months, Hits 40 Mn." *TechCrunch.* Retrieved from http://techcrunch.com/2014/04/09/prezi-puts-on-10m-users-inside-5-months-hits-40m/

Carlson, N. (2013). "Lyft, a Year-Old Startup that Helps Strangers Share Car Rides, Just Raised $60 Million from Andreessen Horowitz and Others." *Business Insider.* Retrieved from http://www.businessinsider.com/lyft-a-startup-that-helps-strangers-share-car-rides-just-raised-60-million-from-andreessen-horowitz-2013-5#ixzz3E9EJLIbZ

Chengappa, B. (2011). "The Origin and Evolution of India's Silicon Valley." *Daily News and Analysis.* Retrieved from http://www.dnaindia.com/analysis/column-the-origin-and-evolution-of-india-s-silicon-valley-1606305

Christensen, C. M. (2011). *The Innovator's Dilemma.* Boston: Harvard Publishing.

Economic Times, The. (2013). "14 Startups to Look Forward to in 2014." *The Economic Times.* Retrieved from http://economictimes.indiatimes.com/slideshows/biz-entrepreneurship/14-startups-to-look-forward-to-in-2014/zomato/slideshow/27778006.cms

Economist, The. (2012). "VC Clone Home." *The Economist.* June 2. Retrieved from http://www.economist.com/node/21556269

Fengler, W. (2012). "How Kenya Became a World Leader for Mobile Money." Retrieved from http://blogs.worldbank.org/africacan/how-kenya-became-a-world-leader-for-mobile-money

Fong, M. (2009). "Clinique, Sony Star in Web Sitcom." *Wall Street Journal.* Retrieved from http://online.wsj.com/articles/SB123810039778551339

Geromel, R. (2012). "Start-up Chile: Importing Entrepreneurs, to Become the Silicon Valley of Latin America." Retrieved from http://www.forbes.com/sites/ricardogeromel/2012/10/05/start-up-chile-importing-entrepreneurs-to-become-the-silicon-valley-of-latin-america/

Gicheru, M. (2012). "Mi-Fone's Mi 3000 is Africa's First Feature Phone with a 60 Day Battery." *TechWeez.* Retrieved from http://www.techweez.com/2012/08/05/mi-fones-mi-3000-is-africas-first-feature-phone-with-a-60-day-battery/

Govindarajan, V. (2012). "The Case for Frugal Thinking." *BusinessWeek.* Retrieved from http://www.businessweek.com/articles/2012-04-17/the-case-for-frugal-thinking

Heim, A. (2011). "How e-Commerce Is Growing in Brazil." *The Next Web.* Retrieved from http://thenextweb.com/la/2011/08/29/how-e-commerce-is-growing-in-brazil/

Hopkins, R. (2014). "EcoBricks and Education: How Plastic Bottle Rubbish Is Helping Build Schools." *The Guardian.* Retrieved from http://www.theguardian.com/lifeandstyle/2014/may/29/ecobricks-and-education-how-plastic-bottle-rubbish-is-helping-build-schools

Johnson, T. (2014). *Seattle Times.* "Mexico on Road to Becoming Carmakers' Nuevo Detroit." Retrieved from http://seattletimes.com/html/businesstechnology/2024604461_mexicomajorcarmakerxml.html

kLab. (2015). Retrieved from http://klab.rw/public/about

KPMG (2012), KPMG Technology Innovation Survey. New York: KPMG.

LeBlanc, R. (2014). "Three Ways Cloud Computing Is Driving Rapid Innovation." *Forbes.* Retrieved from http://www.forbes.com/sites/ibm/2014/09/02/three-ways-cloud-computing-is-driving-rapid-innovation/

Luhn, A. (2013). "Not Just Oil and Oligarchs." *Slate.* Retrieved from http://www.slate.com/articles/technology/the_next_silicon_valley/2013/12/russia_s_innovation_city_skolkovo_plagued_by_doubts_but_it_continues_to.html

Madhavan, N. (2010). "Made in India, for the World." *Business Today*. Retrieved from http://businesstoday.intoday.in/story/made-in-india,-for-the-world/1/5601.html

Markides, C. (2012). "How Disruptive Will Innovations from Emerging Markets Be?" *MIT Sloan Management Review*. Retrieved from http://sloanreview.mit.edu/article/how-disruptive-will-innovations-from-emerging-markets-be/

McNeil Jr., D. (2013). "Car Mechanic Dreams Up a Tool to Ease Births." *NY Times*. Retrieved from http://www.nytimes.com/2013/11/14/health/new-tool-to-ease-difficult-births-a-plastic-bag.html?_r=0

Menezes, B. (2013). "Telecom and IT are Key to Cisco's India Goal of No.1 by 2016." *Daily News and Analysis*. Retrieved from http://www.dnaindia.com/money/report-telecom-and-it-are-key-to-ciscos-india-goal-of-no1-by-2016-1932546

MIT Technology Review. (2013). "World Innovation Clusters." *MIT Technology Review*. Retrieved from http://www.technologyreview.com/news/517626/infographic-the-worlds-technology-hubs/

MobilImza, GSMA Mobile Identity Team and Turkcell. (2012). "Turkcell 'MobilImza,' Mobile Signature in Turkey, a Case Study of Turkcell." Retrieved from http://www.gsma.com/personaldata/wp-content/uploads/2012/09/MI_TurkcellReport_print_FINAL.pdf

Mozur, P. (2014). "China Blocks Google Ahead of Tiananmen Anniversary." *Wall Street Journal*. Retrieved from http://online.wsj.com/articles/ahead-of-tiananmen-anniversary-many-of-googles-services-blocked-in-china-1401773338

O'Connor, C. (2012). "Reverse Innovation's Big Impact for Consumers." Retrieved from http://www.boston.com/business/blogs/global-business-hub/2012/12/reverse_innovat.html

Pilling, D. (2010). "Mismanaging China's Rural Exodus." *Financial Times*. Retrieved from http://www.ft.com/intl/cms/s/0/c6ed2e24-2c78-11df-be45-00144feabdc0.html#axzz3EQwHnYLo

Prabhu, J. and Ahuja, S. (2012). *Jugaad Innovation*. San Francisco: Jossey-Bass.

Radjou, N. (2009). "Polycentric Innovation: The New Global Innovation Agenda for MNCs." *Harvard Business Review*. Retrieved from http://blogs.hbr.org/2009/11/polycentric-innovation-the-new/

Radjou, N. Prabhy, J. Kaipa, P. and Ahuja, S. (2010). "How Xerox Innovates with Emerging Markets' Brainpower." *Harvard Business Review*. Retrieved from http://blogs.hbr.org/2010/08/how-xerox-innovates-with/

Rai, S. (2007). "In India, Grandma Cooks, They Deliver." *NY Times*. Retrieved from http://www.nytimes.com/2007/05/29/business/worldbusiness/29lunch.html?pagewanted=all&_r=0

Togo, Y. and Wartman, W. (1993). *Against All Odds: The Story of the Toyota Motor Corporation and the Family that Created It*. New York: St Martins Press.

Walker, A. (2009). "Balancing Tradeoffs: The Evolution of the Hippo Roller." *Fast Company*. Retrieved from http://www.fastcompany.com/1309505/balancing-tradeoffs-evolution-hippo-roller

Webster, G. (2011). "How Human Waste Could Power Nigeria's Slums." Retrieved from http://www.cnn.com/2011/09/26/world/africa/nigeria-sewage-biogas/

Wee, W. (2012). "Golden Gate Ventures: $10 Million Fund for SEA Startups Launches Today." *Tech in Asia*. Retrieved from http://www.techinasia.com/golden-gate-ventures/

World Bank, The. (2014). "Investing across Borders Database." Retrieved from www.iab.worldbank.org

2
Releasing Trapped Value: The Coming Challenge of Innovation in the Context of Emerging Markets

Bhaskar Chakravorti, Graham Macmillan, and Tony Siesfeld

Introduction

It is commonly accepted that to succeed in emerging markets requires extraordinary focus on the growing ranks of new consumers and their idiosyncratic needs. Innovators with an eye towards serving this growing market will naturally place a strong emphasis on tailoring products, processes, and business models and will be guided by the three mantras of affordability, accessibility, and appropriateness – to make products cheaper, easier to distribute, and suitable for the local environments. In a sense, we can describe such initiatives as innovation for the core, i.e. for the attractive market segments being created because of the new consuming class. In this chapter, we argue that there is an urgent need to focus, in parallel, on innovating in a different direction. Specifically, we argue that it is essential to innovate, not just to serve the core, but to fill gaps in the broader context that surrounds the core opportunity as well.

The term "context" encompasses many elements, from institutions to infrastructure and other public goods, to the state of natural resources, to factors that affect the quality and state of the human condition: health, education, demographics, talent, skill development, etc. The core and the context are connected. The many incomplete elements of the context, we argue, trap a significant portion of the potential value represented by the core consumer opportunity – estimated to be as high as $30 trillion by 2025 by McKinsey & Company. Innovative initiatives, aimed at closing the gaps in the context, hold the potential to release three sources of "trapped value" in emerging markets.

While the precise extent of trapped value is difficult to quantify, the realities of the context in emerging markets indicate that it could have a substantial impact. To explore how core value gets trapped by the context, consider the primary sources, as follows.

Weak or missing institutions

The developing world context is characterized by myriad institutional gaps: missing political and legal systems, and civil society organizations; unavailability of adequate mechanisms for securing financial services; serious deficiencies in the state of the human condition: health, education and skill-building, living conditions, security (internal or with their neighbors or with respect to conflict with nonstate actors or terror groups), and human and property rights; the presence of corruption; and inadequate systems of governance and transparency. In the absence of such institutions, value is lost because of various inefficiencies and eventually in many emerging markets the pace of growth itself is likely to stall. (Our use of the term "institutions" is more specific than the usage initiated by Khanna and Palepu (2010) in the emerging markets literature.)

Without institutions, there are insufficient "guardrails" which maintain business operations within globally acceptable standards of strategic conduct and exercise of market power by companies. The lack of such guardrails can give rise to situations where firms with vastly greater market power have negotiated potentially unsafe and unviable terms with their suppliers or have done so under nontransparent conditions. The collapse of the Rana Plaza garment factory in Bangladesh in 2013, for example, was a particularly poignant and grim reminder of the risks associated with underinvestment in institutions and the associated guardrails.

From the perspective of business, the absence of institutions translates into the prevalence of chronic challenges, such as corruption, difficulty of doing business, heightened political risk, etc. Widely accepted indices and ratings indicate that the emerging markets, in general, fare poorly on all of these dimensions (International Chamber of Commerce et al., 2008). As an example, consider what two widely used indices – the Corruption Perceptions Index (Transparency International) and the Ease of Doing Business Index (The World Bank) – suggest about the quality of institutions in five of the most prominent emerging markets that have recently agreed to establish a development bank as an alternative to the World Bank. These are the BRICS countries: Brazil, Russia, India, China, and South Africa:

Corruption Perceptions Index Rank:

Brazil: 69, Russia: 136, India: 85, China: 100, South Africa: 67 out of 175 countries.

Ease of Doing Business Index Rank:

Brazil: 120, Russia: 62, India: 142, China: 90, South Africa: 43 out of 189 countries.

These countries fare poorly in a segmentation of political risk as well. According to the Economist Intelligence Unit (*The Economist*, 2013), Brazil, China, and South Africa were at high risk of social unrest in 2014, while India and Russia were rated medium-risk.

In addition to the challenges posed by the myriad institutional voids, most emerging markets are reliant on a limited number of industries as their economic engines. Of these, many are disproportionately reliant on natural resources, such as oil or minerals, sectors that are vulnerable to cyclical variations. The absence of policy-led mechanisms to smooth out the inevitable swings in fortune when these industries face a downturn and the associated problems of a "Dutch disease" and a resource curse suggest that natural resource dependency in a great many of these countries can stunt the development of institutions and other job-creating sectors, such as manufacturing.

Insufficient investment in sustainable practices

The rapid pace of growth in emerging markets will inevitably result in pressures on natural resources, raw materials, and the environment. Consider the case of Yum! brands. It derived about half of its revenues from China and faced a sharp decline in 2013 in the Chinese market because of the inability of its suppliers to keep up with demand, leading to an overuse of antibiotics in the local poultry supply chain. The company's ability to grow in additional markets also ran into sustainability challenges. For example, the growth of its KFC franchise in sub-Saharan Africa was limited by the local lack of modernized poultry farming practices.

There are more traditional sustainability challenges as well, as they relate to parts of the supply chain that rely on natural resources. Consider Coca-Cola and the impact on its business of water-related issues. Droughts, unpredictable weather patterns, and more frequent major floods are perennial threats to Coca-Cola's supply of key ingredients – such as sugar cane, sugar beet, and citrus for its fruit juices – which are highly dependent on water supplies. Moreover, it is equally pertinent to keep in mind that the company's core product is fundamentally reliant on clean water supplies and is vulnerable to water shortages around the world.

Poor environmental quality is, frequently, an outcome of underregulated and "dirty" industries being overrepresented in the emerging markets. Pollution alone killed 8.9 million in 2012, making it the single largest cause of unnatural death on the planet, with 94 percent of the people who become ill because of toxic air, soil, and water each year living in these poorer countries (WorstPolluted, 2014). Such losses alone are devastating from a humanitarian perspective, but they also serve as a reminder that growth without investment in sustainability of the resources used can be self-limiting.

Insufficient inclusion of the wider society

Taking a rational approach to organizing business activity, growth in emerging economies is generally concentrated in certain sectors – by industry, by region, and even by cities and pockets within cities. In fact, in an age of data analytics, the targeting of market segments has become increasingly efficient and pointed. As noted earlier, growth in some countries was sharply focused on a few industries, such as natural resource extraction or exports in certain sectors. Besides such unevenness in growth profiles, consider the supporting systems, for example infrastructure development and essential investments in complementary sectors, which were focused on selective market segments, product development, and marketing, and the organization of commercial activity which was focused on certain industries, e.g. real estate, consumer products, and electronics, targeted at the emerging middle class. Much of this focused investment benefited a sliver of the overall population.

Over time, the value of such concentrated growth must be weighed against concerns about the inherent inequity of such lopsided development and the social and political risks that would naturally accumulate. In 2013 alone, protests erupted not only in such chronically troubled cities as Cairo or Kiev but also in cities such as Istanbul and São Paulo that had been celebrating high growth for an extended period of time.

In the paragraph above, we speak, of course, of disruption in a political sense. There is a form of disruption more familiar to a business readership that can cause growth to slow down from the perspective of a single firm. In every case where the strategy of the firm demands a choice regarding where not to play, there is an opportunity for an entrant to offer a stripped-down product that is "good enough," in the sense popularized by the work of Clay Christensen, aimed at the market that has been excluded (Christensen, 1997). Once the entrant has a foothold, experience, scale, and some revenues that help establish a war chest, it can steadily move into the segments that are adjacent to the core and eventually the core itself. Of course, the irony is that as the disrupter moves upmarket into the core, it faces the same challenges because it, too, must be exclusionary in order to pursue rational strategic practices; just as Samsung edged aside Apple in the mobile handset market in many emerging regions by moving upmarket from lower-income segments, Samsung itself is being challenged by Xiaomi.

All of this discussion suggests a key lesson: it is essential to strike a balance between attractive markets and those segments that would normally be excluded in accordance with traditional business strategy metrics. Consider, for example, identifying innovative ways of offering credit and other financial services to the unbanked populace in areas beyond the largest cities, or providing computing technologies to children in rural schools, or life-saving drugs to those afflicted by disease. These markets would appear unattractive at first glance – or even second glance. Yet they represent segments that will be important to serve because they represent prime opportunities for future growth.

Thus, we have three principal sources of trapped value because of insufficient investment in institutions, sustainability, and inclusion. Each area of trapped value represents an innovation challenge for the manager. In each case, an extraordinary new form of value release must occur and it must be done in an extraordinary new way because there are fundamental structural barriers that prevent sufficient levels of investment by businesses. The higher the barriers, the greater the level of creativity and innovation needed to overcome them.

In this chapter, we focus on the fundamental drivers and barriers that make it so difficult to invest adequately in institutions and sustainable and inclusive business activities (SIBA). As we investigate these underlying governing forces, we offer some of the emerging approaches that firms at the vanguard are adopting to address the issues and overcome some of the obstacles. Taken together, these approaches offer an early framework for others to follow.

Investing in institutions: challenges and potential remedies

All firms operating in a market feel the impact of weak and nonexistent institutions. The World Bank's estimates of the cost of corruption alone suggest that it adds up to more than 5 percent of global GDP, with over $1 trillion paid in bribes annually. Corruption adds up to 10 percent to the total cost of doing business globally and up to 25 percent to the cost of procuring contracts in the developing world. While the natural working assumption would be that the onus for establishing these institutions is on the public sector, it is clear that in many emerging markets the political actors do not have the will and, in many cases, the necessary capacity and resources to do so. Moreover, from the perspective of any single firm the fixing of institutional gaps is subject to a classic "tragedy of the commons" problem: the benefits of improving institutions accrue to the collective, whereas no firm can hope to solve the problem by itself; as a result, every firm has an incentive to free ride on the investments of others and a rational incentive to underinvest in institutions. The public goods nature of the problem of weak and missing institutions makes it one of the most difficult forms of trapped value to release. The natural attitude of most executives would be that these are the areas that ought to be handled by local governments, nongovernmental organizations, international agencies, and even by civil society. Otherwise, businesses would run the risk of getting involved in issues that are distracting, consume resources, and are far from their zone of competence and comfort.

The dilemma for the same executives is, of course, that in many emerging markets, there is a deficiency of institutions because the other sectors cannot or will not fill the gaps. The loss of value is felt by the society at large, but by the businesses as well, as their growth potential is adversely affected. Fortunately, all is not lost. Despite the many challenges, there are ways in which businesses can be creative and play an active role in helping close institutional gaps. The opportunity to do so depends largely on the nature of the context – and often with circumstances and levers that are predetermined and beyond a firm's unilateral control. We outline several approaches below; each of them offers some ideas about how a firm may participate in a mechanism to bypass the "tragedy of the commons" problem.

Engaging in public–private partnerships

There are some emerging markets that are sufficiently small or concentrated in terms of the growth industries that a single firm has a disproportionate amount of power to effect change. When the firm is de facto one of the more dominant entities in the country, it has a degree of political leverage that it can put to good use. Consider, for example, the operations of the agricultural trading company, Olam, in Gabon. Olam's business interests in Gabon include palm oil, rubber, and fertilizers, among others. Given its strong interest in overcoming the gaps created by the missing infrastructure, the company has a joint venture with the Gabonese government to build parts of the infrastructure and the associated ecosystems. Specifically, the Nkok special economic zone (SEZ) is a 60–40 partnership between the Republic of Gabon and Olam (for instance, see http://olamgroup.com/locations/west-central-africa/gabon/). Such public–private partnerships can prove to be potential solutions to closing gaps when public institutions have failed. For many firms such as Olam, filling such voids by integrating their response to the upstream challenges has become an important piece of core business strategy.

Using laws as guardrails and leverage

In many situations, there are legal parameters that can act as effective frameworks for firms to self-regulate and negotiate with partners. Such frameworks act both as constraints and as leverage to implement institutional change. Consider the impact of laws such as the US Foreign Corrupt Practices Act (FCPA) or the Dodd–Frank Wall Street Reform and Consumer Protection Act that determine how US listed firms legally engage in business transactions abroad or monitor their use of supplies from politically fragile areas. Many American firms routinely cite the FCPA as a shield against bribery and other forms of corruption in their negotiations and operations in many developing countries, where bribery is a norm in business transactions. Newmont Mining is one such firm and it has used the FCPA as a means to transform the way it conducts transactions in Africa. A telling quote is the following from the company's Director of Corporate and External Affairs Africa (Andersen, 2011):

Newmont's experience, particularly in Africa, has been that FCPA has been an enormously valuable protective device for us ... when you have a government person saying ... "we'll give you that license if you buy us a car or something" ... it's not about look "I'm a mean guy and I don't value our relationship, and therefore I'm not going to give it to you," you say "look, there's a law out there that means I'm going to go to jail if I do that, I'm not going to go to jail for you or anybody else."

Given the multifaceted and complex nature of the institutional gaps and their long histories, it is important to keep in mind that no legal lever provides a silver bullet.

Abiding by consensus-led principles

A third approach is offered by a broad set of principles orchestrated by a mutually respected body and agreed to by consensus among a wide set of firms and other key actors. The public awareness of such agreements and the reputational consequences of violating them help make them binding and reduce the incentive to be a free rider. As an example of such consensus-driven principles, consider the UN Guiding Principles on Business and Human Rights (often referred to as the Ruggie framework, after John Ruggie, the special representative to the UN and their principal author). These principles were unanimously approved by the UN Human Rights Council and agreed to by 387 company human rights policies, along with several other key institutions. These additional signatories include the World Bank, the OECD, the International Chamber of Commerce, and the European Commission.

Responding to crises through collective action

In some situations, there is need for more specific agreements to address targeted institutional gaps. These have often emerged after a crisis that acts as a call to action and simultaneously brings multiple industry players to the table. In many of these cases, the players are unwilling participants, but agree to a set of principles because of the pressures brought on senior management by external stakeholders, the media, and the force of public opinion.

We noted earlier the case of the Rana Plaza garment factory building collapse in 2013 that killed more than 1,100 Bangladeshi garment workers. After the international outrage over the poor factory safety and oversight conditions that led to the disaster and the awareness created about the widespread hazards and unenforced labor and safety laws in poorly regulated manufacturing facilities in Bangladesh and other developing nations, multiple consortia were formed to address the situation. Over 90 retailers, mostly non-US based, along with two labor unions, agreed to a five-year legally binding plan called the Bangladesh Fire and Safety Accord, requiring regular inspections and audits, compensation for repairs, and training by the signatories.

Complying with local government mandates

Requirements made by the public sector and legal systems in the emerging markets can help steer investments towards the closing of institutional gaps. India, for example, became the first country to mandate investments in corporate social responsibility (CSR) through the India Companies Act of 2013, which requires that companies set up CSR board committees, which must, in turn, ensure that the company spends at least 2 percent of the average net profits of the company made during the three immediately preceding financial years on "CSR" activities. Alternatively, there are certain industries in which operating practices can be subject to such mandates. In the aerospace and defense industry, firms often have to make large promises by way of "offset agreements" through transfer of technology, training local suppliers, or other forms of support to local industry or help with the recipient country's export sector.

Participating in precompetitive collaborations

Neutral third-party organizations that have convening power, such as the Gates Foundation or the World Bank, have often been successful in orchestrating precompetitive collaborations in agriculture, nutrition, and health and other sectors in countries where public sector services, guarantees, and regulations are missing. In addition to developing sustainable means of production and promoting products that might otherwise be ignored by the industry, such collaborations can play a critical role in closing institutional gaps by ensuring economies of scale, scope, and standards. In effect, industry players work together to help establish missing parts of industry value chains, capital, knowledge, essential resources, and some basic guarantees that might not otherwise exist in some regions.

In each of the situations outlined above, firms experience the challenge of unilaterally justifying an investment in addressing problems that are public goods. However, in each situation, there is a mechanism that militates against the natural incentive to free ride and offers hope for breaking out of the tragedy of the commons and releasing some part of the trapped value of weak or missing institutions.

Investing in sustainable and inclusive business activities: challenges and potential remedies

As businesses continue to enhance their presence in emerging markets, there is a growing realization that sustainable and inclusive activities are essential to future growth. Undertaking SIBA, and doing so with impact, requires innovation – creating extraordinary new value over the longer term, but requiring investment in extraordinary new ways of doing business in the near term.

As in the case of addressing the institutional gaps, investing in SIBA is complex and multifaceted. To pursue such activities consistently often involves a combination of new products and business models that are competitive in the marketplace while addressing wider contextual gaps associated with sustainability and inclusion. This, however, is difficult. As a result, most organizations have struggled to invest enough and have forgone opportunities, which is why underinvestment in SIBA represents a significant measure of trapped value.

To get a sense of what a combination of SIBA might consist of and why it is central to the innovation agenda for corporations doing business in emerging markets, consider the case of Mars Incorporated, a leading food manufacturer. The company innovates in both product and business model in order to pursue SIBA. To promote environmental sustainability and ensure access to future supply, it is applying genomics to enhance the productivity of small farmers, while moving towards 100 percent certified sustainable sourcing for most of its key agricultural raw materials. Moreover, it is working to ensure that certification does not constitute an economic burden for the farmers. Mars has also expanded its business model to include several key partnerships that give it added leverage to meet its objectives. For example, it has teamed up with the International Finance Corporation in Indonesia to set up the Cocoa Sustainability Partnership, a multi-stakeholder forum on cocoa collaboration, and with the Rainforest Alliance to train farmers in sustainable cocoa practices (Shapiro, 2014).

Yet, not all companies invest in SIBA or do so consistently. This raises the natural question: why do some companies invest in SIBA, while others do not, and yet others simply do not invest enough? The value of identifying the key drivers and barriers within firms is that it provides a basis for seeking solutions and actions to be recommended to release trapped value through SIBA.

To gain some insight into the drivers and barriers, we conducted primary research with 42 large companies with interests in emerging markets, which was complemented by secondary research of existing case studies and the literature. Of the companies that were surveyed:

- 50 percent reported revenue >$5bn
- 55 percent engaged in small, so-called "bottom of the pyramid (BoP)" enterprises as suppliers
- They operated across the value chain, and were multi-industry (63 percent operated in agriculture, finance and banking, or consumer products)
- Their operations covered eight regions around the world

In our investigations into the questions, *why* companies invest in SIBA initiatives and *how* they move from inception to implementation, we also set out to answer several questions that delve into managerial incentives, specifically the key enablers and barriers, and the managers' decision-making environments:

- *Language frameworks*: How do companies define SIBA? What language do they use to talk about these issues internally?
- *Motivation and analysis of the business case*: What rationale(s) do companies cite for developing these activities? How do they articulate the business case to gain approval and budgets?
- *Internal structuring and organization*: How do companies structure and support such initiatives internally? How do they address: investment time horizons, payoff uncertainty, profitability, reporting, etc.?
- *External structures, partnerships, and execution*: What additional capacity do companies have to develop to successfully execute and scale these initiatives? What varieties of partnerships enable them to succeed? What is the role of philanthropic and public support?

Our findings suggest that even as more companies invest in SIBA, this is still, at best, an evolving and emergent factor within the typical corporation. Though there are notable pockets of innovation, and perhaps even distinct segments (more on this below) among companies engaging in SIBA, as corporate practice, the space currently lacks cohesion, consistency, and sustained commitment from management.

In many instances, SIBA has had impact in a limited setting, but has not been deployed at scale. In other instances, commitment to SIBA varies with the firm's performance during upturns and downturns of the business cycle. In other cases, SIBA has been consigned to a CSR or sustainability department without integration into core business strategy and the profit and loss (P&L) calculations of the business units. As a result, SIBA investments and the returns associated with them are not considered to be business-driven, which automatically places them in a lower position in the organization's strategic and financial priorities.

The primary findings of our research are threefold: (1) language matters; (2) there are too many ways to make the business case within the organization; and (3) there are many barriers to investing in SIBA at scale, with corresponding strategies to address them. Let us consider each of these in turn.

Language matters

One significant insight gleaned from our research is the lack of common vocabulary among actors in this field. There has been a proliferation of buzzwords and terminologies – a clear sign of growing momentum, but no common and clearly bounded set of concepts have gained dominance. Academics, consultants, and development actors often use terms such as "inclusive business" and "shared value"; however our research shows that the business community has not embraced this language and there is, as yet, no alternative terminology or standard definition among companies. In fact, under one in five companies interviewed reported using these terms, and in some cases managers claimed that terms such as *inclusive business* or *social business* risk diminishing the strength of the business case for investment, and they avoid using them.

Despite these challenges, there is a growing recognition of the need for incorporating SIBA in a systematic way across a wide range of companies that are not only major players in their respective markets but also help set the tone across their industry.

Perhaps the most widely cited example is Unilever, the Anglo-Dutch consumer products giant. According to Unilever literature (see Unilever, 2015): "We believe that businesses like ours can and should play an important role in generating wealth and jobs around the world, improving skills and offering access to markets. We can also help by sharing technology and best practice." Even in firms in an industry such as high technology that compete to serve high-margin customers – primarily other businesses – and invest in R&D and acquisitions to retain competitive positions, SIBA appears to be making inroads. Consider the case of Cisco. According to its corporate literature (see Cisco, 2015), "Organic trickle down benefits take an inordinately long time to the larger population. Hence, inorganic means are required to accelerate the elevation of the capabilities of the population to integrate and partake in the overall economic development. This approach creates economic, political, and social stability and further accelerates the growth of the nation."

Given the breadth of industries and their very different orientations and motivations to engage in SIBA, clearly, a standard definition is needed. Despite the lack of a common vocabulary, our research shows that businesses are beginning to coalesce around *core elements* that ought to be included in a definition:

- *Strategic alignment*: Alignment with core business strategy and/or moral purpose of business (rather than ancillary philanthropy)
- *Commercial viability*: Importance of, at a minimum, cost recovery
- *Positive social outcomes*: Clear benefit for society and/or environment
- *Scalability*: Potential to scale within or across geographies
- *Execution through collaborative models*: Need to leverage partnerships, including those with private, public, and social sector players, wherever reasonable and feasible

Within these common elements, there is still room for debate and variation depending on each company's context. They include acceptable profit margins relative to other investment opportunities; the time horizon to cost recovery; the definition of what is "socially" or "environmentally" beneficial; and impact expectations, measurement, and reporting. Each business is different in the trade-offs it makes.

A key conclusion that can be drawn is that alignment on standardized terminology and language frameworks will improve communications within organizations, across functions as well as across sectors, enabling more effective collaboration and partnerships that accelerate the scaling of SIBA activities.

Too many ways to make the business case

Our research revealed the different journeys that companies have taken in initiating their investments in SIBA. Usually, the decision originated with an internal champion. One company narrated how theirs was a direct result of their CEO being questioned by Kofi Annan at Davos. Others pointed to the success of M-PESA and how the availability of public funds emboldened key managers to take on experiments they were reluctant to on their own. Other managers cited the need to manage supply chain risks in fragile regions. Not only were the motives for engaging in SIBA varied, but companies also tended to cite multiple motives. In our survey, companies cited an average of 4.6 motivations for engaging in SIBA, ranging from social impact to business case rationales. Three in five companies (61 percent) cited generating social or environmental impact among their top three motivations, while nearly all (95 percent) cited at least one business case rationale.

There is a wide variance among the business case rationales. Even though the majority of the motivations cited may be characterized as "defensive" (avoiding loss/mitigating risk) or "maintaining" (staying competitive/keeping up), there were frequent references to "affirmative" rationales as well (seeking growth/opening new markets).

The top business case motivations for SIBA included:

1. Preserve license to operate;
2. Avoid reputational damage;
3. Avoid future supply disruptions;
4. Maintain competitive position;
5. Respond to internal demand;
6. Differentiate products;
7. Capture revenues and build loyalty.

The relative importance of each motivation across the companies surveyed is indicated by Figure 2.1.

These results have several implications. First, there can be a fragmentation of the interests to promote SIBA, thereby making it difficult to build a critical mass of support behind it. Second, different organizations are galvanized by different kinds of motivations to place initiatives at the top of their investment priorities. SIBA advocacy by internal champions would have to

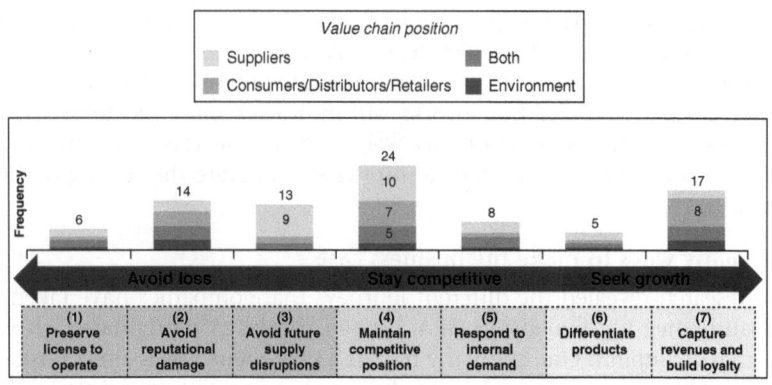

Figure 2.1   Frequency of citation of business case rationales
Source: Chakravorti et al. (2014).

dovetail with corporate imperatives in order for SIBA to garner the requisite levels of managerial and financial support. Third, depending on the primary motivation, the "ownership" of SIBA could be located in different places in the organization. This location could influence the degree of corporate support and visibility that SIBA enjoys. Finally, some motivations, more than others, may make it easier to quantify and communicate the impact of SIBA; for example, avoiding loss is easier to explain and quantify than one that involves a longer-term and more diffuse objective, such as growth. The overarching implication is that for a given company, senior management must make a choice – pick the primary business case logic and organize around it to get SIBA launched.

For some companies, picking the business case that will resonate within the organization and with its various stakeholders is easier than for others. For a chocolate producer, such as Mondelez for example, ensuring that cocoa farmers do well, are part of a flourishing community, and find it worth their while to grow cocoa are essential to the well-being of Mondelez itself. The case to take action to ensure that cocoa farmers are well-compensated is not hard to make for a company that relies so heavily on it as an essential raw material. However, many shareholders complained that the company management was overinvesting in SIBA and paying less attention to its sales performance (see Daneshkhu and Oakley, 2015). The "return on SIBA" is harder to demonstrate in this latter instance.

Barriers to investing in SIBA at scale with corresponding remedies

There are three primary barriers to pursuing SIBA at scale.

Absence of organizational home

The first of these barriers is the absence of a natural, long-term organizational "home" for SIBA. The majority of companies interviewed had not settled on how and where best to house SIBA over the long term, as it required new internal organizational structures, including compensation, talent management, and measurement systems, along with new business models and new partnerships.

Among the companies interviewed, there were different organizational models. One, a major cosmetics skin and hair care company, employed an "isolate and replicate" model, initiated by regional offices, with funding from the CSR budget at headquarters to enable piloting and successful deployment in one regional market followed by replication in other regions. Alternatively, a large diversified supplier of construction materials invested in an affordable housing program that was run on a "let a thousand flowers bloom" approach – a portfolio of experiments funded through a combination of a central innovation budget and budgets of local microfinance institutions. The projects were implemented in a highly decentralized manner. The "right" model varies with the nature of the company, the industry, and the core areas of SIBA activity; however, other than the CSR budgets and departments kept separate from the business units, SIBA integrated with core business strategy was still an activity in search of a permanent home.

A potential remedy is to create the "space" for SIBA colocated within the business units, with distinct decision rules, management incentives, budgets, and metrics. Two of the most important areas in which such SIBA-focused units need their own yardsticks are:

- *Differential investment time horizons*: Over half (53 percent) of companies reported longer time horizons to profitability for SIBA relative to commercial businesses, with only 17 percent of initiatives reaching profitability within three years.
- *Relief from quarterly P&L targets*: Three-quarters of SIBA receive funding from commercial budgets, clearly signaling returns expectations, but potentially limiting the ability to take on longer time horizons to prove the business case given the higher degrees of uncertainty associated with SIBA. There is a clear need to set up separate funds and even external sources of funds to derisk SIBA, enabling companies and business unit P&L managers to accept differential returns and time horizons.

Companies must continue to boldly experiment with new funding structures, reporting structures, success metrics (e.g. time horizons and profit expectations), and measurement schemas. Borrowing from the models used to incubate and scale up new venture units – along the lines of creating "ambidextrous" organizations – might offer a possible route to institutionalizing SIBA within traditional corporate structures.

For inspiration as well as to get a feel for alternative models that have worked, consider the advice from three CEOs (see McKinsey & Company, 2014) on how to foster innovative units within large organizations and manage ambidexterity. At Intuit, Scott Cook established a series of systems and a culture where if there was an idea that someone was passionate about, the company had a system that made it easy, fast, and cheap to run an experiment. According to Idealab's Bill Gross, the key was in equitization and autonomy, which provide managers the greatest incentives to go off the beaten path. Autodesk's Carl Bass placed great value in external threats and crises. "The threat of somebody doing something is one of the biggest tools you have to motivate, encourage, scare people into taking risks they wouldn't otherwise do," according to Bass.

Local constraints

Implementing SIBA is frequently constrained by inadequate local infrastructure, ill-equipped and fragmented suppliers and partners, and the difficulties of monitoring and measuring results in unfamiliar territory. In addition, a key difficulty on the ground for most companies is the inadequacy of supportive public services, human capital, and institutions. It is notable that two of the five most commonly cited constraints were weak local infrastructure and limited consumer education; both are traditionally areas addressed by the public sector. These are, again, areas that are vulnerable to the free-rider problem. Since these enabling systems are public goods, everyone stands to benefit from them, which inhibits individual companies from stepping up and being willing to invest unilaterally in these areas.

A potential solution to these challenges is to engage partners with knowledge, credible relationships, and access to the local territory. Such partnerships can be drawn from a spectrum of players, e.g. NGOs and development organizations that can provide on-the-ground support ranging from training of suppliers and distributors to consumer education. Private sector partners provide technologies and services tailored to on-the-ground realities, such as transport/logistics and production/processing. In select cases, even competitors can be effective partners in precompetitive collaborations to develop smallholder farmers or to certify supply chains.

In addition to developing innovative internal structures that nurture SIBA and enable it to succeed, external partners will continue to be essential to overcoming constraints. SIBA calls for heightened forms of collaboration that cross public–private boundaries, as well as, in some cases, even competitive boundaries. Consider the case of Lafarge, the French industrial company, with businesses in cement, construction aggregates, and concrete. It has a long tradition of working in partnership with multiple agencies and players with local knowledge, including NGOs, such as Wildlife Habitat Council, Habitat for Humanity, and Room to Read, as well as with government ministries.

Difficulties in measuring impact

For SIBA to scale up, a critical obstacle is that of impact measurement. Currently, managers who wish to make the case for greater investment in SIBA within their organizations are hampered by the limited tools available to measure the impact of such investments. About a quarter of our respondents indicated impact measurement challenges as a key barrier.

Impact measurement and communication of outcomes are increasingly essential for multinational companies with clear impact agendas: 57 percent of Fortune 500 companies report on environmental, social, and governance impacts. Consumers are demanding clear, comparable impact information. Governments and donors supporting multinational companies require more than periodic reporting; they are seeking assurances that interventions are creating the kind of social change promised by the companies. Managers require information to make informed investment decisions. Where impact efficiency and effectiveness are tied to financial performance, regular assessment is a key management tool. Managers need to go beyond periodic sustainability reporting to assess their long-term social and environmental impacts.

This is another area where partnerships can be valuable in overcoming a barrier. Impact measurement has been a common challenge faced by NGOs and development agencies historically. Many such organizations have invested in tools and have a track record in innovative measurement and evaluation techniques. Partnering with such organizations can improve a company's ability to credibly and (cost-)effectively assess the impact of SIBA.

In addition, it is often useful to set targets that are measurable and readily associated with impact on the surrounding context and on the company's business. For example, SABMiller, the beer producer, has a target as follows: by 2020, they plan to directly support over half a million small enterprises to enhance their business growth and family livelihoods. Given that these small businesses employ at least 1.5 million people with benefits accruing to a further 6 million of their family members, it is clear that these populations with increasing purchasing power can contribute directly to SABMiller's top line by becoming consumers of beer. However, many of these small enterprises win the company's support because of their participation in SABMiller's value chain as suppliers and distributors and, thereby, contribute to its long-term sustainability.

To summarize, there are several barriers to scaling up SIBA. Surprisingly, there are other barriers that would, intuitively, be potential hindrances, but were not cited as significant obstacles to SIBA in our research. The top barriers – as well as the less frequently cited ones – revealed by our research are summarized by Figure 2.2. For each of the key barriers, there is a potential solution that can be put in place. Much as in the case of our earlier discussions, these "point solutions" need to comprise a system of mutually reinforcing activities.

In closing: escaping the core competency trap to release trapped value

We noted earlier that the pursuit of innovation in the context that surrounds the core opportunities in emerging markets may demand that the

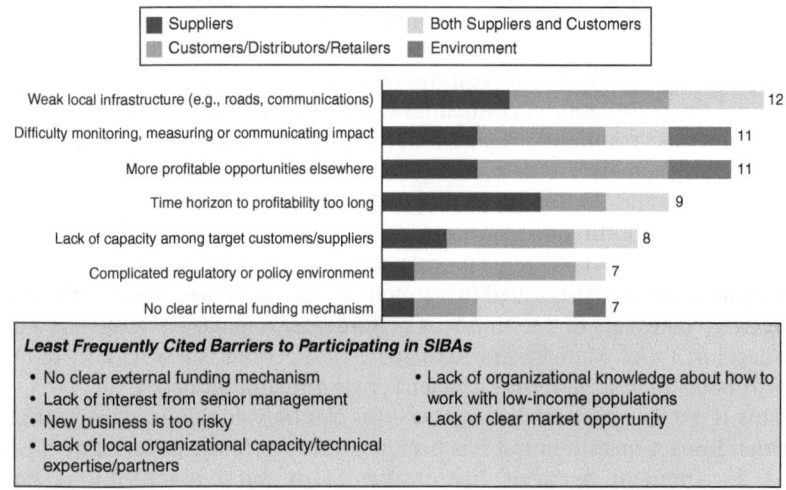

Figure 2.2   Barriers to participating in sustainable and inclusive business
Source: Chakravorti et al. (2014).

firm's managers find innovative ways to break away from one of the basic axioms of modern business: the need to focus, in order to optimize the use of scarce resources. In doing so, a manager may run afoul of a second foundational axiom: sticking to the organization's core competencies. In many emerging markets, closing the contextual gaps and effectively investing in the necessary institutions cannot be reliably delegated to others. In these situations, the firm needs to step into the breach and extend beyond its core competencies and consciously add to its system of activities.

Companies that rigidly adhere to the second axiom fall prey to such a "core competency trap." Those that have more well-rounded strategies for innovating in the emerging markets context have found ways around the trap. However, such action takes leadership and foresight, and the willingness to judiciously extend beyond the natural position in the value chain and find the right "hooks" and interfaces for complementary support from external partners. We offer two examples for illustrative purposes.

Olam, the leading agri-business operator, has moved outside its core competency even while collaborating with local governments to develop public–private partnerships for filling infrastructure gaps, with other government agencies such as USAID, along with NGOs such as TechnoServe, to support farmer organizations. Similarly, Novartis, the leading health care, agri-business, and consumer health company, conducts health education in rural and periurban cells to raise awareness of local disease and disease prevention techniques, organizes health camps, and deploys health supervisors to build the capacity of local medical professionals.

Unlike traditional CSR, investments in releasing trapped value are intended to address market inefficiencies and contextual gaps as part of the natural process of seeking profitable growth, rather than for purely philanthropic reasons or to build social and political capital or brand equity in international markets. This represents, to our mind, a new innovation challenge for companies – to transform the context that surrounds the core opportunities in emerging markets.

It is clear that this form of innovation in elements that are separate from the core business is a complex task. For the companies that innovate in such integration between the core and the context, it will be well worth the sacrifices in the near term. Given that the hope for corporate growth rests with the rising middle class in emerging markets, if done well such innovation may prove to be the most enduring sources of competitive advantage in the years to come.

In a widely circulated map produced by McKinsey (the same folks who gave us advance notice of the $30 trillion estimate of the potential that awaits businesses in 2025) about the migration of the world's center of economic gravity, the center is moving steadily southeastwards from around Reykjavik to hover around Novosibirsk by 2025. Innovation in releasing trapped value is hard, will require collective will, and it will require managers to develop a degree of "contextual intelligence." However, developing this new form of intelligence and incorporating it into innovation practices within the organization are essential. Otherwise, there is a risk that the center of economic gravity will never reach its 2025 destination in Novosibirsk and will be lost in a Siberian wilderness for several decades to come.

Acknowledgements

The authors are grateful to Hui Wen Chan, Anamitra Deb, Jianwei Dong, Sara Eshelman, Kate Fedosova, Jamilah Welch, and Jessica Zhao for their research, analysis, integration, and other numerous forms of support in the preparation of this study.

References

Andersen, C. (2011). "New and Emerging Financial Reporting Requirements and the EITI." Extractive Industries Transparency Initiative Global Conference. Retrieved from http://icar.ngo/analysis/why-the-foreign-corrupt-practices-act-is-good-for-business/ [accessed 11/26/2015].

Chakravorti, B., Macmillan, G., and Siesfeld, T. (2014). "Good for Growth or Growth for Good." The Fletcher School, Monitor Institute, and Citi Foundation. Retrieved from http://fletcher.tufts.edu/Inclusive-Business/SIBA-report-2014

Christensen, C. M. (1997). *The Innovator's Dilemma: When New Technologies Cause Great Firms to Fail.* Boston, MA: Harvard Business School Press.

Cisco. (2015). Retrieved from http://www.cisco.com/web/IN/about/inclusive_growth. html
Daneshkhu, S. and Oakley, D. (2015). "Unilever under Pressure to Step Up Growth Rate." *Financial Times*. Retrieved from http://www.ft.com/cms/s/0/7c79452e-ae5c-11e4-8188-00144feab7de.html#axzz3sczJSOaW [accessed 11/26/2015].
International Chamber of Commerce, Transparency International, United Nations Global Compact, and the World Economic Forum Partnering against Corruption Initiative (PACI) (2008). *The Business Case against Corruption*. 15 July. Retrieved from http://www.iccwbo.org/Advocacy-Codes-and-Rules/Document-centre/2008/ The-Business-Case-against-Corruption/ [accessed 11/26/2015].
Khanna, T. and Palepu, K. (2010). *Winning in Emerging Markets*. Boston, MA: Harvard Business School Press.
McKinsey & Company. (2014). "How Big Companies Can Innovate." Retrieved from http://www.mckinsey.com/insights/innovation/how_big_companies_can_innovate
Shapiro, H. Y. (2014). "Unleashing the Power of Genomics." *Sustain: Cutting-Edge Business Solutions*. Retrieved from http://www.sustainbusiness.org/ unleashing-the-power-of-genomics
The Economist (2013). "Protesting Predictions." December 23. Retrieved from http:// www.economist.com/blogs/theworldin2014/2013/12/social-unrest-2014 [accessed 11/26/2015].
Unilever. (2015). "Sustainable Living." Retrieved from http://www.unilever.com/ sustainable-living-2014/enhancing-livelihoods/inclusive-business/
WorstPolluted. (2014). "Top Ten Countries Turning the Corner on Toxic Pollution." Blacksmith Institute, Green Cross (Global Alliance on Health and Pollution). Retrieved from http://worstpolluted.org/docs/TopTen2014.pdf

3
Innovation in Emerging Markets: Asia

Rebecca A. Fannin

Just a decade ago, Asian markets were not on the global map of innovation hotspots. The focus was on Silicon Valley and Israel. But today, that has all changed. China, once known primarily as the factory of the world, and India, long a global capital for outsourcing, have moved into the spotlight as start-up nations.

Entrepreneurial talent, venture capital, government policies, start-up ecosystems, improved infrastructure, tech hotspots, free trade zones, economic growth, and urban consumer markets have led this change. From large multinationals Lenovo, Huawei, Infosys, Samsung, and Acer to emerging players Baidu, Alibaba, Tencent, Xiaomi, and Flipkart, Asian innovation is no longer in the shadows. The region's large and fast-growing digital economy is fine-tuned for the latest gadgets from smartphones to Internet-connected wearable devices, bypassing the personal computer era and e-mail.

Already, Asia counts the most Internet users (Table 3.1) and smartphone subscribers (Table 3.2) in the world. Moreover, four China websites – Alibaba, Baidu, Tencent, and Sohu – rank among the top ten largest internationally (Internet Trends, 2014, p. 131). Considering that mobile Internet

Table 3.1 Leading Internet markets in Asia's emerging countries, 2013, US comparison

	# of users	% Change from 2012	Population penetration (%)
China	618 million	+10	46
India	154 million	+27	13
Indonesia	71 million	+13	28
Philippines	38 million	+27	36
Vietnam	37 million	+14	39
US	263 million	+2	83

Source: China Internet Network Information Center, United Nations/Internet Telecommunications Union, US Census Bureau, Indonesia – APJII (Indonesian Internet Service Providers Association).

Table 3.2 Leading smartphone markets in Asia's emerging countries, 2013, US comparison

	# of subscribers	% Change from 2012	Population penetration (%)
China	422 million	+26	31
India	117 million	+55	10
Indonesia	48 million	+42	19
Philippines	20 million	+43	19
Vietnam	17 million	+156	19
US	188 million	+21	59

Source: Informa, IDC Vietnam, Gartner, Morgan Stanley Research.

penetration is still relatively low in some of Asia's developing nations, the only direction is up.

Throughout Asia's switched-on markets, mobile communications is where the action is for texts, chats, shopping, banking, and payments. Take Tencent's WeChat in China. This mobile messaging app zoomed to more than 300 million users within just two years after its 2011 launch. Rival service Whatsapp has emerged as tops in India.

From transportation to financial services to education, many business sectors are being disrupted by mobile communications advances. OlaCabs, an Indian version of taxi-calling mobile app Uber, has ramped up in India. DidiTaxi, its counterpart in China, claims more than 100 million users. In traffic-jammed Beijing, passengers bid by mobile phone for a cab ride. Alipay's money market fund Yu'E Bao, accessed by a mobile app, skyrocketed to $89 billion in assets within 10 months after its May 2013 launch (Internet Trends, 2014, p. 136).

Necessity being the mother of invention, cleantech has become a focal point of innovation in Asia too, with advancements for cleaning up water and air pollution. Tesla cars are being sold in China, with multiple electric charging stations in Shanghai and Beijing. Speaking at a Silicon Dragon forum in Shanghai, Tesla manager Dan Hsu predicted "China will be our biggest market globally" (Fannin, 2014b).

Asia's leaps in innovation have been led by private enterprise start-ups that did not exist just 15 years ago. Today, these start-ups have become tech titans and are competing aggressively. Baidu dominates online search in China while Tencent towers in chat and social messaging. Kakao Talk from Korea and Line from Japan are forces in social messaging and gaming. The e-commerce giant Alibaba envelops everything from online shopping to logistics to payments.

Tech innovation clusters have sprung up throughout Asia and are spinning out products that can be commercialized. Like in Silicon Valley, these clusters have formed from collective resources for start-ups: nearby universities, incubators and accelerators, software parks, talent, and financing. Beijing's Haidian tech district is home to Tsinghua University (China's Massachusetts Institute of Technology), venture firms at Tsinghua Software Park, as well as Baidu and Microsoft. Taiwan's expansive Hinschu Science Park is headquarters for semiconductor leaders TSMC (Taiwan Semiconductor Manufacturing Company) and UMC (United Microelectronics Corporation) and start-ups in LEDs. Bangalore's campuses are dominated by outsourcing leaders Wipro and Infosys but also start-ups in mobile and e-commerce. Singapore's Fusionopolis houses digital media start-ups while Hong Kong's Kennedy Town and Sheung Wan are home to fledgling businesses.

Asia's start-up boom has its roots in the late-1990s dotcom boom when the "returnee" entrepreneurs – so-called sea turtles – journeyed home from overseas to launch Internet start-ups that copied successful Western business models. With Ivy League graduate degrees, professional experience, and Silicon Valley savvy, they established search, e-commerce, and social media businesses, raised capital from Sand Hill Road, grabbed market share, and built publicly traded tech winners. Baidu emerged as China's Google, Weibo as China's Twitter, Renren as China's Facebook, Dangdang as China's Amazon, Ctrip as China's Expedia while Alibaba morphed into an eBay, Yahoo, and Amazon combined. Chinese leaders Jack Ma of Alibaba and Robin Li of Baidu became icons in China while Xiaomi co-founder Lei Jun was compared to Steve Jobs.

Today, in Beijing, Shanghai, Shenzhen, Hangzhou, and Chengdu, "the tech ecosystem is buzzing with unprecedented activity, a wealth of talent, as well as considerable innovation," points out Chris Evdemon, a partner at incubator and venture fund Innovation Works (Evdemon, 2014). He notes that "Chinese developers are already beginning to leapfrog their foreign counterparts in consumer mobile Internet products, and are catching up fast in pretty much every other tech sector. Chinese entrepreneurs have already been innovating in terms of new business models and processes for several years." It is a trend that bodes well for increased protection of intellectual property in China.

India is closing the gap with China

India lags behind China by at least five years in developing its own start-up and innovation ecosystem, notes *Startup Asia* (2012), but could catch up some day. Over the past decade, Flipkart has emerged as India's Amazon, NASDAQ-listed MakeMyTrip developed as India's Travelocity, and Snapdeal launched as India's eBay and GroupOn. The trend is gaining momentum: in 2014, Snapdeal.com picked up $627 million from Japanese tech giant Softbank, OlaCabs raised $210 million from Softbank, and there is talk that mobile marketing firm inMobi may be acquired by Google.

From copying to microinnovating

While the copycats defined the first generation of Asian start-ups, the region's entrepreneurs have gained confidence. While not yet developing disruptive breakthroughs, they have become adept at microinnovations for the local market and skilled at money-making business models, points out Kai-Fu Lee, the founder of Beijing-based Innovation Works and a former head of Google in China (Silicon Dragon Video, 2011).

A leading indicator of change is that innovations developed in Asia are now being copied by Western markets. The "freemium" model of monetization from video games – free games but charges for premium added-on services and virtual merchandise – originated in Korea, Japan, and China where there was a reluctance to pay upfront for games. That idea has now spread to the West. The combination of all-in-one gaming, social sharing, and messaging services that started in China is now being copied by US mobile communications start-ups such as Curse Inc. in Huntsville, Alabama (Fannin, 2014b).

Homegrown Asian brands are gaining an edge over overseas competitors that have failed to tweak offerings for local tastes. Chinese smartphone maker Xiaomi, formed in 2010, is now China's top-selling smartphone, and ranks third globally. Xiaomi's success is due to building affordable Android devices, designed by Chinese, for Chinese, points out venture capitalist Evdemon. Poaching senior executive Hugo Barra from Google's Android team has helped too (Evdemon, 2014). As Xiaomi has expanded into Asia and branched out into TVs, fitness monitors, and tablets, its December 2014 financing of $1 billion put Xiaomi's market valuation at a staggering $46 billion.

Flames of innovation

Ideas are developing at a fast pace within this start-up ecosystem, fed by cloud computing for online collaboration, speedier Internet connections, and tech get-togethers for networking and information exchange. Eric Feng, chief technology officer of Flipboard Inc., recalls that when he worked at Microsoft Asia in Beijing a decade ago, "it took 900 staffers and $100 million in investment for a major product or service launch" (Silicon Dragon Talk, 2014). Now projects for Flipboard in China can be completed within months by a small team, he says.

Sand Hill Road's impact on Asian innovations

A main spark for Asia's entrepreneurship has been venture capital from Silicon Valley's Sand Hill Road firms (see Table 3.3). Venture partners often fund promising start-ups together, fostering faster cycles of innovation. Crowd-funding is another impetus, notably the launch of Israel's OurCrowd and

Table 3.3 Leading venture investors in China and India

China	India
Charles River Ventures	Helion Venture
Draper Fisher Jurvetson	IDG Capital
GGV Capital	Inventus Capital Partners
Kleiner Perkins	Kae Capital
IDG Capital	Lightbox Ventures
Lightspeed Venture	Mumbai Angels
Matrix	Nexus Venture Partners
Mayfield Fund	Norwest Venture Partners
NEA	SAIF
Redpoint Ventures	Sequoia Capital
Sequoia Capital	
Sutter Hill	

Source: Silicon Dragon Ventures.

Table 3.4 Venture capital in Asia nearly triples

	2006	2009	2010	2011	2012	2013
Global	$42.7 Bn	$35.2 Bn	$46.4 Bn	$54.7 Bn	$47.6 Bn	$48.5 Bn
China	$2.5 Bn	$2.8 Bn	$6.1 Bn	$6.5 Bn	$5.0 Bn	$3.5 Bn
India	$0.6 Bn	$0.8 Bn	$0.9 Bn	$1.5 Bn	$1.6 Bn	$1.8 Bn
U.S.	$31.1 Bn	$24.5 Bn	$29.3 Bn	$36.2 Bn	$32.8 Bn	$33.1 Bn

Source: DowJones VentureSource (2013), Silicon Dragon Ventures (2013).

Hong Kong's Investable.vc. Accelerating regional innovation, local venture capital firms have set up shop in Asia to invest in homegrown start-ups.

Venture capital in the region has been highly cyclical. In the initial rush of start-up financing to the region more than a decade ago, pioneering venture investors made fortunes from bets on Baidu, Alibaba, Tencent, and MakeMyTrip. Venture investment in China and India – the two beacons – peaked in 2011, nearly tripling the amount seen in 2006. Funding then leveled off with the burst of the bubble economy, the global financial crisis, and accounting scandals at a few publicly traded Chinese companies (see Table 3.4). Now, as the issues have been worked through, venture funding is on the rebound.

In proof of its impact, over the last five years, some $0.5 trillion in market capitalization has been created by 36 publicly traded e-commerce and media companies from key emerging markets, led by China (see Table 3.5).

Angel investors are another spark in Asian innovation – just like in Silicon Valley. Founders of Asia's most successful emerging companies and start-ups are pouring their new-found wealth into *more* start-ups. Networks of angel investors have sprung up, including Mumbai Angels and AngelVest

Table 3.5 Top ten online companies in emerging markets by market capitalization

	Market Cap	Country
Alibaba	$219bn	China
Tencent	$138.6bn	China
Baidu	$76.6bn	China
JD.com	$35.3bn	China
Netease	$11.2bn	China
VIPShop	$10.7bn	China
Qihoo 360	$8.8bn	China
Yandex	$8.8bn	Russia
Ctrip	$7.3bn	China
Mail.ru	$5.9bn	Russia

Source: Rise Capital with data drawn from Bloomberg, Yahoo Finance, company data, September 2014.

Table 3.6 Arrival of angel investors in Asia

	Total rounds 2013	% of Total Angel $
U.S.	3480	12%
Europe	1395	11%
China	314	5%
India	222	17%
Israel	166	3%

Source: Dow Jones VentureSource, Silicon Dragon Ventures.

in Shanghai (see Table 3.6), to provide funding and mentorship to the next group of entrepreneurs. Billionaire Lei Jun is the angel force behind Xiaomi and social entertainment service YY.

Asia climbs up the rungs for new patents

It is little surprise that Asia is climbing the ladder for new patents (see Table 3.7). China ascended from eighth in the world for new patent applications in 2006 to sixth place in 2008, and to third place in 2013. Japan has been a perennial top player on the patent scales, as has South Korea.

This trend is also illustrated by geographic shifts (see Table 3.8).

Two China tech titans – ZTE and Huawei – rank tops while Japan scores with Panasonic, Sharp, and Toyota (see Table 3.9).

The next step for Asian start-ups is international expansion. Chinese tech titans are investing heavily in US tech start-ups (see Table 3.10) to gain market share and knowhow, with Alibaba at the forefront.

Table 3.7 Shift in new patents to China (%)

	2008		2013
1. U.S.	32.70%	1. U.S.	27.80%
2. Japan	17.50%	2. Japan	21.40%
3. Germany	11.30%	3. China	10.50%
4. S. Korea	4.80%	4. Germany	8.70%
5. France	4.20%	5. S. Korea	5%
6. China	3.70%	6. France	3.80%
7. U.K.	3.40%	7. U.K.	2.40%

Source: World Intellectual Property Organization.

Table 3.8 Patent applications trends – China's rise, Japan's stronghold

	2008	2012	2013	YTY % change
U.S.	52,280	51,207	57,239	+ 10.8%
Japan	28,760	43,660	43,918	+ 0.6%
China	1,706	18,267	21,516	+ 15.6%

Source: World Intellectual Property Organization.

Table 3.9 Top patent filers by company and origin, 2013

	# of patents	Country of Origin
Panasonic	2,881	Japan
ZTE	2,309	China
Huawei	2,094	China
Qualcom	2,036	U.S.
Intel	1,852	U.S.
Sharp	1,840	Japan
Bosch	1,786	Germany
Toyota	1,696	Japan
Ericsson	1,467	Sweden
Phillips	1,423	Netherlands

Source: World Intellectual Property Organization.

"These are strategic deals. Asian companies are paying their tuition by buying into U.S. tech startups," says Hany Nada, a managing partner at GGV Capital in Menlo Park, California, and an early investor in Alibaba (Silicon Dragon Talk, 2014).

Expect Asian innovation to continue to unfold as Shanghai, Tokyo, and Seoul vie with Silicon Valley for tech innovation leadership. The trend of Start-up Asia is not even into its third decade but already is having a profound influence on the shape of innovations for tomorrow.

Table 3.10 China buys into US tech start-ups

Date		Transaction
Feb. 2011	Tencent $400M	Game developer Riot Games
Apr-12	Tencent $400M	Game developer Epic Games
Oct. 2012	Renren $77M	College lender Social Finance
Feb. 2013	Baidu $30M	Mobile safety firm Trust Go
Oct. 2013	Alibaba $50M	App search engine Quixey
Jan. 2014	Alibaba $15M	Luxury e-commerce site 1stdibs
Mar-14	Alibaba $215M	Mobile messaging app Tango
Apr-14	Alibaba $250M	Ride-sharing company Lyft
Jun-14	Alibaba $50M	Remote control app Peel
Jul-14	Alibaba $120M	Mobile gaming startup Kabam
Oct. 2014	Fosun $200M	Movie producer Studio 8
	Alibaba $206M	Subscription service ShopRunner
	Baidu $10M	Mapping company Indoor Atlas

Source: Silicon Dragon Ventures.

Case studies of Asian innovators

Huawei Technologies Company, China: where core R&D is the champion

A visit to the sprawling headquarters of Huawei Technologies Co. in the southern Chinese city of Shenzhen offers an example of China's climb up the innovation ladder. This telecom-plus company has moved up as a leading-edge industry leader with a global footprint as new innovation management processes have been adopted, patents granted, and a series of nifty smartphones and ultraspeed communications services have launched.

The trend of "made in China" to "invented in China" has been seen most clearly among the Chinese tech start-ups that have scaled over the past decade and continue to microinnovate, acquire, and expand. Now this thread is increasingly prominent among China's larger multinationals.

It is ironic that this manufacturing-heavy Shenzhen, a hub for low-cost knock-offs, is emerging as a center of corporate innovation. The city is home to the telecom-plus companies Huawei and ZTE, social messaging player Tencent, supply chain manager PCH International, electronics maker Foxconn, and Warren Buffet-invested electric car maker BYD.

This Pearl River delta city still has a brusque style and rough feel, and with its hot and humid climate, cannot compare to the lifestyle and upbeat, creative energy of Silicon Valley. Yet that innate Chinese inventive knack is increasingly evident among Shenzhen's large conglomerates and exemplifies China's ascendancy in the value chain from a production-only base to a stronger foundation in science and technology.

To take one example, Huawei is leaping forward with innovations after long being seen as a copier of technologies from Cisco. The vast headquarters of Huawei, founded in 1988 by engineer Ren Zhengfei, a former engineer with the People's Liberation Army, looks like a leafy college campus. English-speaking guides take corporate visitors around several sleek showrooms that exhibit a vast array of Huawei technical breakthroughs from networking infrastructure equipment to its own nifty smartphones under the brand name Ascend, an Android phone that has been compared favorably with the latest Apple iPhone makes.

Li Yingtao is the president of Huawei's innovation division named (oddly) the 2012 Laboratories. During an hour-long interview while being served tea, Li elaborated how Huawei is keeping its innovation locomotive on track.

He pointed to the launch of superthin Ascend brand name smartphones and its crash- and waterproof handsets. He cited Huawei's work on seamless mobile broadband connections on the Shanghai–Beijing bullet trains. Then, he mentioned the company's pioneering work to offer in-flight Wi-Fi service for airline passengers. Li emphasized too, Huawei's long history of pioneering new telecom systems, such as the introduction of 3G service to European customers (Fannin, 2013).

Huawei keeps its innovation pipeline pumped with its 2012 Laboratories, comprised of 15,000 engineers and researchers, 20 percent of them outside China and a substantial percentage from India. The unit is under Li's direction and, significantly, reports directly to the CEO.

Huawei's integrated team of marketing, technical, and R&D managers aim to make sure products match requirements, working with suppliers and customers. Joint innovation centers with customers and collaborations with universities and research institutes work on breakthroughs.

"You can't close the door and work on innovation on your own," says Li, who holds a doctorate degree from Harbin Institute of Technology in China and has worked at Huawei since 1997 in various research and technology managerial roles (Fannin, 2014a). Sounding like any clued-in Silicon Valley tech entrepreneur, Li says that Huawei aims to create some solutions that anticipate customer needs. "Not every time will we be successful, but if you are innovative, you have to accept that." A perennial challenge is getting the right balance of investment for success and failure in the future. "Without sufficient failure, we might not have tested enough in the future," Li adds (Fannin, 2014a).

MisFit Wearables, Vietnam: following the Steve Jobs model

Misfit Wearables represents the new breed of world-class innovative start-ups with Vietnamese roots. The company is spearheaded by Vietnamese refugee Sonny Vu, working with a team in Ho Chi Minh City, Seoul, and San Francisco. Its products are positioned squarely in one of the "sweet spots" of technology – wearable computing products.

Its main product is Misfit Shine, a device that can be worn like a watch to measure sports activities, track progress, and share results. Shine is the only wearable activity device that is waterproof, fashionable, and requires no battery charge – it works through a connection to a smartphone. The product launched in mid-2013 and within two quarters sold 300,000 units. Shine is already available in 35 countries.

An idea of the company's culture can be found in its name, drawing upon a Steve Jobs quote honoring "the misfits, the rebels, the troublemakers" who "think different" and create the next, new thing (ABC News, 2011). Misift Wearables was formed in 2011 on the day that Jobs died, and former Apple leader John Sculley is a cofounder of Misfit Wearables and a mentor to CEO Vu.

After getting its start from a crowd-funding campaign, Misfit Wearables raised a substantial amount of venture capital from a who's who of start-up investors – PayPal cofounder Peter Thiel's Founder Fund, billionaire Vinod Khosla of Khosla Ventures, plus Li Ka-shing's Horizon Ventures from Hong Kong.

This start-up's originality is a contrast to Vietnam's homegrown start-ups that have copied Internet and e-commerce brands such as VinaBook, which is like Amazon, and Peacesoft, like eBay, and Tencent, like VNG. Most Vietnamese start-ups have marketed their wares within their home market and received funding from local firms, primarily IDG Ventures Vietnam, one of three main funds from the Boston-based parent company that invests in Chinese, Vietnamese, and Indian start-ups.

How did Misfit Wearables break out of this mold? For one thing, the company is truly operating in an international structure, selecting the best resources from the best sources in the world. The company's software engineers, graphic designers, and data scientists are in Ho Chi Minh City, the main base of operations where 55 of its 90 employees work. Sensor development and manufacturing are done in Seoul. Hardware engineering and design of the product's interface for users are from San Francisco.

This international footprint stems in part from Vu's own career trajectory. He arrived in the US in 1979 with his parents as one of many "boat people" who fled North Vietnam and settled in America. Bright and hard-working, he got a mathematics degree from the University of Illinois and a PhD from the top-notch Massachusetts Institute of Technology (MIT). He also worked as a researcher at Microsoft, refining Chinese natural language processing for speech and handwriting recognition. He formed his first start-up AgaMatrix in 2001 with Sridhar Iyengar, a second-generation Indian entrepreneur he met as an undergrad. Their background as immigrants no doubt fueled their ambition and drive to get ahead. The two ran and built up AgaMatrix for nearly a decade into a $20 million business, and its core technology – an iPhone-connected medical device – became the basis for Misfit.

CEO Vu credits the start-up's innovation stripes to its small team, which gives it an advantage in moving fast and staying agile – the lean start-up model. Teams get together more or less at random with no real structure for managing innovation – it just happens. And it happens sometimes through day-long jams for designing new products. The goal is to dream and think of new products that might be developed several years into the future. "We don't ask the customer what products to make. That's our job," says Vu, sounding a theme from Steve Jobs' method of designing (Silicon Dragon SF, 2014).

Even so, Vu says product innovation is incremental, not one big lightbulb moment. "We get to base two and then base three and eventually climb the mountain," he says. "We don't really know what the next killer app will be, and honestly if we did know it, we would be plowing all our money into it" (Silicon Dragon SF, 2014).

One area that Vu does see as promising is in a related field, wearable payment devices. Wearing a bracelet with a built-in sensor, consumers can swipe to make payments, much like credit cards are swiped through a reader.

Ultimately, where Vu wants to see Misfit Wearables go is into a much broader array of ambient computing products for the home, the body, and the car (Silicon Dragon SF, 2014). "We want technology to blend into the background and serve us," he emphasizes, "rather than us staring at tiny dots on a screen."

inMobi, India: where mobile meets advertising globally

Mobile communications is the rage today, and one of the companies that is best at capitalizing on this trend is inMobi, an Indian business that specializes in mobile advertising services. With the rise of smartphones and technologies to reach consumers through mobile ads rather than old-fashioned TV sets or even personal computers, inMobi is well-positioned in a large global market of $13.1 billion that technology research firm Gartner estimates will reach $42 billion by 2017 (Gartner, 2014).

Founded in 2007, inMobi is at the forefront of Indian tech start-ups with a global footprint. This mobile advertising company expanded from India to nearby Southeast Asian countries first in 2008, and then stepped up the pace over the next few years by moving into the more developed advertising markets of Europe, Japan, and the US. The next epicenter for mobile advertising and for inMobi is the emerging markets of Asia, driven by a surge in India, China, and Indonesia.

A culture of thinking and acting globally has carried the company forward, starting with its worldly sounding name, which was changed early on from mKhoj, meaning "search" in Hindi. Today, inMobi has 17 offices worldwide and its ad network spans 165 countries – and nearly all of its revenues are from outside India. It is the largest independently owned mobile ad network globally, with Google AdWords as the leader, according to a ranking by Seattle-based online marketing firm HasOffers. inMobi claims 795 million monthly active users (Byrne, 2014).

What has put this Indian company in the lead is a bundle of factors – its global perspective, innovative culture modeled after Silicon Valley, quick pivots, laser-like focus on one of the fastest-growing sectors of the new digital economy, and continual technology upgrades to stay ahead of competitors.

Credit goes to its leader, Naveen Tewari, an MBA graduate of Harvard and ex-McKinsey consultant who despite many start-up hurdles never gave up pursuing his dream and has managed to be in the right place at the right time with the right connections. A mechanical engineering graduate from the Indian Institute of Technology Kanpur in 2000, Tewari landed a job as a business analyst for McKinsey in Mumbai, where he had the opportunity to observe how Indian business magnate Mukesh Ambani launched Reliance Infocomm in 2002 and ushered in digital communications to India. While an MBA student at Harvard from 2003 to 2005, he did a stint in Boston with venture firm Charles River Ventures, and that increased his thirst for entrepreneurship. He moved to Silicon Valley and worked for a year in small teams to get a business off the ground. When that failed, Tewari went on vacation to India for one month and started to work on his own business plan – and he has not looked back since.

Tewari was able to put inMobi on the extreme fast track by using his smarts to attract significant sums of funding from some of the best venture capital and technology firms. He raised $500,000 from the well-connected angel investing group Mumbai Angels in 2006 for his first business idea, a text-based mobile search engine. When that business did not gain traction quickly enough, he did not waste time in pivoting the business model to mobile advertising. Living off maxed-out credit cards and with no cash coming in, he and his cofounders – two former classmates Abhay Singhal and Amit Gupta – almost did not make it through to the next funding round. Luckily, the big break came in mid-2007, when Tewari was able to connect with and nab $7.1 million from the well-regarded Valley venture firm Kleiner Perkins and its India investing partner Sherpalo Ventures.

With a decision to move from Mumbai to Bangalore – India's Silicon Valley – the start-up team gained access to the city's large pool of talented software coders. As the start-up scaled, top engineering hires were recruited from Google, Yahoo, and MSN. Many of these new hires were ambitious returnees to India like Tewari. Borrowing a highly effective tactic from the Valley, the new employees were lured in with stock options.

inMobi's contemporary office in a central business corporate park has an upbeat energy not uncommon in techie upstarts. The fast pace of scaling up meant that inMobi soon had to return to the venture coffers for money. The start-up's original two backers put in another $8 million in 2010. Then, Tewari was able to convince Japan's leading technology company Softbank to invest $200 million in 2010 – a daring and well-timed pitch as Tewari was plotting an entry strategy for the large Japanese market.

Throughout its start-up journey, the inMobi team has benefited from mentorship by industry-leading tech titans. Technology industry insider and inMobi backer Ram Shriram is one of the first investors in Google and a key exec in building online bookseller Amazon and Internet browser Netscape. Softbank leader and inMobi investor Masayoshi Son was an early backer of Alibaba.

Pressures are on the seven-year-old inMobi to go public or be acquired by one of the tech giants. So far, inMobi's founding team have resisted the pull to go public. CEO Tewari says he will look to make that move when the company reaches a significant scale. The long-term aim is $1 billion in revenues, double its 2013 revenues.

Aiming to build up its mobile ecosystem and compete with new entrants such as Google and Apple, inMobi has made several overseas acquisitions: San Francisco-based MMTG Labs, a creator of customized app stores; Metaflow Solutions, a UK-based specialist in mobile app management and distribution; and Overlay Media, a mobile data analytics firm in the UK.

A constant churn of new products is also part of inMobi's winning business model, where revenue is earned every time a user views or clicks on an ad. Early in 2014, the company debuted a platform for publishers and advertisers to create customized ad layouts and interfaces within the inMobi mobile ad network to attract higher viewership and monetization. Around the same time, inMobi launched a state-of-the-art video ad platform for delivering interactive content and promotions.

The next step for inMobi is to become a player in data analytics. This leading mobile start-up has developed technology to help advertisers to better place relevant messages by analyzing hundreds of terabytes of data daily and predicting consumer behavior. inMobi counts Japan's Yamaha Motor Co., Microsoft, sporting goods maker Adidas, fashion brand Lancôme, and retailer Macy's as customers.

For inMobi, the challenge in the fast-moving mobile advertising market will be to continue embracing and innovating technology ad formats that blur the boundaries between content and advertising without irritating the consumer.

Alibaba, China: a truly innovative leader on a mission

In China, Alibaba is synonymous with e-commerce and this could someday be true globally.

Started in 1999 in the eastern Chinese city of Hangzhou by former English translator Jack Ma and a small team, Alibaba has emerged as a powerhouse in China's online economy. Alibaba's breadth of offerings across the e-commerce ecosystem gives it a commanding lead in China and could give it an edge against rivals in Western markets.

Alibaba aims to build the future of commerce and intends to be around a long time doing just that – in Ma's words, at least 102 years to span three centuries. Its success is due to the management team's sense of mission, long-term focus, commitment to guiding values, and organic growth of several leading businesses, all in e-commerce or related fields.

This once little-known Chinese e-commerce giant that Ma named after the globally recognized Ali Baba – the character who discovers hidden treasures in a thieves' cave with the magical phrase "Open Sesame" – already has an international footprint as a facilitator of cross-border e-commerce between

China and the West. In 2013, Alibaba claimed 12 percent of overall revenues from international commerce. Interestingly, the company is fast extending its global outreach with numerous investments in seemingly unrelated and innovative Silicon Valley tech start-ups.

Alibaba's expanse across multiple e-commerce sectors has been compared to eBay, Amazon, GroupOn, and PayPal combined. Alibaba handles $248 billion in merchandise transactions, eclipsing both Amazon, which handles half that amount, and eBay, which handles one-third that total (see Table 3.11).

Its multiple retail and wholesale marketplaces are designed for both global and Chinese markets. The original site, Alibaba.com, is its global wholesale marketplace that connects Chinese manufacturers to small businesses globally, while 1688.com is the company's China wholesale marketplace. Taobao, its Mandarin-language business-to-consumer site, links buyers and sellers looking for bargain-priced jewelry, games, and the types of gadgets found in street fairs. Tmall is a virtual shopping center, giving global brands a way to sell their goods directly to Chinese consumers. Juhuasuan is Alibaba's popular group-buying marketplace. AliExpress is its global consumer marketplace, linking global consumers to Chinese wholesalers and exporters. Then there's Alipay, an online payment system for processing online purchases. Rounding out the offerings, Alibaba has cloud computing services, logistics systems, and online marketing services. Alibaba generates the bulk of its revenues from seller fees for online marketing services and commissions on merchandise sold using Alipay, counting $5.5 billion in revenues for fiscal year 2013.

The market opportunities for Alibaba are the rising spending power of Chinese consumers, increased penetration of online shopping in China, growth in the types of products and services bought over the Internet, and expanded use of mobile devices for purchases (see Table 3.12). Additionally, logistics infrastructure and delivery systems are improving nationwide, which bodes well for e-commerce in China.

So does the trend toward increased online shopping in China (see Table 3.13). Chinese Internet market research firm iResearch projects e-commerce will grow to 27.2 percent of China's total consumption in 2016 from 7.9 percent in 2013.

Table 3.11 How Alibaba stacks up

	Gross Merchandise Value 2013
Alibaba	$248 billion
Amazon	$100 billion
eBay	$75 billion

Source: Silicon Dragon Ventures.

Table 3.12 Online shopping penetration in China (%)

Year	Penetration
2008	1.10%
2009	2.00%
2010	2.90%
2011	4.30%
2012	6.30%
2013	7.90%
2014E	9.10%
2015E	10.40%
2016E	11.50%

Source: iResearch.
(E = estimated).

Table 3.13 Market potential for online shopping in China

Online shoppers as percent of total Internet users	
China	48.90%
U.S.	74.20%
U.K.	77.60%
Japan	55.80%
Germany	79.50%

Source: CNNIC for China, IDG for other countries.

The challenges of China's retail marketplace mean that online shopping could leapfrog the offline market. In smaller cities, there is not a strong base of brick and mortar retail outlets, product selection is limited, and quality is inconsistent. Chinese shoppers are going online for convenience, lower prices, ease of payment, and greater merchandise availability, not just for low-end products but for luxury goods too. New safeguards against counterfeit products have additionally given online shopping a boost.

As Western-style consumerism catches on in China, the e-commerce market is forecasted by 2020 to be larger than the US, UK, Japan, Germany, and France combined, according to a KPMG report, "China's Connected Consumers" (KPMG, 2014, p. 6). By 2015, the report projects e-commerce transactions in China to reach $540 billion or 7.5 percent of total global retail transactions.

Within this fast-growing market, Alibaba claims an enviable position – an 80 percent market share of the e-commerce market in the world's most populated country, with a middle class of 500 million. Its only holiday promotion Single's Day is the equivalent to the shopping frenzy in the US on the day after Thanksgiving, Black Friday. In mobile commerce, Alibaba accounts for 76.2 percent of total mobile retail merchandise sold, aside from virtual items from gaming sites, according to iResearch. Plus, Alibaba has developed a surround-the-consumer strategy with multiple platforms (see Table 3.14).

Alibaba is well-positioned in the instant messaging and microblogging markets, two of the more heated sectors of China's increasingly advanced digital economy. In 2013, Alibaba led a $1.2 billion investment in the YouTube of China (Youku) and injected $586 million into the Twitter of China (Weibo). Plus, Alibaba has developed its own mobile messaging app Laiwang to compete with Tencent's popular WeChat, and launched the money market fund Yu'E Bao, which now ranks as one of the larger funds worldwide.

Alibaba's position in e-commerce in China has been compared with WalMart in the US. It is a point made by Nazar Yasin, managing partner of emerging markets investing firm Rise Capital (Silicon Dragon Talk, 2014). Alibaba has a 7.4 percent share of China's retail market, handling merchandise transactions worth $248 billion. WalMart has a 7.7 percent share of US retail sales with $336 billion in gross sales.

Now, Alibaba is stepping out in the world. Its initial public offering on the New York Stock Exchange in September 2014 was a watershed moment for the brand and for China tech companies as a whole. In a record-breaking IPO (initial public offering) globally, Alibaba raised $25 billion with a market valuation of $228.5 billion, topping by far Facebook's $16 billion IPO in 2012 and market valuation of $104 billion then.

Expansion in Western markets is the next step in Alibaba's strategy. Alibaba has been investing in American tech brands. In a 12-month period ending August 2014, the Chinese e-commerce company invested $968 million in nine US tech start-ups, according to investment banking firm

Table 3.14   Timeline of Alibaba's innovative new businesses in the e-commerce marketplace

1999	Alibaba.com	Global wholesale
2000	1688.com	China wholesale
2003	Taobao	China online auction
2003	Aliwangwang	Instant messaging
2004	Alipay	Online payment
2007	Alimama	Online ad exchange
2008	TMall	Global brands portal
2009	Aliyuan	Cloud computing
2010	Juhuasuan	China group buying
2010	AliExpress	Consumer wholesale
2011	Laiwang	Phone messaging app
2013	Mobile Taobao	Shopping app
2013	China Smart Logistics	24-hour deliveries
2014	11 Main	US shopping site

Source: Alibaba Group, Silicon Dragon Ventures.

Williams Capital Advisors in Palo Alto, California. The investments included messaging app Tango, video game start-up Kabam, ride-sharing company Lyft, remote control app Peel, subscription service ShopRunner, search engine app Quixey, and online sports memorabilia retailer Fanatics.

Jack Ma's influence at the helm

An important factor in Alibaba's success is its culture, set by its ambitious and visionary leader Jack Ma (see Table 3.15).

A tightly centralized and controlled management structure built around 28 senior management executives who have a big hand in board member nominees is part of the formula. This unusual – some might say innovative – structure paves the way for the founding team to continue having a major influence on its culture.

Throughout Alibaba's 15-year history, aggressive marketing and promotion – even publicity stunts such as a top salesman diving into Hangzhou's West Lake to celebrate sales successes – have driven Alibaba's success. Alibaba trounced eBay in China by introducing rival Taobao in 2002 and undercutting the American brand on its standard 15 percent commission on sales. Instead of charging sellers a fee for listing their goods, Alibaba charged for extras such as personalized web pages and banner ads. Alibaba further won favor with local consumers by offering an escrow system where buyers could return items to sellers up to six weeks after the sale.

By 2005, the underdog Taobao had jumped ahead of an overconfident eBay and claimed a 59 percent market share. A year later, as Taobao continued to gain, eBay shut down its China site, merged with Chinese portal Tom Online, and effectively withdrew from the market.

Connections to leading American tech innovator Yahoo have played a major role in Alibaba's triumphs. Management gained valuable insights into

Table 3.15 Alibaba value system

Customer first – The interests of our community of buyers and sellers must be our first priority.
Teamwork – We believe teamwork enables ordinary people to achieve extraordinary things.
Embrace change – In this fast-changing world, we must be flexible, innovative and ready to adapt to new business conditions in order to survive.
Integrity – We expect our people to uphold the highest standards of honesty and to deliver on their commitments.
Passion – We expect our people to approach everything with fire in their belly and never give up on doing what they believe is right.
Commitment – Employees who demonstrate perseverance and excellence are richly rewarded. Nothing should be taken lightly as we encourage our people to "work happily, and live seriously."

Source: Alibaba Group (2014).

online advertising techniques after Yahoo invested $1 billion in 2005 for a 40 percent stake in Alibaba.

All this is a long way from when Ma and 17 cofounders cobbled together $60,000 to form Alibaba and when Ma called on unimpressed venture capitalists along Sand Hill Road. To get a start in 1999, Alibaba raised $5 million from Goldman Sachs and some Chinese venture investors. The breaks came when Alibaba drew $20 million from Japan's Softbank in 2000, and then in 2005 pulled in $82 million from a group of well-connected US–China investors, among them GGV Capital.

A cult-like loyalty among staffers at Alibaba is central to the company's culture. An annual sales force rally called AliFest led by the charismatic Ma in Hangzhou brings together sellers, journalists, investors, and celebrities (Arnold Schwarzenegger among them) to build positive energy and position it for the big leagues.

Not that it was always so easy. Alibaba almost failed. In fact, Ma's first start-up in 2005, China Pages, did fail, mostly because China's small businesses were not ready for the Web.

His business-to-business site, Alibaba.com, came close to going under during the dotcom crash of 2001 but recovered in 2003. Alibaba was able to engineer a turnaround orchestrated by former General Electric executive Savio Kwan through cost-cutting, layoffs, management development training, and a renewed focus on its core business in China. That foundation has served Alibaba well as it doubled revenues annually and reached $200 million in revenues in 2007 and $8 billion in 2013.

Not that Alibaba has been without controversy. There was a corruption scandal in 2011 over sellers cheating thousands of foreign merchants with fake storefronts on its China web pages. The CEO and COO were let go and 100 salespeople were fired. There was Alibaba's ownership shift of online payment unit Alipay in late 2010 from shareholders including a miffed Yahoo to an entity owned mostly by Ma – a move he said was made to get around foreign ownership restrictions and obtain a banking license in China.

Such a move may not be so surprising to Alibaba observers who know that the visionary strategist Ma believes in "customers first, employees second, and shareholders third." Alibaba is representative of the new generation of Chinese tech titans who are boldly venturing out as business leaders and having a global impact.

National policy dictates the flavor of innovations

The Silicon Valley style of entrepreneurship is spreading throughout Asia, thanks to the region's business talent, capital, and resources plus government-run research and innovation initiatives.

The best example is Singapore, which has relied for the past few decades on government investment and programs to build a knowledge-based economy and stimulate innovation. Singapore has become a hub for finance and commerce, logistics, and research and development, and a headquarters for multinational biotech companies such as Merck, Novartis, Sanofi, and Johnson & Johnson. Yet Singapore lacks a high-profile tech star.

Much of Singapore's push to upgrade its innovation credentials revolves around its National Research Foundation (NRF), which was set up in 2006 to develop policies to foster the Lion City's development technologies ranging from cybersecurity to the environment.

As part of its outreach, the NRF funds 15 fellowships annually for young scientists and researchers over a five-year period, supports 11 venture funds that invest in Singapore tech start-ups, and coinvests with several local technology incubators. Other initiatives include a government-backed start-up fund, Vertex Ventures, and seed funding group Spring. The government additionally supports the state-of-the-art digital facility Fusionopolis, which houses entrepreneurial projects in mobile games, rich media, e-publishing, and social media. Animation company Sparky Animation is a successful offshoot.

Offering a backdrop to this push, the National University of Singapore (NUS) and an INSEAD campus offer a good talent base and springboard for start-ups. Borrowing a model of innovation from Stanford University, NUS provides space for start-ups and invests in them for an equity stake. One recent notable success is Tencube, a mobile phone security system acquired by US tech leader McAfee in a multimillion deal.

Inspired by the Singaporean innovation drive, Nestlé Singapore has embarked on a new program. The huge company encouraged innovation, creativity, and entrepreneurship with teams that relied on mentors and benchmarking with competitors to design and dream up big ideas for a better tomorrow (Business Innovation Culture, 2013). This cross-functional approach of fast, effective, and efficient innovation has since been introduced to Nestlé markets globally.

Turning to mainland China, entire business markets such as outsourcing have developed thanks to government subsidies, incentives, and training. For instance, China crafted a $1 billion program to boost its outsourcing market, providing subsidies, incentives, and training with the goal of developing 1000 vendors and 10 outsourcing hubs. The e-commerce sector is getting a boost, too, underscored by investments in greater broadband coverage and improved mobile networks.

Overshadowed by China's entrepreneurial surge, Hong Kong's budding start-up ecosystem is now on steroids as this finance and trading hub seeks to anchor its economy to high-growth creative enterprises. A major issue has been a shortage of venture capital and angel investment for start-ups. Now, the government is establishing a technology innovation bureau to help Hong Kong reach that tipping point, while incubators and accelerators such as Nest.vc have sprung up by the dozen.

Likewise, the government-supported initiative StartmeupHK, launched in late 2013, is helping to fund and groom champion start-ups with the goal of encouraging them to base their operations in Hong Kong.

Hong Kong's resources to stimulate start-ups include Cyberport and the Hong Kong Science and Technology Park, which run incubator programs, and provide office space, financial assistance, technology support, professional service subsidies, and meet-ups. Outblaze, a leading developer of mobile games and entertainment apps, grew up at Cyberport. Feeding Hong Kong's start-up ecosystem are the Hong Kong University of Science and Technology and the Chinese University of Hong Kong.

Succeeding because of the government or in spite of it

The saying goes that businesses in China succeed because of the government. In India, the opposite is said – entrepreneurs succeed *despite* the government. That has been true for decades as innovators have struggled with bureaucratic quagmires and the country's poor infrastructure.

Recently elected Prime Minister Narendra Modi is showing he is serious about boosting new businesses in India. The government has created a Ministry of Entrepreneurship to work on supporting small and medium-sized enterprises and upgrade worker skills to be more globally competitive.

Ultimately, what India needs to foster a culture of innovation are more entrepreneurial heroes who can motivate the next generation. India may yet close the gap with China on tech innovation with its free-thinking style and culture steeped in 50-plus years of democracy.

Tigers can learn from dragons

Judging from the growing number of world-leading tech titans that are emerging from Asia in the form of both start-ups and big corporations, it is clear that the balance of innovation is shifting from West to East. China – and increasingly India – are the leaders in this parade, and no other region in the world can compare to their progress in recent decades with urbanization, an uplift from poverty and illiteracy, and the creation of thousands of emerging and innovative small businesses.

The tigers in Asia's developing markets can learn from China's mixture of government support and free enterprise. While not the ideal standard for every market, there is little doubt of China's leap on the innovation ladder.

What successful innovation nations have in common are a risk-taking culture, a global outlook, a huge talent pool, a start-up ecosystem, and an advanced technological infrastructure. Only a few nations in the world have all these factors. The US ranks top, and about the only other market that can compare with its heft is China, although arguably Israel is a start-up nation in its own right.

Many of Asia's markets still struggle to change a culture that is widely seen as authoritarian, where conformity is rewarded rather than the out-of-the-box thinking that Silicon Valley thrives on.

It took 30 years for Silicon Valley to fully develop, but with lightning-speed communications today, the pace will be a lot quicker for the Silicon Dragon and Tiger markets. For years, the words "innovation" and "Asia" were never spoken in the same sentence. That is certainly not the case today – nor likely tomorrow.

References

ABC News. (2011). "Steve Jobs: 20 Best Quotes." Retrieved from http://abcnews. go.com/Technology/steve-jobs-death-20-best-quotes/story?id=14681795

Alibaba Group (2014). http://www.alibabagroup.com/en/about/culture

Business Innovation Culture. (2013). "Nestle Singapore Driving Innovation Culture." Retrieved from http://www.bic.sg/inspiration/2014/2/5/nestle-singapore-driving-innovation-culture

Byrne, R. (2014). "Top 25 Mobile Advertising Ecosystems." *Venturebeat.* Retrieved from http://venturebeat.com/2014/02/25/the-top-25-mobile-advertising-ecosystems-beyond-facebook-report/

Dow Jones Venture Source (2013). Silicon Dragon Ventures analysis http://www.ey.com/Publication/vwLUAssets/EY-venture-capital-insights-2013-year-end/$FILE/EY-venture-capital-insights-2013-year-end.pdf page 5 www.silicondragonventures.com

Evdemon, C. (2014). "The Coming Out of China's Tech Companies." Retrieved from https://medium.com/@evdemon

Fannin, R. (2013). "Huawei Pumps Up the Innovation Engines from Shenzhen." *Forbes.* Retrieved from http://www.forbes.com/sites/rebeccafannin/2013/06/26/huawei-pumps-up-the-innovation-engines-from-shenzhen/

Fannin, R. (2014a). "Copying China Business Models in the U.S. Catches On." *Forbes.* Retrieved from http://www.forbes.com/sites/rebeccafannin/2014/07/08/copying-china-business-models-in-the-u-s-catches-on-as-a-new-tech-startup-trend-quite-the-reverse/

Fannin, R. (2014b). "Tesla Eyes China as Biggest Market." *Forbes.* Retrieved from http://www.forbes.com/sites/rebeccafannin/2014/09/26/tesla-eyes-china-as-biggest-market-not-just-a-toy-for-rich-kids/

Gartner (2014). Van der Meulen, R and www.gartner.com/newsroom/id/2653121

Internet Trends (2014). Kleiner Perkins Caufield Byers, Mary Meeker, p. 131 http://kpcbweb2.s3.amazonaws.com/files/85/Internet_Trends_2014_vFINAL_-_05_28_14-_PDF.pdf?1401286773

Internet Trends (2014). Kleiner Perkins Caufield Byers, Mary Meeker, p. 136 http://kpcbweb2.s3.amazonaws.com/files/85/Internet_Trends_2014_vFINAL_-_05_28_14-_PDF.pdf?1401286773

KPMG. (2014). "China's Connected Consumers." Retrieved from http://www.kpmg.com/CN/en/IssuesAndInsights/ArticlesPublications/Documents/China-Connected-Consumers-201402-v2.pdf

Silicon Dragon Video: Nov. 14, 2011 Kai-Fu Lee chats with Startup Asia author https://www.youtube.com/watch?v=nK6Lw1bXaBk

Silicon Dragon SF. (2014). "Tech Chat – Sonny Vu, Misfit Wearables." Retrieved from https://www.youtube.com/watch?v=taxqeHNy6qQ

Silicon Dragon Talk. (2014a). "Chinovation." Retrieved from http://www.youtube.com/watch?v=gN-NRikRu7c

Silicon Dragon Talk. (2014b). "Sizing up Alibaba." Retrieved from https://www.youtube.com/watch?v=u3opIREg4T0

Silicon Dragon Talk. (2014c). "The BAT are buying!" Retrieved from http://www.youtube.com/watch?v=NoNPhwcoO_8&feature=youtube

Van der Meule, R. and Rivera, J. (2014). "Mobile Advertising Spending Will Reach $18 Billion in 2014." *Gartner.* Retrieved from http://www.gartner.com/newsroom/id/2653121

4
Innovation in Emerging Markets: The Case of Latin America

Lourdes Casanova, Jeff Dayton-Johnson, Nils Olaya Fonstad, and Sukriti Jain

Introduction

With economies the world over realizing that scarcity of resources will be the reality of the future, with green growth becoming a necessity instead of a choice, and social development assuming an importance equal to economic growth – with all these challenges and more, policy makers worldwide have realized the importance of innovation in their rendition of good governance. For a long time, innovation has been largely associated with the developed world. However, in the past three decades, developing economies have realized its importance and are embracing it as a critical mission.

There are structural and socioeconomic differences between countries, which translate into differences in how they approach, employ, accept, and benefit from the myriad forms of innovation. In this chapter, we will dwell upon this further and will aim to understand innovation from the perspective of Latin America. We will discuss where these countries stand today in terms of their innovation quotient, the challenges they endure, and recommendations that can facilitate the way forward. With this backdrop, we will also discuss how there are consequential aspects of innovation specific to these countries and how there is a need for an evolved assessment matrix, modified to correctly capture innovation in the context of these underlined differences. In the future, this will be imperative for these countries if they are to channel their resources and policy framework in the correct direction and in timely fashion make their way out of unproductive ventures. This will also provide the world with a more calibrated measure to understand and learn from the Latin American way of innovation.

To achieve the aforementioned objectives, scholars from INSEAD and the Development Center of the Organization for Economic Cooperation and Development (OECD) published a study – *InnovaLatino*, funded by the Telefónica Foundation.

The case for innovation in Latin America

Innovation is conventionally assessed in terms of measures like, but not limited to, R&D expenditure as a share of gross domestic product (GDP), patent applications, and high-tech exports. However, if we look at Latin American economies, more than half of their GDP churn comes from natural resources like agriculture, mining, and petroleum extraction. Such sectors generally have a lower R&D intensity and call for different forms of innovation that have their origins beyond laboratory research. Duly recognizing the importance of the basic innovation indicators and the fact that Latin America faces gross underinvestment and underrepresentation in them relative to the rest of the world, thereby warranting humungous efforts for improvement, we still need to identify additional yardsticks that can provide us the complete picture of the Latin American innovation scene.

A step in this regard is the broadened definition of innovation having its roots in the concept of novelty. The *Oslo Manual* (OECD, 2005) – first published by the OECD and Eurostat in 1992, and currently in its third edition – defines innovation as "the implementation of a new or significantly improved product (good or service), or process, a new marketing method, or a new organizational method in business practices, workplace organization or external relations."

Measuring innovation the conventional way – where does Latin America stand?[1]

There are many yardsticks by which we gauge the measure of innovation-led development in an economy. The performance of Latin American economies vis-à-vis OECD countries across some of these traditional measures such as public and private R&D expenditure as a share of GDP, number of patent applications, high-technology exports as share of total exports, and quality of education is discussed below.

- *Public and private R&D expenditure as a share of GDP* is probably the most basic innovation indicator. All Latin American economies are well below the OECD average expenditure of 2.40 percent (see Figure 4.1). The regional average of 0.31 percent is barely 7.5 percent of the R&D expenditure of South Korea (4.04 percent). Barring Brazil (1.21 percent), all other Latin American countries fall short of even Greece (0.69 percent), which is supposedly the laggard of the OECD nations. There are also wide differences among the Latin American economies, the range being as high as 1.17 percent, indicative of the gulf of development within the region.

With Latin American countries' GDP structure skewed towards natural resource-based agriculture, mining, and petroleum extraction sectors, some

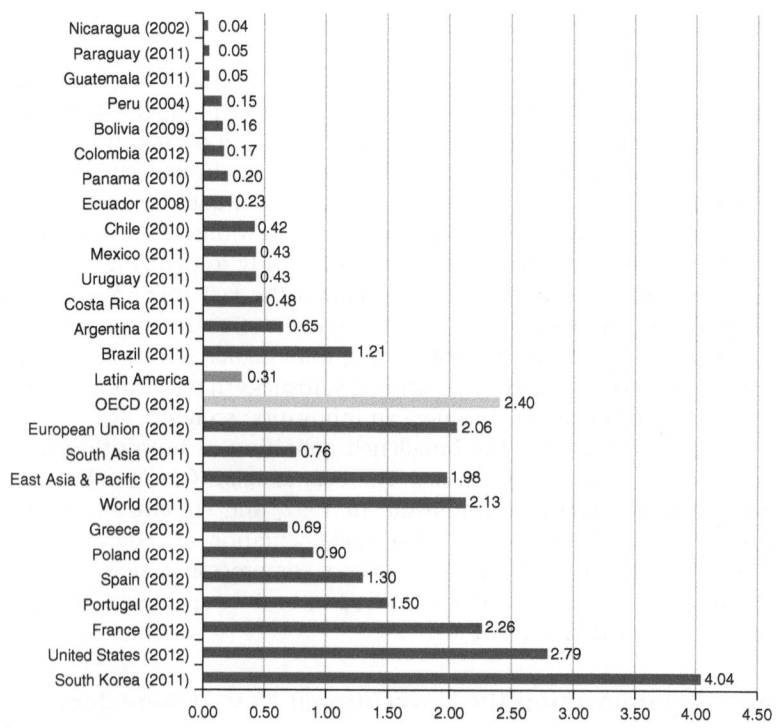

Figure 4.1   R&D expenditure as a percentage of GDP, 2012
Source: The World Bank, World Development Indicators (2014a).
Notes: The average for Latin America is computed for the Latin American countries in the figure including Mexico and Chile.

level of disparity in R&D investments is expected. However, clearly the difference is large enough for only structural differences to account for it completely. Further, not discounting the importance of R&D expenditure, evidently the Latin American countries have a huge amount of catching up to do.

• *Number of patent applications* is another favored indicator of the innovation intensity of an economy. The gap is even more glaring on this front. According to the Science and Technology Indicators of the World Bank, the OECD averaged a whopping 24,414 patent applications filed by residents, as compared to Latin America, which averaged a mere 425. Patents are negligible to nonexistent in most Latin American nations, and even the top performer Brazil (4,804) falls short of even 3.5 percent of the applications filed by South Korea (148,136).

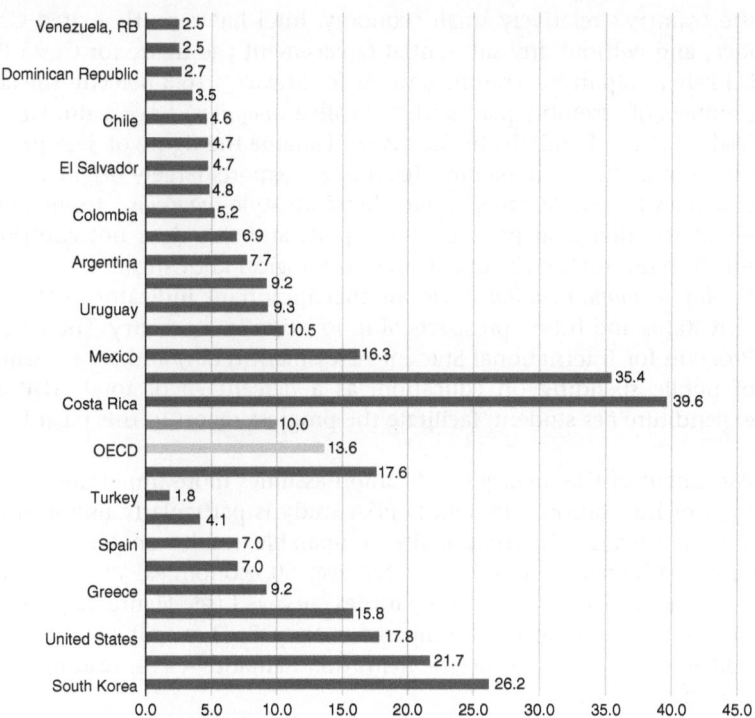

Figure 4.2 High-technology exports as a percentage of total exports, 2012
Source: The World Bank, World Development Indicators, Latest Science and Technology Indicators (2014c).
Notes: These statistics are simple counts based on the date of application and the applicant's country of residence. The OECD average includes Mexico and Chile, which are also included in the average for Latin America. The average for Latin America is computed for the Latin American countries included in the graph. Most recent values (MRV) taken in the cases of Venezuela, Panama, and Honduras as data for the specified year or full period are not available.

- *High-technology exports* as a share of all exports (see Figure 4.2), which is often regarded as the degree of technological specialization of production structures, is another traditional benchmark of innovation. On average, 10.0 percent of Latin American exports are high-tech in nature, relative to 13.6 percent of OECD exports and 17.6 percent of world exports. Even though the Latin American average compares well with the OECD average with respect to this indicator, there is still more to this than meets the eye. Latin America's average is bundled upwards owing to the unusually high contributions from Costa Rica, Panama, and Mexico. Costa Rica's 39.6 percent share is on account of Intel's share of its total exports in the country's relatively small economy. Intel has recently exited Costa Rica, and without any substantial replacement the figure for Costa Rica is likely to dip in the coming year. As for Mexico's 16.3 percent, the large numbers of assembly plants (the so-called *maquilas*) have a similar misleading effect. Similarly, in the case of Panama the figure of 35.4 percent does not paint a true picture. Technology exports largely originate from Panama's free trade zone, where "Mexican style *maquilas*" assemble different technological products for export, and this does not contribute much to improving the country's technological know-how.

• *Quality of human capital* is yet another important indicator of the current status and future prospects of innovation in a country. The OECD's Program for International Student Assessment (PISA) and an assessment of public spending on education, as a percentage of total GDP and expenditure per student, facilitate the progress report in this regard.

Assessment of the quality of education assumes more importance in the context of innovation. The OECD PISA study is particularly useful in this respect. It provides internationally comparable results on test scores in science, mathematics, and reading for over 50 economies. We can clearly see that the six Latin American countries surveyed (see Figure 4.3) are well behind in their test scores as compared to the OECD overall average of 497 as well as the averages in specific fields (mathematics – 494, reading – 496, and science – 501).

While lower scores can be partly explained by lower public spending on education and expenditure incurred per pupil, it is not the entire explanation since Latin American countries rank below the average performance expected by their expenditure level. Even Brazil, the front-runner of Latin America in expenditure on education, really falls behind in comparison to all countries when we look at the PISA scores achieved. There are economies in the PISA study that spend similar amounts as a whole and per pupil to Latin American countries; 15-year-olds in Lithuania, Germany, or Spain substantially outperform their Latin American counterparts, even with similar education expenditure patterns.

Do Latin American 15-year-olds aspire to careers in science and technology? A recent report based on the OECD PISA study concludes that the quality of science education is but one part of developing such an interest. Other factors are required that complement the educational system, including better content on the Internet on science or entrepreneurship, prizes for student projects as well as cultural events that kindle such interest. The presence or absence of these other factors, in turn, likely derives from more general social attitudes towards both innovation and entrepreneurship.

Regarding quality of human talent and based on the findings of a new report *Manpower's Ninth Annual Talent Shortage Survey* (2014), we can state

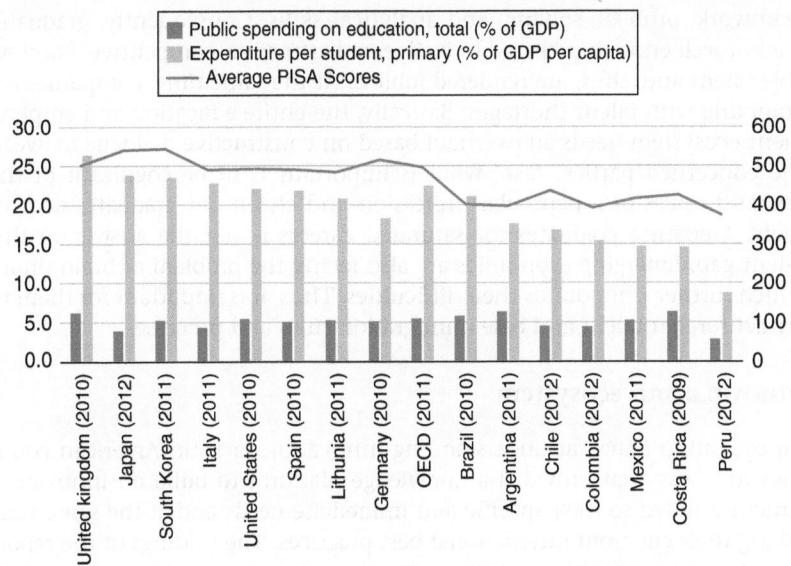

Figure 4.3 Assessment of PISA scores and spending on education, 2012
Source: OECD Statistics (2014) – Education Indicator-Mean Scores in PISA, 2012.
The World Bank, World Development Indicators (2014d) – Public Spending on Education, total (%
of GDP) and (2014b) Expenditure per Student, Primary (% of GDP per capita).
Notes: OECD averages are essentially based on 2011 numbers, but incorporate the latest possible
figures for some OECD nations for which 2011 numbers were not available. Shanghai and Macau
are amongst the highest performers in the PISA study but expenditure data for China are not
available and hence not included in this graph.

that Latin America faces severe talent shortages, and the problem has been
getting worse through the years. Five Latin American countries – Peru,
Brazil, Argentina, Panama, and Colombia – featured in the top ten countries
where employers faced difficulty in filling jobs in 2014. According to the
report, and regarding shortages, Brazil was in the top ten in 2013 and thus,
this is an alarming situation for the region. Further, Japan has been reported
to be the top nation facing talent shortages for the third year in a row. With
36 percent of employers worldwide facing this issue, it is clearly a problem
across the globe. The eurozone, which has been on the receiving end of the
recession, seems to be the only one that is faring well, possibly on account
of a recent relative improvement in the labor market situation.

For Latin America, a key reason for this striking imbalance can be the lack
of quality shown above in the education systems of these countries. There
seemingly is a disconnect between universities and industry in terms of market requirements like vocational training, technical skills, entrepreneurship,

teamwork, problem-solving, and analytical skills. Consequently, graduates are not well enough prepared to be deemed attractive, competitive, employable talent and, thus, are rendered jobless. At the same time, companies are grappling with talent shortages. Basically, the entire education and employment ecosystem needs an overhaul based on constructive dialogue between the concerned parties. Also, what is important is to be cognizant of the demand forces of a particular profession and churn out graduates accordingly. Preparing graduates for saturated careers is not the answer to this talent gap. Emerging economies are also facing the problem of brain drain, which further compounds their difficulties. Thus, it is important for them to attract foreign talent and ease immigration rules and processes.

InnovaLatino ecosystem

InnovaLatino aimed at understanding innovation in Latin American countries in a way that provided a knowledge platform to build an innovation structure suited to their specific and immediate needs and at the same time taking their cue from international best practices. The findings of the report were based on primary research undertaken in eight countries – Argentina, Brazil, Chile, Colombia, Costa Rica, Mexico, Peru, and Uruguay – and it compiled a list of 55 model innovation case studies – "vignettes," from which the region can learn and get inspiration. Accordingly, innovators can be differentiated into six varieties along two dimensions. The first dimension is the size of the organization (based on revenues). Size both enables and constrains how effectively and efficiently an organization engages in innovation activities. It also influences the kinds of resources – such as credit – it can access.

The second dimension distinguishes between organizations driven primarily by maximizing profits and those driven by maximizing social benefits at large, such as poverty reduction, health care for the poor, social justice, and improved literacy among others. Understanding the benefits that an innovator seeks enables us to assess more accurately what critical success factors correlate with different outputs. Taken together, these help define five different types of innovators, which together constitute an innovation ecosystem and they are: public institutions, big companies, small and medium-size companies, and social innovators. There is another category of innovation, which is social, and ranges from corporate social responsibility (CSR) practices to social entrepreneurs.

Based on the concept of novelty, we could also distinguish four types of innovation: product innovation, process innovation, marketing and branding innovation, and business model innovation. In spite of the innovation gap in terms of traditional R&D and patents shown in the previous sections, the broad spectrum of innovation in Latin America lends considerable hope.

Innovation through the Latin American prism

As discussed earlier, Latin American economies are witnessing unconventional innovation, presenting us with a reason for hope and even rejoicing. Some foremost illustrations of this are that Latin American governments are leading the way in innovation, that the region is charting the green innovation frontiers, technological innovation is happening for a brighter future, and collaboration between universities, research institutions, and the private sector is starting to happen. All this is explained below.

- *Latin American governments leading the way in innovation*: Policy makers in Latin America time and again have been saddled with the dilemma of whether to promote innovation in the natural resource-based sectors that drive the economy now or to work towards building the innovation framework essentially based on R&D, patents, and technology. The answer lies in embracing both simultaneously as together they complete the innovation puzzle for these economies. Firms are rising to the challenge of boosting innovation in natural resource-intensive sectors; an example of this is provided by *EMBRAPA*, the Brazilian Agricultural Research Corporation (Empresa Brasileira de Pesquisa Agropecuária, in Portuguese), a government-owned research institute. EMBRAPA, created in 1973, is behind the transformation of Brazil into an agricultural powerhouse and a 150 percent increase of productivity with an increase of agricultural land of only 20 percent in the sector in the last 30 years. In the field of policy initiatives, *Chile´s Development Agency* (CORFO, Corporación de Fomento de la Producción, in Spanish) has launched focused programs to promote process innovations in the mining sector and introduce new species of fish in the aquaculture sector. Similarly, in Argentina the development of dynamic clusters linked to natural resource-intensive sectors has received public funding from *FONTAR* (Fondo Tecnológico Argentino, in Spanish) to execute both individual and associative innovation projects. This has been the case, for example, for the agricultural machinery cluster.
- *Charting the green innovation frontiers*: Innovation is of central importance in combating environmental degradation and can be a key factor in making green growth possible through the development and deployment of environmental technologies. Some Latin American governments and firms are already shifting to greener growth models. Latin America has been very successful with ethanol. Other than biofuels, another example of green innovation is Grupo Islita, which leads a group of Costa Rican enterprises with the common goal of promoting responsible tourism practices that foster cultural authenticity, economic opportunity, and optimal environmental standards. As a member of the World Heritage

Alliance for Sustainable Tourism, Grupo Islita is committed to promoting and preserving World Heritage Sites through sustainable tourism.

- *Technological innovation for a brighter future*: Another particularly noteworthy example is the successful penetration of *mobile technology and Internet users in Latin America*. It is a perfect example of interaction of firms external to the region, providers located in Latin America, customer take-up of new applications, and public policies, particularly deregulation, liberalization, and the presence of an adequate regulatory framework. Many Latin American countries have more Internet and mobile technology users than the global average. Latin America's mobile subscribers' average at 120 percent (see Figure 4.4) is way ahead of the world (93 percent), United States and Asia-Pacific (86 percent) averages and just short of the European average (123 percent). Most countries like Panama, Uruguay, Guatemala, Argentina, El Salvador, Chile, Brazil, Costa Rica, Nicaragua, Ecuador, Colombia, Venezuela, and Paraguay have surpassed the 100% threshold. Peru, Bolivia, and Honduras are very close to full coverage. With innovative business models of "pay per second billing" and prepaid packages, mobile operators have fueled this growth

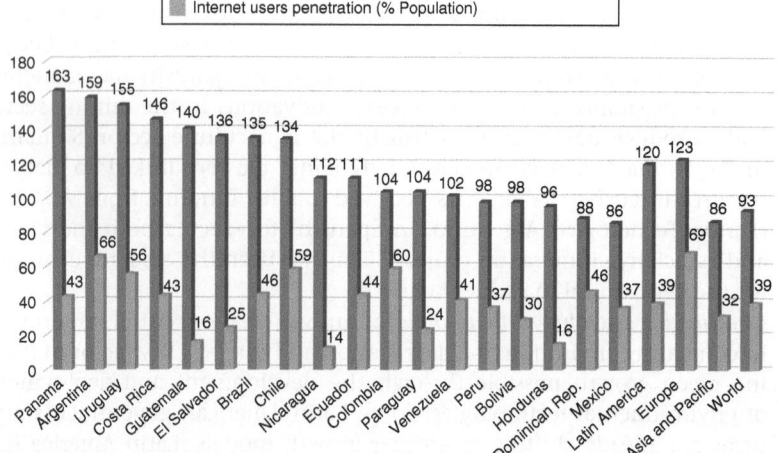

Figure 4.4   Mobile technology and Internet penetration

Source: ITU World Telecommunication (2014), ICT Indicators database; Internet World Stats (2014), Internet Users Penetration 2013.

Notes: Internet users' penetration data for Latin American countries are as of June 30, 2012. The average for Latin America is computed for the Latin American countries in the figure including Mexico and Chile; Internet users' penetration for the Asia-Pacific pertains to Asia only.

further, making services accessible to all sections of society including the so-called "bottom of the pyramid," those still considered poor. Mobile technology allows those citizens below the poverty level to connect with their families and business partners as well as organize their work better. Cleaning ladies all over Latin America can organize their working days in such a way as to reduce traveling time. In terms of Internet users, Latin America's average at 39 percent coincides with the world average and is more than the Asia-Pacific average (32 percent), though it is just about half the European average (69 percent).

With increased competition in the handset manufacturing industry and economies of scale seeping in, it is expected that smartphones will soon become accessible to the masses, which will most likely further drive the Internet usage upwards. This will be an important precursor to harnessing the compelling power of mobile and Internet technology. High-speed broadband connections, in particular, promise to provide an important platform for greater and more productive entrepreneurial activity in many countries of the region, but also for the provision of basic public services like health and education to disadvantaged sectors of the population. Broadband is furthermore a critical platform for businesses, social health delivery, e-governance, and use of ICT for development. Two examples illustrating this are the Intelligent Coffee Identification Card in Colombia where the new minister Diego Molano has implemented an award-winning policy *Vive Digital*[2] (which could be translated as "Live digitally"), and Fundación País Digital in Chile. The Intelligent Coffee ID, accessed via mobile telephones, works as a cash-in and cash-out tool. Growers' profits are deposited by the coffee growers' federation in the ID along with transfers provided by the national government (e.g., crop renovation and protection). Once the money is credited to the ID, the farmer can use it to withdraw cash or buy at selected stores. As for Fundación País Digital, it is a nonprofit foundation aimed at serving as a bridge between the public and private sectors in areas such as smart cities, education and digital development, and the use of ICT in productive processes.

- *Fostering innovation through collaboration between universities/research institutions and the private sector* is another important measure that can lay the foundations of innovation. Unless there is an open dialogue between these parties, innovation making commercial and economic sense for business will not be adequately taken up by universities and research institutions. This linkage is important for both parties to appreciate and cater to each other's needs and motivations. The Casanova et al. (2011) survey highlights that cooperation with education institutions and firms more generally is increasingly being recognized as important. According to the survey, for more than two in five firms (44 percent), cooperation is very important for the development of their innovation activities, and about the same proportion (41 percent) actually engage in some form of cooperation. Universities in Latin America – such as Tec de Monterrey (partnering with technology companies) in Mexico, University of São Paulo in Brazil, Universidad de los Andes in Colombia, Universidad del Pacífico in Perú, and Universidad Católica in Chile – can and are starting to play an important role in this area.

Key take-aways

Based on our analysis above and findings from Casanova et al. (2011), we have arrived at key recommendations, which can facilitate the strengthening of the innovation capacity of the Latin American nations and other emerging economies facing similar challenges and paradigms. These recommendations explained below are: a national vision and a drive for partnerships, innovation related to natural resources, human capital as the catalyst of innovation, supporting small and medium-size enterprises (SMEs) through cluster policies, innovation information ecosystems beyond traditional measures, and support for innovation leading to social inclusion and sustainability.

- *A national vision and a drive for partnerships*: It would be fair to say that government holds the top job, mainly in emerging countries, as far as the innovation corporation of any nation is concerned. The impetus has to come from central planning in terms of ensuring the funding, the infrastructure, the synergy between all stakeholders, the readiness of human capital in terms of skills as well as the innovation culture. Thus, clearly a national vision marked by a drive for partnerships where all parties contribute is a prerequisite. A mutually reinforcing nexus between the government, research institutions and universities, and the private sector will go a long way in channeling intention and funding into commercially viable and acceptable innovations that can advance the development of a nation. *Vive Digital*, an initiative championed by the Colombian government and mentioned above, is one such initiative that has produced tangible results in the right direction.

Further, there is a need to promote private participation. In Latin America, R&D financing is largely centered around public institutions. Typically 60 percent of overall innovation expenditure is publicly financed. The private sector must be engaged to invest in innovation on their own as well as through the more collaborative and, thus, relatively easier means of public–private partnerships.

- *Innovation fueled by natural resources*: Prudence lies in building on your strengths and core areas before venturing into the unknown. Thus, we feel that Latin American economies will go a long way if they work around their structural onus and focus on specific sectors like natural resources including mining and petroleum extraction, agriculture, and tourism/gastronomy as there is already some kind of momentum and it would be best to benefit from and build on it further. Investments in ICT are prevalent all over the world and in the region as well and should also be a priority.

- *Human capital as the catalyst of innovation*: At the core of the innovation infrastructure is the development of human capital. From leadership that fosters innovation policy, which in turn breeds innovation culture and innovation initiative in an educated and skilled workforce that renders innovation implementation – everything hinges on the people in the economy. Thus, it is indispensable that government invests in the education of its citizens and undertakes measures to kindle their interest and acceptance of these novel notions. People must be empowered to take up the futuristic cause of innovation and entrepreneurship and in effect the socioeconomic development and sustainability of their nations. Amidst infrastructure constraints, the growing technology prowess of the continent must be harnessed to impart education through different initiatives using technology and advances in eLearning (Universidad Virtual from the Tec de Monterrey in Mexico was a pioneer in this field). Also, since quality of education is paramount, it is imperative that skill-based vocational training that promotes employability should be encouraged.

- *Supporting micro and SMEs through cluster policies*: Because of their relatively small size, lack of funding support, and inability to deal with the costs, gestation period, and uncertainty associated with innovation, more support is needed for micro and SMEs. Promoting cluster developments, where costs, risks, and human capital can be shared, moving from tax incentives for increased R&D (which is seldom undertaken by SMEs) to loans and grants for SMEs, pro-innovation regulations and policies, and public financing for innovation projects in defined clusters (like in Chile where funding from mining tax is channeled through Innovation for Competitiveness) are some measures that can pave the way to some extent. During the last decade, a number of Latin American governments implemented policies to promote clusters for different purposes: fostering SMEs, such as the Arranjo Productivo Local (APL), a program carried out by SEBRAE (Brazilian Service of Support for Micro and Small Companies, Serviço Brasileiro de Apoio à Micro e Pequenas Empresas, in Portuguese) in Brazil; promoting regional development as in the case of the cluster program of the city of Antioquia, Colombia; or looking for innovative solutions to challenges faced by a sector or group of companies, as in the case of the Technology Consortia Program implemented by CORFO in Chile.

- *Innovation information ecosystem beyond traditional measures*: As discussed earlier, current innovation indicators are not equipped to capture the innovation landscape of emerging economies like that of Latin America, where nontechnological new-to-market or new-to-firm innovations are more prevalent. Thus, as advocated by the OECD Innovation Strategy, Latin American countries need to develop a comprehensive matrix for innovation measurement specific to these economies. A progressive stride in this regard is the *Global Innovation Index (GII) 2014*, which in its 7th edition was copublished by Cornell University, INSEAD, and the World Intellectual Property Organization (WIPO, an agency of the United Nations). It is a comprehensive ranking of world economies, based on a framework of *innovation inputs* like institutions, human capital and research, infrastructure, market sophistication, and business sophistication and *innovation outputs* like knowledge and technology outputs and creative outputs. Such a measure is a step in the right direction and can help capture the true innovation scene of these economies, which can then be translated into well-directed government policies aimed at economic and social development.
- *Innovation leading to social inclusion and sustainability*: Social entrepreneurs, nongovernmental organizations (NGOs), enterprises with their CSR agenda along with governments and public institutions must ensure that there is a balance between economic innovation and social inclusion and sustainability. This is especially crucial in the context of the emerging and developing economies of Latin America. The region has numerous examples, as detailed in this chapter, which are upholding this end of the innovation balance and the idea is to continue in this direction.

Thus, clearly from the perspective of Latin American countries, grappling with increased pressures on per capita resource availability, innovation can help improve productivity, economic growth, social inclusion, and sustainability. Needless to say, government has a pivotal role to play if this innovation dream is to come anywhere close to fruition. Comprehensive, symbiotic, and coherent policy-making at the center creating synergies between all stakeholders, increased R&D expenditure, and pro-innovation incentives and regulations have improved what has been called hard and soft (such as education) infrastructure with the availability of educated and skilled human capital – all this and much more lie in the hands of the government. To complement these efforts, the private sector also needs to play its part in terms of taking up the cause of innovation risk taking, especially against the backdrop of high costs and long gestation periods associated with these initiatives. The private sector, individually and in partnership with the public sector, needs to foster an investment and innovation culture. It is important to celebrate local best practices that can be replicated and will be useful in the context of local requirements, rather than expending energy importing international ways and means that are eventually not relevant or useful in the Latin American context.

In conclusion, Latin America is innovating, albeit in its own unique way. It is redefining innovation, giving new paradigms and facets to innovation and, thus, charting its growth story, tailor-made to tackle its weaknesses and build on its core strengths. Thus, the economies are becoming better equipped to seize their opportunities, even with the backdrop of scarcity of resources and low R&D expenditure. There is no doubt that Latin America needs to continue building upon this foundation, and at the same time strive to improve its performance across the traditional indicators.

Notes

This chapter builds upon research published in *InnovaLatino: Fostering Innovation in Latin America* (2011), the result of an INSEAD/OECD research project funded by Fundación Telefónica about innovation in Latin America. The contribution of Anna Pietikäinen, Counsellor, OECD Sahel and West Africa Club, as well as all InnovaLatino co-authors is gratefully acknowledged.

1. A commonly accepted measure for innovation, albeit a broader one, is productivity. We have not covered the concept of total factor productivity as it is beyond the scope of this chapter.
2. Vive Digital won the world's Best Technology Policy award at the Mobile Congress in Barcelona in 2012.

References

Casanova, L., Castellani, F., Dayton-Johnson, J., Dutta, S., Olaya-Fonstad, N., and Paunov, C. (2011). "InnovaLatino: Fostering Innovation in Latin America." Retrieved from http://www.innovalatino.org/
Internet World Stats. (2014). "Internet Users Penetration 2013." Retrieved from http://www.internetworldstats.com/stats.htm
ITU World Telecommunication. (2014). "ICT Indicators Database 2014." Retrieved from http://www.itu.int/en/ITU-D/Statistics/Pages/stat/default.aspx
OECD Statistics. (2014). "Education Indicator-Mean Scores in PISA 2012." Retrieved from http://www.oecd.org/statistics/
Organization for Economic Cooperation and Development. (2005). *The Oslo Manual: Guidelines for Collecting and Interpreting Innovation Data. The Measurement of Scientific and Technological Activities*, 3rd edn. Retrieved from http://www.oecd.org/
The World Bank. (2014a). "World Development Indicators." Retrieved from http://data.worldbank.org/indicator/GB.XPD.RSDV.GD.ZS/countries/1W?order=wbapi_data_value_2012%20wbapi_data_value%20wbapi_data_value-last&sort=asc&display=default

The World Bank. (2014b). "World Development Indicators – Expenditure per Student, Primary (% of GDP per Capita)." Retrieved from http://data.worldbank.org/indicator/SE.XPD.PRIM.PC.ZS?display=default

The World Bank. (2014c). "World Development Indicators, Latest Science and Technology Indicators." Retrieved from http://wdi.worldbank.org/table/5.13

The World Bank. (2014d). "World Development Indicators – Public Spending on Education, Total (% of GDP)." Retrieved from http://data.worldbank.org/indicator/SE.XPD.TOTL.GD.ZS?display=default

5
Entrepreneurship and Innovation in the Middle East: Current Challenges and Recommended Policies

Norean R. Sharpe and Christopher M. Schroeder

Introduction

Despite an unstable political climate, innovation in the Middle East has enormous potential if regulatory reforms can be implemented. While Israel may have been at the forefront of entrepreneurial activity, other nations in the region are showing signs of growth in innovative start-ups and services. In this chapter, we will focus our discussion on those countries that have implemented business reforms and exhibited notable financial investment, organizational support, and individual start-ups. These same economies have also experienced transitions and shifts in access to technology. It is not difficult to find individual examples of high-tech start-ups, incubator labs, and serious capital investments – but such innovation may be surprising, given the continued level of unrest and uncertainty in the region. While it may be tempting to treat the Middle East as one homogeneous region, it is important not to aggregate all the countries and cultures, as each has its own set of advantages and challenges.

We and other scholars have observed that, by all metrics, such as unemployment, educational attainment, investment, and female participation, the level of innovation varies across national borders. These economic, political, and legal differences contribute to distinct attitudes toward change and firm-level innovation across cultures. While there is much in common among the countries in northern Africa, there are also variations in religions, regulations, public policy, access to information technology, and logistical and legal infrastructure. In this chapter, we will identify organizations that currently support innovation, share examples of successful entrepreneurial activity in the region, and recommend national policies needed for innovation to be sustainable and scalable.

Current national policies and regulations

The World Bank reports titled *Doing Business* have studied regulations that encourage business and innovation since 2004. It is clear that innovation requires regulations and rules that support business owners and investors. These regulations need to be accessible and easy to implement. Regulations impact our ability to start a business, apply for credit, raise capital, protect investors, move goods across borders, hire employees, and even declare bankruptcy. The *Doing Business 2011* report showed that 61 percent of economies in the Middle East and North Africa reformed regulations facilitating businesses in 2009 or 2010. In addition, about 20 percent of the reforms that facilitated trade were in the Middle East and North Africa (MENA). Out of the 18 economies examined in MENA, 11 were found to have reformed business regulations such as customs procedures, including Bahrain, Egypt, and the United Arab Emirates (World Bank, 2010).

Since then, however, little progress has been made. The *Doing Business 2015* report showed that the regions where the most regulatory reforms have taken place are Europe, Asia, and sub-Saharan Africa. The one exception was the United Arab Emirates (UAE), which ranked 22 out of the 189 economies studied. In fact, the UAE was ranked in the top ten economies for improving the most, though in its three core areas of success – registering property, getting credit, and protecting minority investors – only the latter has direct impact on start-ups. Overall, the UAE enacted a total of 30 reforms in 2013 and 2014 that protected minority investors (World Bank, 2014).

Outside the UAE, the Middle East ranks among the lowest regions for new businesses; according to the World Bank, Saudi Arabia ranks at 49, Qatar ranks at 50, Bahrain ranks at 53, and Tunisia at 60. The remaining economies studied by the World Bank were ranked between 66 and 188 (see Figure 5.1 for the complete rankings).[1]

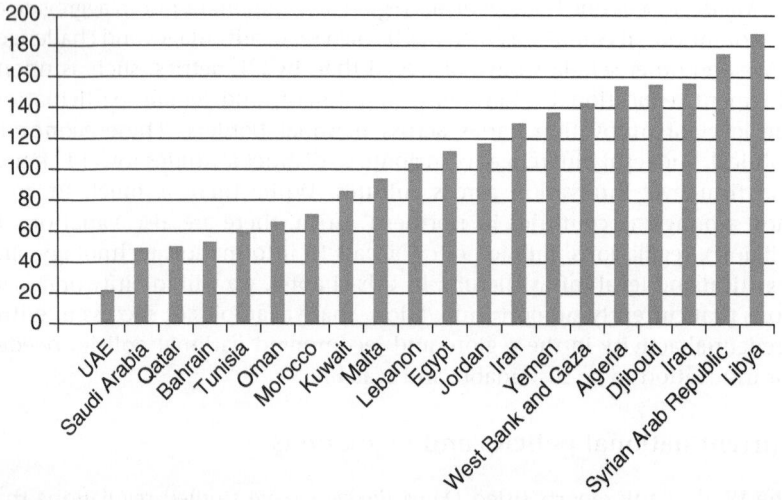

Figure 5.1   Ease of doing business rank
Source: World Bank (2014).

Many of these economies lack the basic services needed to support entrepreneurs. For example, new firms need access to credit. Thus credit bureaus and credit registries are needed to help banks assess the credit risk of micro or small start-ups (Brown et al., 2009). In fact, a recent study showed that access to credit increased faster for small businesses than large firms after credit system reform (Brown and Zehnder, 2007). The fact that the MENA region has among the lowest number of economies that provide credit scores and online access to credit data, may help explain the low amount of entrepreneurial activity in that region.[2]

The importance of access to credit, as well as other measures, such as the ability to recover from bankruptcy, can be seen in the example of Saudi Arabia. Although this country is rated highly by the World Bank in the areas of dealing with construction permits (21), access to electricity (22), registering property (20), and taxes (3), it ranks far lower in getting credit (71) and resolving insolvency (163), which hurts its overall ranking. An important measure of a nation's ability to assist firms out of bankruptcy is the recovery rate, as measured by cents on the dollar. In Figure 5.2, we can see that Saudi Arabia is underperforming in the area of resolving insolvency compared to other nations (see Figure 5.2).

Besides the challenges of recovering from bankruptcy, getting funding, and obtaining licenses to start a business, many cite national policies governing the moving of products across borders as an issue. For example, according to the World Bank, in Egypt, Jordan, and Lebanon, it takes 8, 12, and 9 days, respectively, to start a business and 12, 12, and 22 days to export

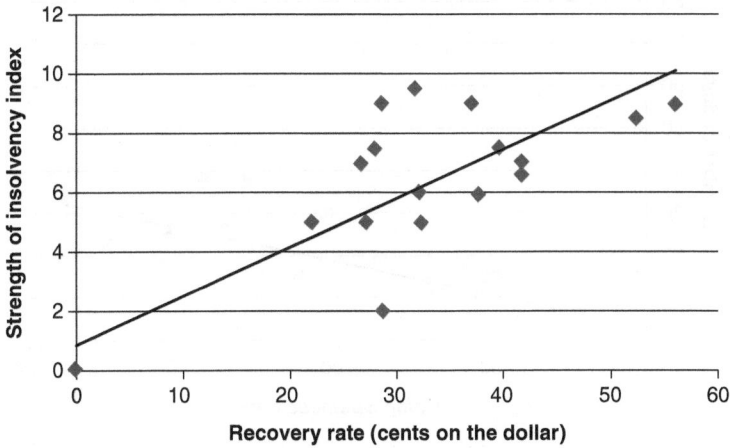

Figure 5.2 Strength of insolvency index
Source: World Bank (2014).
Note: The correlation between the strength of insolvency index and the recovery rate in the MENA nations is 0.82 ($p < 0.01$) (Doing Business 2015 database).

products. To put this in perspective, this compares to 2.5 days to start a business and 6 days to export in Singapore (ranked 1) and 5.6 days to start a business and 6 days to export products in the US (ranked 7). While Jordan has made recent reforms in trading regulations (by improving infrastructure at the port of Aqaba), neither Egypt nor Lebanon have made any progress in trading efficiency. Figure 5.3 shows the relationship between the *Doing Business 2015* ranking and days to export and import.

Due to the lack of critical regulatory reform, the economic climate in the Middle East remains challenging for all those who want to take risks, embrace change, raise capital, and capitalize on new access to the Internet and technology. Compared to other regions, MENA ranks far below OECD high-income countries in providing a "business friendly environment" (World Bank, 2014). As one might expect, those economies with a higher gross national income (GNI) per capita in the Middle East in general had a more favorable business rating (see Figure 5.4).

In confirmation of these challenges, another recent report from the World Bank (2013) analyzed data from over 5000 firms in Egypt, Jordan, Lebanon, Morocco, Saudi Arabia, Syria, Gaza and the West Bank, and Yemen and reported similar regulatory obstacles. All firms in the World Bank survey (both male- and female-owned firms) shared their perceived struggles with regulations (including taxes, customs, trade, and licensing) and access to

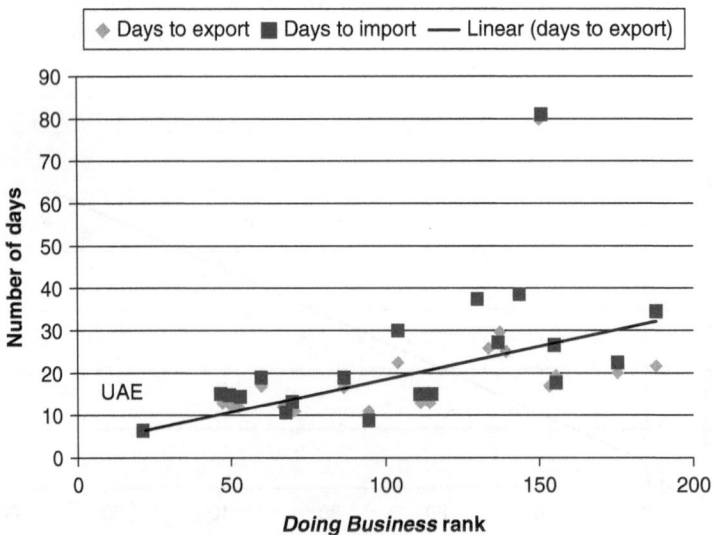

Figure 5.3   Days to export and import vs rank
Source: World Bank (2014).

Note: the correlation between the Doing Business rank and the days to export in the MENA nations is 0.48 ($p < 0.05$) (Doing Business 2015 database).

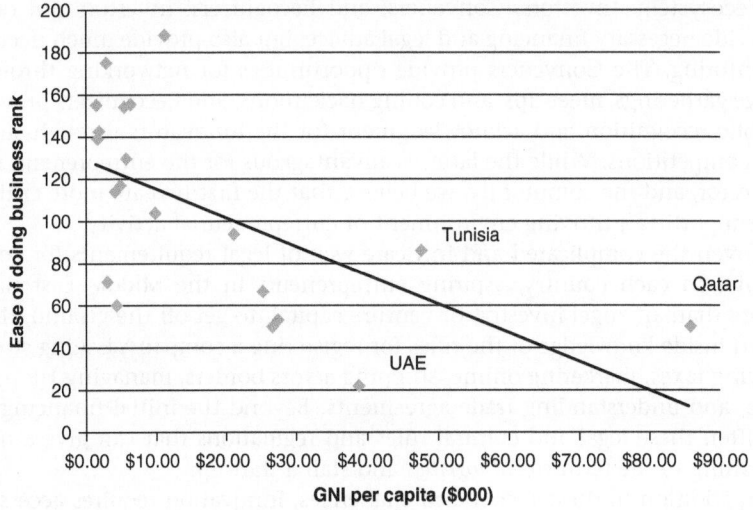

Figure 5.4 *Doing business* rank vs GNI per capita
Source: World Bank (2014).
Note: the correlation between the Doing Business rank and the GNI per capita in the MENA nations is –0.60 ($p < 0.01$) (Doing Business 2015 database).

financing. In these categories, the perceived barriers were higher among firms in the Middle East than among firms in East and Central Asia, Latin America and the Caribbean, and Europe.

Can we show that changes in regulations impact existing firms, employment, and new businesses? Research confirms the relationship between regulatory and legal reforms and measures of growth in economies and entrepreneurial activity (Djankov et al., 2002, 2007, 2010). Other research shows the connection between making business registration easier and increased firm formation in Portugal (Branstetter et al., 2013), Mexico (Bruhn, 2011; Kaplan et al., 2011), and Brazil (Monteiro and Assuncao, 2012) and between trade reforms and increased economic growth (Chang et al., 2009). Additionally, recent research suggests that an increase in ability to recover debt can improve the performance of credit markets (Love et al., 2013) and that an increase in registration procedures is a deterrent to entrepreneurship (Dreher and Gassebner, 2013). One reason that Israel has such an entrepreneurial environment is due to its financial and regulatory structures. Israeli laws regarding bankruptcy and new businesses help to facilitate innovation – perhaps explaining Israel's higher ranking. Clearly, access to financing, at all stages of the process, is critical for any start-up.

In addition to regulations and policies, individuals and organizations are critical to supporting an ecosystem of innovation. Schroeder (2013) outlined three types of individuals or organizations that are required to support such an ecosystem: Investors, Conveners, and Recognizers. Investors not only provide necessary financing and legal advice, but also provide much needed mentoring. The Conveners provide opportunities for networking through large gatherings, meet-ups, and coding hackathons, and Recognizers provide public recognition and acknowledgment for the innovators through start-up competitions. While the latter is advantageous for the entrepreneur, the investor, and the community, we believe that the first two are more critical to supporting a thriving environment of entrepreneurial activity.

Given the complicated and intricate web of legal requirements for companies in each country, aspiring entrepreneurs in the Middle East need more than an angel investor, or venture capital, to get off the ground; they need inside knowledge of the rules for registering a company, issuing stock, paying taxes, marketing online, shipping across borders, managing HR policies, and understanding trade agreements. Beyond the initial financing, it is often these legal and cultural rules and regulations that can give a new meaning to the term *global logistics* and stall a start-up.

In addition to these recognized challenges, innovation requires access to technology. In today's high-tech world, a new business would find it next to impossible to thrive without access to advanced technology – and the education and/or personnel to know how to take advantage of it. This latter necessity can partly explain the growth in financial, technical, management, and computer science training; access to education and professional mentoring/coaching has become essential to encourage and empower entrepreneurs.

Evidence and examples of innovation

Beyond providing seed funding, Investors include those people or organizations who provide support and incubation space. One such organization, Oasis500 in Jordan, provides training, mentoring, and funding for local start-ups. The advantage of Oasis500 is that they provide customized regional training and help to leverage strategies and opportunities already proven in the West.

Relatively new on the scene (since 2011), Flat6Labs was cofounded by Ahmed El Alfi and Hany Sonbaty and provides space for young businesses to work, have access to technology, and receive training in HR, finance, marketing, and accounting for up to three months (a cycle).

With over 80 mentors, Flat6Labs is having a real impact. In the original Cairo office alone, Flat6Labs has spent about $1.5 million to support nearly 50 firms. Over 50% of these young firms have received external funding and examples include a start-up that recycles trash bags into furniture coverings (ReformStudio, a 2014 Cartier Women's Initiative Award Winner); a tech start-up that uses thermal imaging to scan cows to diagnose disease and infections to maximize milk production; and a firm that takes cooking oil from restaurants and uses if for biodiesel fuel. Perhaps the most successful firm is a nonprofit that provides online education. The firm Nafham (meaning "we understand" in Arabic) has produced and crowd-sourced video lessons for the entire Egyptian K-12 curriculum for free delivery on YouTube. They are the most widely used online education site in the Arab region, delivering 100,000 daily video lessons worldwide.[3]

Flat6Labs has recently expanded to two new locations – Jeddah and Abu Dhabi. The Jeddah location is a collaboration with a young start-up named Qotuf (Arabic for "low-hanging fruit"), and already has success stories to share. While only in their third cycle of projects, this incubator has been able to support several young entrepreneurs in launching their businesses. Examples include a young woman who is establishing her own e-business for signs, a team of innovators who are developing their own educational "brain teaser" online computer games for children in Arabic, and a husband and wife team who are launching their own Arabic e-grocery business (similar to Peapod.com in the US).[4] In speaking with each of these young innovators, it is apparent that the financial and logistical support provided by Flat6Labs and Qotuf is critical to their success. Given the challenge of moving actual products, and the easy access to technology, it is no accident that the businesses supported by this incubator all have a technical/Internet focus.

Historically, one of the first industries to gain a foothold in the Middle East was telecommunications. The first mobile spectrum auction in Morocco in 1998 was expected to raise $50 to $70 million; Moroccan officials were stunned when the first bid came in at $1 billion. Government officials had assumed that mobile telephony was a luxury toy – rather than a revolutionary means of communications and data sharing. Meanwhile, businesses had the foresight to envision a world where everyone used a mobile device. Today, most countries in the Arab world have more than 100 percent mobile penetration (since many citizens own more than one device), thus revolutionizing people's ability to communicate, connect, and transact. With greater ramifications, mobile operators in the region expect that by the end of the decade smartphone penetration will surpass 60 percent, which has, in fact, already happened in most of the Gulf.

Smartphones enable more than connection and communication; they provide accessibility and affordability for many businesses and consumers. This means that those in the Middle East now have an unprecedented ability to collaborate, experiment, and market their goods and services directly to their customers. Users have access to knowledge, competitors, and consumers. Thus, it should not be surprising that many of our examples of thriving start-ups in the Middle East involve e-commerce, e-marketing, e-education, or e-trade.

According to Schroeder (2013), new firms are either Improvisers – those who adopt known and proven Western models, Problem Solvers – those who work to solve regional challenges, or Global Players – those innovators who are launching unique firms that are scalable outside the local region. Early Improvisers relied heavily on frameworks that were already successful in the US. A classic example is Aramex, the Middle East's largest logistics and transport company, which was largely based on the successful structure of FedEx. It was the first Middle East firm to go public in 1997 and its founder Fadi Ghandour is now an investor in other area start-ups. The firm continues to grow and reported an increase in revenues of 12 percent (to AED 922M) and in net profits of 16 percent (to AED 69.5M) in the third quarter of 2014 compared to Q3 in 2013.[5]

Another early start-up, Maktoob, an all-Arabic Internet portal founded in 1998 in Jordan, was referred to as the "Yahoo of the Middle East" even before Yahoo bought them in 2008 for $175M. The private equity firm Abraaj, founded in 2002, manages $7.5B in assets and has 25 offices around the globe with over 300 employees. Their six major hubs are Dubai, Istanbul, Mexico City, Mumbai, Nairobi, and Singapore and the firm's founder Arif Naqvi was recently appointed to the Global Impact Board by the UN Secretary General. Other examples of firms that "borrowed" innovative ideas from the West include Altibbi.com, an Arab medical site similar to WebMD.com, ArabMatrimony.com, a popular marriage site in the Middle East similar to Match.com, and Namshi, an online retail site, similar to Zappos.

In addition to the firms above that have focused on the retail or financial space, other start-ups have chosen to tackle real problems in the region. Examples include RecycloBekia, founded in 2011, which resells and recycles computers and electronics; Bey2ollak, launched in 2010, which boasts over 400,000 users who share updates on real-time traffic patterns; and Emokhalfa, the first platform in Egypt that enables users to report bad drivers instantaneously. Each of these firms has pioneered approaches to solving regional issues and providing a service to consumers. While none of these have grown their reach outside the MENA region, probably the best example of a company that has become a "Global Player" worldwide is WeatherHD. Created by Amir Ramadan in 2010, this site allows users to track weather in cities around the globe, comparing multiple forecasts for the same cities and combining information with a unique graphic interface. It is now called ClearDay with enhanced radar and weather data and is easily downloaded from the iTunes App Store.

Newer firms have continued to use the Internet, either to provide services, or to sell their products using e-commerce. Hind Hobeika, a Lebanese female and former professional swimmer, used a combination of crowd funding and institutional funding in 2010 to launch Instabeat. The firm's primary product is smart swimming goggles that track and display information, including heart rate, calories burned, and laps completed, to help competitive swimmers track their performance. Her firm receives orders from over 50 countries and the goggles were a top pick for wearable technology at the

2014 International Computer Electronics Show. Other small firms in the Middle East use technology to stream video in Arabic (Istikana, Ltd), provide a smartphone app that replaces loyalty cards (Cashbury), or offer group buying savings, similar to Groupon (Offerna.com). Finally, Genieo Innovation, founded in 2008, is an Israeli firm that provides a personalized homepage and other options. A modern success story, Genieo Innovation was acquired in 2014 for $34M by Somoto.

Besides e-commerce and e-marketing, a growing area of innovation in the MENA region is in education. Most likely in response to a poor public education system in the region, which ranks among the lowest in the world, both in outcomes and in skills, online education is growing. As much as 15 percent of the start-ups in the region are offering online or mobile platforms and videos to supplement learning and classes to millions of users. One of the reasons that Saudi Arabia is number one in per capita usage of YouTube in the world is the demand by women for education videos and online courses. In fact, with approximately half of the Middle East's population of 350 million being women, this demographic provides an untapped resource – both for entrepreneurs and consumers.

Women rising

While the entrepreneurial climate for women may be particularly challenging, they are making advances. A recent report from the World Bank (2012) analyzed data from over 5,000 firms in the Middle East and found that women in the region own about 13 percent of the firms, and of these female-owned firms, only 8 percent are micro firms (with <10 employees), while over 30 percent have more than 250 employees. The countries with the greatest percentage of "large" female-owned firms are Egypt, Morocco, and Saudi Arabia. The World Bank also found that female-owned firms hire more workers in Egypt, Jordan, Saudi Arabia, and Gaza and the West Bank – with women constituting a larger proportion of the employees at female-owned firms.

According to the recent Global Entrepreneurship Monitor (GEM) 2012 Women's Report, for every woman entrepreneur, six women intend to start a business in the MENA/Mid-Asia region. In fact, this ratio compares to just 2.5 men intending to begin a start-up for every male entrepreneur in the same region. In addition, women in the MENA region are more likely to generate businesses in the consumer sector, while men are more likely to begin start-ups in the transformative and extractive sectors. Sole ownership among entrepreneurs is similar for women and men in the MENA region (about two thirds of start-ups), while nearly 90 percent of female-owned businesses in Israel are solely owned (Kelly et al., 2013).

To understand the potential of women in the MENA region, consider the results of the GEM Report, which showed that the MENA/Mid-Asia region had the lowest level of female total entrepreneurial activity (TEA) of any region at about 4 percent – and the highest gender disparity, since the male TEA level for that region is approximately 15 percent.[6] In addition, the World Bank Enterprise Survey data revealed that while women own fewer firms than men in every region examined, they are a smaller minority in the MENA region than in East Asia, Latin America, Europe, and Central Asia.

One theory for the increase in female entrepreneurial activity is that inspiration and innovation are born out of necessity. It is the lack of employment, struggle to earn a living, and desire to have a better living that drive the passion for change and willingness to take risks. A prior publication summarizes cases of many women-driven initiatives (Syeed and Zafar, 2012). Their examples include entrepreneurs across all industries, all stages of development, and all socioeconomic statuses. This is consistent with the GEM 2012 Report, which found that female entrepreneurs in the MENA/Mid-Asia region were distributed equally across three levels of income: lowest, middle, and highest (Kelly et al., 2013).

Challenges and recommended reforms

This connection between necessity and inspiration crosses both genders. Schroeder (2013) provides many examples of businesses built out of "passion and necessity." In fact, several entrepreneurs he interviewed made the case that the bursting of the tech bubble in 2001 and the worldwide crash in 2008 actually promoted ecosystems of entrepreneurial activity. Thus ongoing challenges that persist in the Middle East may, in fact, provide the impetus and stimulation that budding innovators need to take risks and enact change.

One remaining challenge is that regional and international officials are accustomed to a top-down world of mass programs and bureaucratic regulations. They are slow to change and unclear what to do in a bottom-up world that would encourage individual innovation and economic growth. Crucially, governments often fail to understand the implications of shifts in technological, legal, and economic regulations for economic growth and job creation in the region.

According to entrepreneurs in the Middle East, there are three primary reasons that innovation is stifled in the region. First, in many countries there is confusion between what is needed by tech start-up companies versus "small and medium-sized enterprises" or SMEs. The former require smaller amounts of capital but flexible legal structures, while the latter tend to be established and profitable enterprises. Governments, especially in the Gulf, have passed procurement laws, such as requiring governments to have 10 percent or more contracts with SMEs, and have granted SMEs exemptions from customs tax for equipment, raw materials, and goods for production purposes. Important as these are – and in most countries in the region well over two thirds of nonoil and nongovernment jobs are in SMEs – they have no impact on tech-enabled start-up opportunities.

Second, governments often focus on creating and enabling businesses with investment capital and support services in "free zones" – areas like the Dubai Media City or Dubai Internet City – with less attention to concrete legal reform. Kuwait, among other countries, is setting up investment funds which, again, are more focused on SMEs and less on tech-enabled start-ups. Bahrain and Dubai's free zones still pose difficult requirements for new companies to simply register and commence business. Egypt and Jordan have supported start-up initiatives like Jordan's Oasis500 – noted earlier as one of the first business incubators in the region set up as a private/public partnership – but the future focus is uncertain as governments and ministers have so often repeatedly changed hands. The UAE's recent announcement of a new category of work visas to be issued allowing people to work without having to rent an office is encouraging, but details of their impact remain unclear.

Third, even where good laws have been put on the books, they have taken an extremely long time to get passed and their enforcement has been less clear. The Gulf Cooperation Council (GCC) recently passed laws to stream-line customs and improve free movement without visas among its union, but they took nearly a decade to pass and acquiring permits can be costly and take months. The ability to freely and cost-effectively move goods and services from outside the union is hampered by the paperwork required and corruption. In fact, the number one issue that held up laws being passed more quickly was the appropriate mechanism for the distribution of customs revenues. One official confided: "The fact is, entrenched status quo inter-ests are unwilling to give up what, in their minds, are special privileges – like openness to foreign ownership and more competitive real estate pricing – which work against startups." A start-up who took nearly six months and had to pay nearly $50,000 in fees, lawyers, and travels – nearly half of his first round of investment – to register in one of the free areas of Dubai added: "You don't feel you're talking to smart people, people who care – they just follow the book exactly and cannot think, don't want to think. They are not customer service oriented. They are afraid to take any risk. It's like dealing with robots."

Based on interactions with current and aspiring entrepreneurs, the follow-ing recommendations are needed to improve the climate for innovation in the Middle East. Their implementation will require those in leadership to make entrepreneurship a priority in their economies:

- *Loosen trade restrictions*: E-commerce has enormous growth potential in the Arab world; consumers in the region now spend more than $1 billion a year online. This figure is expected to double within two years. To give a sense of the opportunity, offline retail sales in the Arab Middle East total about $425 billion a year. However, each country in the region has its own obscure, cumbersome, and costly shipping regulations. Moving goods between Arab countries is analogous to the untenable situation of each state in the US having its own customs laws that require checking each package and adding fees and taxes at every border. Note that the indicators measured by the World Bank in *Doing Business 2015* included the documents, cost, and time required for imports and exports by seaport.

None of the economies in the Middle East showed notable improvements in the area identified as "trading across borders." Although the economies in the lowest quartile in terms of days required to import and export products have improved since 2005, the average number of days to import in these worst-performing economies is nearly 50 days, and to export it is about 40 days. In fact, only Jordan and Morocco in the Middle East were cited as making improvements in trading across borders in 2013–14 – and in Yemen, trading actually became more challenging due to inefficiencies in port operations and logistics (World Bank, 2014). An improvement in the ability of the nations in the Middle East to move goods and services across borders would likely result in vastly improved access to consumers and revenue.

- *Encourage free circulation of knowledge and ideas*: The Arab world urgently needs a system to transfer promising technology developed in universities and other research institutions into the start-up ecosystem. For example, universities in Gulf countries are developing advanced solar-power and desalinization technology. However, there is no web of connections among research labs and the start-up community such as one finds at MIT or Stanford. One bright light is MASDAR City. This recent project in Abu Dhabi aims to build a tech-focused, solar-based city with partners around the world. However, skeptics are concerned that it will be a walled enclave for the elite and that technology parks generally tend to isolate technological innovation, as opposed to embracing and enhancing societal impact.

Another problem is that Arab countries generally lack investment laws designed to encourage tech-based innovation. The legal picture is not entirely bleak. For example, it is easier to enforce a contract in parts of the Arab world than it is in many East Asian countries. Yet most countries in the region lack clear and coherent legal protections for real and intellectual property rights. Most Arab countries restrict financial flows, making it difficult for businesses to repatriate capital. Laws that criminalize bankruptcy tend to stifle the ability of entrepreneurs to fail quickly and then restart. Such restrictions deter entrepreneurs and investors from pursuing new ventures in the Middle East.

- *Modernize regulatory regimes*: Some of the most interesting mobile payment start-ups exist in the Middle East. However, Arab entrepreneurs have complained that government institutions that issue permits cannot move fast enough to keep up with them. In most Arab countries, separate bureaucracies regulate telephony, banking, domestic transactions, and foreign transactions. Yet mobile payment technology touches all these sectors. In addition, in most countries, bankruptcy laws are nascent or, for example, even treated as a criminal violation. Acceptance of failure is a necessity for a vibrant inovation ecosystem, and efficiency in shutting down a business that does not work is essential. By updating regulatory regimes so that they reflect twenty-first-century economic realities, Arab governments can help accelerate the growth and impact of innovative tech companies.

- *Facilitate risk protection for early-stage investors*: In an effort to encourage foreign direct investment in emerging markets, large aid institutions, such as the US Overseas Private Investment Corporation and the World Bank, have long offered debt financing and political-risk insurance for companies that invest in these markets. While not everyone agrees about the overall efficacy of these programs, risk-mitigation products have certainly helped a number of Western companies reach middle-class consumers in the Middle East and other emerging markets that combine economic potential and political volatility. Start-ups rarely qualify for debt financing, and multilateral aid organizations are often ill-equipped to support equity investment. Moreover, the amounts required to launch a typical tech start-up are much smaller than these institutions are used to disbursing. Aid organizations should support the emerging Arab start-up economy by focusing on "bottom up" investment products and strategies.

- *Expand opportunities for women*: Data generated by the GEM Report and the World Bank lead us to believe that the critical path toward more women-run businesses is expanding access to financial and technical educational and training opportunities – particularly in disciplines that teach and support innovation, risk-taking, financial investment, and launching new ventures. As evidence that training and education will spur innovation, we cite the current relationship between education and entrepreneurial activity; in most regions, women entrepreneurs are more likely to have post-secondary education than women who are not entrepreneurs (30 vs 26 percent for MENA/Mid-Asia) and more likely than male entrepreneurs (30 vs 26 percent for MENA/Mid-Asia). For a comparison, 70 percent of female entrepreneurs in the US and 55 percent in Israel have a post-secondary degree (Kelly et al., 2013).

Our analysis of the World Bank data showed that unemployment among women in the Middle East is relatively high, although it differs by country; it has been lower over the past five years in Lebanon, Israel, and Qatar (2–12 percent), compared to Egypt, Saudi Arabia, and Jordan (14–28 percent). These employment patterns stem from numerous challenges that women face in the Middle East. It is well-documented that the perception and tradition of women outside the home vary by nation and culture, and are different in other regions. These attitudes are compounded by the restrictions for women in areas that make working and starting companies more difficult, such as regulations on transportation, ownership, banking, and even daily interactions with men.

Given this persistence in unemployment among women, women in many countries in the Middle East are finding entrepreneurship a necessary path – a way to create their own opportunity. Thus it is critical that nations in this region find ways to provide focused training and skills in working with technical advisors, lawyers, incubators, and investors. This can be accomplished by (1) creating more public–private partnerships; (2) publicizing and expanding accelerator and incubator organizations; and (3) developing focused curricula at universities.

Conclusion

In the near future, there will be many more Arab consumers and innovators with access to technology. As a result, the nature of problem solving in the Arab world will change dramatically. We will see more bottom-up innovation that challenges the technical and regulatory capacities of governments and other large institutions. If these economies do not enact regulatory and financial reforms, they will be missing a generational opportunity to embrace new drivers of entrepreneurial activity and job creation. For individual, corporate, and national economic growth, Arab officials and the international community should listen and adapt to the needs of start-up entrepreneurs across the Middle East. In addition, any changes in the business environment that help promote female employment and leadership will reap vast rewards and benefits to the region.

Notes

1. Israel was ranked at 40, but is not included in MENA, since it is considered an OECD nation by the World Bank.
2. Note that the other economy in MENA that also enacted credit reforms was Bahrain (World Bank, 2014).
3. Information obtained from interview with Flat6Labs cofounder Ahmed El Alfi on December 2, 2014.
4. Interview with Majid Bin Ayed Al-Ayed, Director of Business Development, Qotuf, in Jeddah on November 26, 2014.
5. AED stands for the UAE's currency: the United Arab Emirates dirham.

6. Note that the economies surveyed in the GEM Report include Algeria, Egypt, Iran, Pakistan, Palestine, and Tunisia. Israel was considered separately in the GEM Report.

References

Branstetter, L., Lima, F., Lowell, J. T., and Venancio, A. (2013). "Do Entry Regulations Deter Entrepreneurship and Job Creation? Evidence from Recent Reforms in Portugal." *The Economic Journal*, 124: 577, 805–32. doi:10.1111/ecoj.12044.

Brown, M., Jappelli, J., and Pagano, M. (2009). "Information Sharing and Credit: Firm-Level Evidence from Transition Countries." *Journal of Financial Intermediation*, 18: 151–72.

Brown, M. and Zehnder, C. (2007). "Credit Reporting, Relationship Banking, and Loan Repayment." *Journal of Money, Credit and Banking*, 39 (8): 1883–918.

Bruhn, M. (2011). "License to Sell: The Effect of Business Registration Reform on Entrepreneurial Activity in Mexico." *The Review of Economics and Statistics*, 93 (1): 382–6.

Chang, R., Kaltani, L., and Loayza, N. (2009). "Openness Can Be Good for Growth: The Role of Policy Complementarities." *Journal of Development Economics*, 90: 33–49.

Djankov, S., Ganser, T., McLiesh, C., Ramalho, R., and Shleifer, A. (2010). "The Effect of Corporate Taxes on Investment and Entrepreneurship." *American Economic Journal: Macroeconomics*, 2 (3): 31–64.

Djankov, S., La Porta, R., Lopez-de-Silanes, F., and Shleifer, A. (2002). "The Regulation of Entry." *The Quarterly Journal of Economics*, 117 (1): 1–37.

Djankov, S., McLiesh, C., and Shleifer, A. (2007). "Private Credit in 129 Countries." *Journal of Financial Economics*, 84 (2): 299–329.

Dreher, A. and Gassebner, M. (2013). "Greasing the Wheels? The Impact of Regulations and Corruption on Firm Entry." *Public Choice*, 155 (3–4): 413–32.

Kaplan, D., Piedra, E., and Seira, E. (2011). "Entry Regulation and Business Start-Ups: Evidence from Mexico." *Journal of Development Economics*, 95: 1501–15.

Kelly, D. J., Brush, C. G., Greene, P., and Litovsky, Y. (2013). *Global Entrepreneurship Monitor: 2012 Women's Report*. Wellesley, MA: Global Entrepreneurship Research Association.

Love, I., Martinez Peria, M. S., and Singh, S. (2013). *Collateral Registries for Movable Assets: Does Their Introduction Spur Firms' Access to Bank Finance?* Washington, DC: World Bank.

Monteiro, C. M. J. and Assunção, J. (2012). "Coming Out of the Shadows? Estimating the Impact of Bureaucracy Simplification and Tax Cut on Formality in Brazilian Microenterprises." *Journal of Development Economics*, 99: 105–15.

Schroeder, C. (2013). *Startup Rising: The Entrepreneurial Revolution Remaking the Middle East*. New York: Palgrave Macmillan.

Syeed, N. and Zafar, R. (2012). *Arab Women Rising*. Philadelphia, PA: Knowledge@Wharton.

World Bank. (2010). "Doing Business 2011: Making a Difference for Entrepreneurs." Retrieved from http://www.worldbank.org/

World Bank. (2012). "The Environment for Women's Entrepreneurship in the Middle East and North Africa Region." Retrieved from http://www.worldbank.org/

World Bank. (2013). "Enterprise Surveys in the Middle East," Conducted in partnership with EBRD and the European Investment Bank. Retrieved from http://www.enterprisesurveys.org/

World Bank. (2014). "Doing Business 2015: Going Beyond Efficiency." Retrieved from http://www.worldbank.org/

6
Innovation in Central Europe

Marina Dabić, Jadranka Švarc, and Emira Bečić

Innovation and research policies in the Central European transition countries

The research and innovation system in Central Europe (CE) has not been studied in sufficient depth, either on a theoretical or on an empirical level. Consequently, it has sometimes been surmised that CE is simply lagging behind the rest of Europe in terms of scientific and technological performance (Archibugi and Coco, 2005). Moreover, although CE comprises 13 countries, there is some disagreement as to which ones should be classified as part of it.[1] For the purposes of this chapter we focus on five emerging economies: the three largest transition economies – Hungary, Poland, and the Czech Republic – and two ex-Yugoslavia countries, now also EU members – the republics of Slovenia and Croatia, whose size, geographic locale, and cultural similarity merit inclusion. Moreover, their economic structure, research level, and technical development are comparable. In terms of innovation and small and medium-sized enterprise (SME) competitiveness, these countries are home to growing numbers of innovative companies which successfully compete on global markets, e.g. Prezi, Ravimed, Rimac car, Invea-tech, and GEA. Presently, these countries are among the most promising emergent markets in CE.

Twenty-five years ago, all these countries were only starting their transition from socialist, centrally planned economies to market-style economies and democracy. For decades trapped behind the Iron Curtain, in the embrace of the Soviet Union (except Croatia and Slovenia), these countries were not very familiar with Schumpeter's concept of innovation as a process of creative destruction initiated by an entrepreneur and with innovation as an essential driver of the "capitalist machine" and economic growth in general. Nor did they have any concept of the national system of innovation (Lundvall, 1992) and its purpose. However, after the major sociopolitical and economic changes of 1990, Hungary, Poland, the Czech Republic, Slovenia, and Croatia made tremendous progress in establishing national innovation systems and policies to foster innovation and catch up with developed European countries not only in concepts and policies but also in competitiveness and growth (Bečić and Dabić, 2012). They turned out to be quick learners under the pressure of tremendous socioeconomic changes that accompanied their transition to a market economy and political freedom.

The most competitive countries in terms of ease of doing business and business environment are Slovenia and Poland; in the global rankings of the Global Entrepreneurship and Development Index (GEDI) Slovenia took 24th place and Poland 27th out of 121 countries. As regards the Doing Business Index, Slovenia is ranked 33rd and Poland 45th out of 189 countries. They are followed by the Czech Republic and Hungary, while Croatia is facing more difficulties in creating a stimulating business environment and appropriate institutions to foster business competitiveness.

As regards economic freedom, the Czech Republic is the highest ranked country with a score of 72.2 in 2014. Second place is shared by Hungary and Poland with a score of 67. Slovenia takes fourth place with a score of 62.7 while Croatia is last with an index score of 60.4.

According to the Globalization Index 2014, Hungary is the ninth most globalized country in the world. Our other analyzed CE countries are ranked among the top 33 in the annual Globalization Index (out of 192 listed countries in 2014).

The trends of data in other composite indices (the Global Innovation Index (GII), the Networked Readiness Index (NRI), and the Global Enabling Trade Index (ETI)) also clearly indicate that the five selected countries in CE need improvements in terms of competitiveness, innovation, and technology development. The countries' competitiveness would be enhanced by improvements in their performance-related technological readiness and innovation ecosystems. The 2013 data show that the business demography[2] of the CE selected countries consisted of some 3.27 million active enterprises (16 percent of the EU-27[3] total) with 15.76 million people employed (12 percent of EU-27).

Enterprise births are often thought to be a key determinant of job creation and economic growth, as newly emerging competition stimulates a country's enterprise population to become more efficient and competitive. On the EU level, enterprise birth and death rates average about 10 percent of the total number of enterprises. Compared to EU aggregate values, the Czech Republic, Poland, and Slovenia recorded higher enterprise birth rates.

According to the type of activity start-ups focus on, Poland is the most production-oriented economy in Europe with 41 percent of start-ups and early-stage enterprises in production industries. It is followed by Slovenia and Hungary, while Croatia has the smallest share of start-up companies in the production sector (18 percent). A decline in the share of enterprises operating in production industries (from 47 percent in 2011 to 41 percent in

2012) characterizes all the countries, chiefly in favor of extraction industries and B2C (business-to-customer) services.[4]

The share of start-ups in B2B (business-to-business) services, however, is considered the true indicator of advanced economic development. It is highest in Slovenia (42 percent) and the figures for Croatia and Hungary are the next highest.[5]

The role of national innovation policy in stimulating innovation and economic development

The innovation policies and systems in the CE countries have mostly been shaped by the theoretical framework of national/regional innovation systems which highlights various innovation actors and the interactions between them at the regional and national levels (Lundvall, 1992). The emerging CE economies embraced the idea that the competitiveness of a nation does not depend so much on the scale of research and other resources but rather on the way the available resources are managed, organized, and governed, both at the enterprise and at the national level. The idea of building up the institutional setup of private and public institutions that would by mutual interaction foster and accelerate the creation and broad commercialization of innovation was rather attractive to policy makers all over the world, and especially so to policy makers in economies which were trying to make a transition from a socialist planned economy to innovation-driven economies in an increasingly globalized world.

The first phase of transition was devastating for all these economies. It was characterized by a huge decline in economic growth, largely as a result of socioeconomic and institutional uncertainty, disrupted production, and the loss of traditional markets. Innovation and technological development were threatened by the sudden transition to a liberal economy and the privatization of companies which were the pillars of technological development as the most applicative and developmental research was carried out at their in-house institutes (Švarc, 2011). As a rule, the governments abruptly withdrew financial support to the majority of industrial (now private) institutes. The process, known as "shock without therapy" (Radošević, 1996), led to the collapse of the majority of industrial institutes since they were unable to find new markets for their research activities. State-owned companies still play a significant role in the economies of these countries, especially the large companies in the sector of energy and resources. For example, four state-controlled Polish firms are listed among the ten largest CE companies in the energy and resources sector in terms of sales revenues (Deloitte, 2013). These are the fuel groups PKN Orlen and Lotos (first and fifth, respectively), the energy group PGE (sixth) and the gas giant PGNiG (ninth). Regardless of ownership, there are 166 Polish companies in the CE Top 500 companies as ranked by Deloitte. Poland is followed by the Czech Republic with 87 companies and Hungary with 62 companies among the CE Top 500. Slovenia, represented by eight companies, is sixth among the 15 countries analyzed, while Croatia, represented by 12 companies, shares seventh place with Lithuania. The top three positions in the CE Top 500 are held by one Polish (PKN Orlen – energy), one Czech (Škoda Auto – automotive), and one Hungarian (MOL – energy) company.

By contrast with the rapid devastation of the long-term accumulation of technology in industry, the public research sector was subjected to a more gradual process of reform and remained a substantial basis for education and recruitment of new generations of researchers and human resources. Human resources in science and technology (HRST) in these countries are nowadays close to the EU average, while the number of science and technology graduates surpasses the EU average in Poland, Slovenia, and Croatia (Figure 6.1). The share of young people educated to upper secondary or tertiary level in all the countries is above the EU average (Croatia tops the list of all the member states). All of the above illustrates the potential in human resources for future development and growth.

Apart from their similar socialist historical legacies and patterns of transition process, the national research and innovation systems of the selected

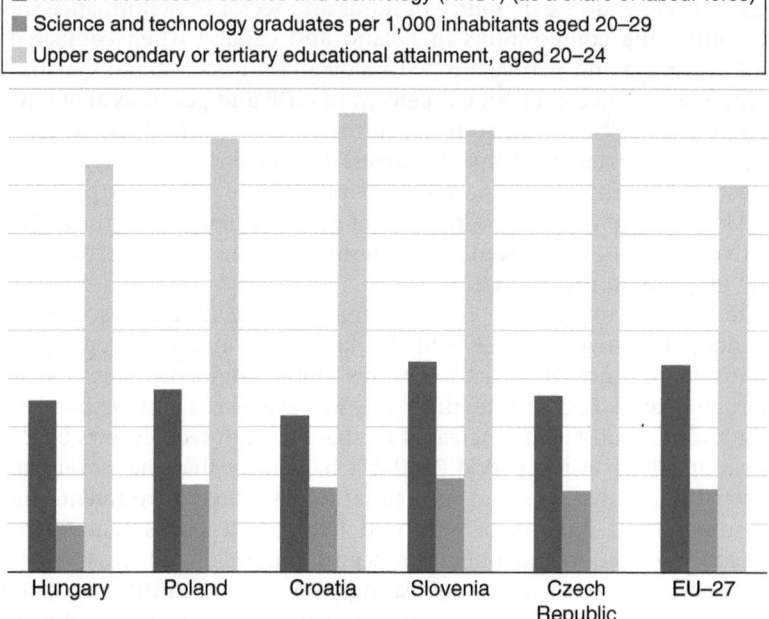

Figure 6.1 Human resources in science and technology
Source: Eurostat (2012).

countries evolved into rather diverse systems in terms of their size, enterprise composition, research intensity, and structural configuration.

There are great differences among the countries regarding R&D investments: while Slovenia surpassed the EU average in R&D investment by quite a way and stood alongside the most innovative members, for example Sweden and Finland, Poland and Croatia's spending on science was lower than in most other European countries, i.e. only around 13–15 percent of the euros per capita spent in Slovenia. The low R&D investments in these countries are conditioned by the low investments of the business sector in R&D. This is due to specialization in more traditional sectors, but also due to the difficulties in access to finance and developing businesses abroad. The weak research capacity of companies in Croatia is also a consequence of the deep and persistent economic crisis that hit Croatia in 2008 and a lack of large companies capable of investing in R&D. The level of investment in R&D fell from above 1 percent in 2004 to 0.75 percent of Croatia's GDP and has stagnated at that level since 2010. Companies in Poland mainly rely on foreign technologies: over 50 percent of R&D investments in Poland cover the purchases of foreign products and services (Erawatch, 2013b).

Underinvestment by the private sector, however, remains the main weakness of the Croatian and Polish research systems. It has far-reaching and adverse effects on economic development and growth. The breakdown of total R&D expenditure by source of funds and sector of performance shows the contrasting compositions in Poland and Croatia when compared to the EU average and particularly with Slovenia. In Poland and Croatia the business sector invests about 0.3 percent of GDP and performs about 40–45 percent of all research, which is not sufficient for knowledge-based growth. By contrast, Slovenian companies invest 1.6 percent of GDP and perform almost 76 percent of all research activities (Figure 6.2). This proves that R&D is a priority for the development of medium-high and high-tech competitive enterprises in Slovenia. As a result Slovenia had the sixth highest R&D intensity in the EU in 2012.

The Polish economy, the seventh largest in the EU-28, has undergone a structural change to achieve higher knowledge intensity (a 28 percent improvement since 2000) and Poland's global competitiveness has been improving at a higher rate than the EU average. Polish exports have been growing and Poland increased its share of high-tech exports by 2 percent annually over the period 2000–10. It is likely that this development reflects the positive effects of substantial foreign direct investment inflows and the related imports of advanced investment goods that upgraded domestic production structures (European Commission, 2013).

European policy has had an increasingly important influence on national innovation and research policies, the dynamics of innovation and technological development, and overall progress of the emerging CE countries. Unlike the countries that became EU members during the fifth round of EU

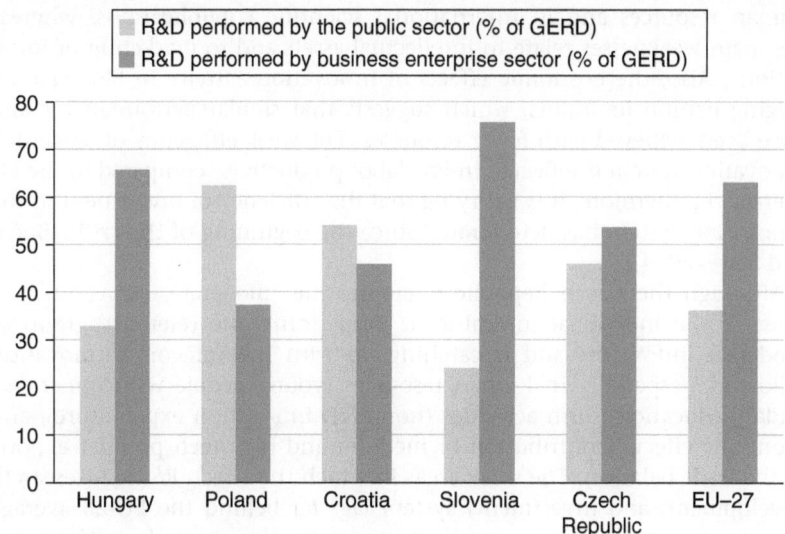

Figure 6.2 R&D performed by public and business sectors, 2012
Note: GERD = gross domestic expenditure on R&D.
Source: Eurostat (2012).

enlargement in 2004, Croatia became a member state almost a decade later, in 2013.

Central Europe has undergone a process of extensive modernization of research and innovation systems through the process of Europeanization. Their research and innovation systems are thus largely determined by the common European strategy of research and innovation which fosters a transition to a knowledge economy to overcome stagnant economic growth.

The progress in achieving common goals in research and innovation policies is regularly monitored (usually once a year) through different analyses and assessments. The Innovation Union Scoreboard (IUS) is a tool which is meant to help monitor the implementation of the Europe 2020 Innovation Union flagship initiative, and provides a comparative assessment of the innovation performance of the member states.

Based on their average innovation performance the member states are classified into four performance groups: "innovation leaders," "innovation followers," "moderate innovators," and "modest innovators." By the IUS 2014 (European Commission, 2014b), all analyzed emerging CE economies belonged to the "moderate innovators" group in which innovation performance is below the EU average, except Slovenia which was an "innovation follower," which means that its aggregate innovation performance was above or close to the EU average. The relative strengths of Slovenia are in human resources and in international scientific copublications, whereas the main weaknesses relate to intellectual assets and to marketing of innovations. Also, the economic effects of innovation activity in Slovenia are lagging behind its inputs, which suggests that similar performance could have been achieved with fewer resources. The weak efficiency of Slovenia's innovation system is reflected in low labor productivity compared to the EU average. Furthermore, it is worrying that the efficiency of investment in the innovation system has deteriorated since the beginning of the crisis (Bučar and Stare, 2014).

Although the Czech Republic is among the "moderate innovators," in most of the individual indicators it outperforms the reference group of moderate innovators and is catching up with the category "innovation followers," especially in human resources (young people with upper secondary education), firm activities (non-R&D innovation expenditure), and economic effects (contribution of medium and high-tech product exports to the trade balance). The weak areas in which the Czech R&D&I (research, development, and investment) system lags far behind the EU-28 average are concentrated in open, excellent, and attractive research systems (top scientific publications and non-EU doctorate students), finance and support (venture capital), intellectual assets (patents, trademarks, and designs), and in license and patent revenues from abroad.

Similarly to the Czech Republic, Croatia outperforms the EU average in the share of young people with upper secondary education and in non-R&D innovation expenditure, but in all other areas it is below the EU average, most prominently in the R&D expenditure of the business sector (around 55 percent of the total resources for R&D is provided by the government), knowledge-intensive service exports, and intellectual assets (patents, trademarks, and designs).

Poland is performing below the average of the EU for most indicators. Its performance is the weakest in terms of business investment in R&D, number of innovative companies, and linkages and entrepreneurship efforts. The Scoreboard, on the other hand, reveals the high quality of Polish human resources and recent growth in intellectual assets (patents, trademarks, and designs).

License and patent revenues from abroad, international scientific copublications, and fast-growing innovative firms are the relative strengths of Hungary. High growth is observed for Community trademarks, R&D expenditure in the business sector, and sales share of new innovations. A large decline in growth is observed for non-R&D innovation expenditure, in R&D expenditure in the public sector, SMEs innovating in-house, and Community designs.

Another recent common initiative focuses on fostering innovation, employment, and economic growth following the idea of comparative business advantages and specialization in innovation commonly known as

Table 6.1 Some identified hot spots and specializations in science and technology

	Hot spots in key technologies
Slovenia	Health, food and agriculture; ICT; materials; new production technologies; environment
Croatia	Health care sector; food processing and agrobusiness; energy technology; electronics and advanced materials and digital techniques
Czech Republic	Automobiles; transport; construction; materials; energy; environment
Hungary	Health; environment; automobiles; biotechnology
Poland	Food, agriculture, and fisheries; energy; environment; security; ICT; materials

Source: European Commission (2013).

the concept of the "smart specialization strategy."[6] It gained political and economic significance in 2011 when it was established by the European Commission's proposal for Cohesion Policy in 2014–20[7] as a precondition for using the European Structural and Investment Funds (ESI). Since ESI represents a significant financial input for all EU member states, the national/regional research and innovation strategy for smart specialization (S3) has become a first-rank policy issue.

An analysis of EU research and innovation performance (European Commission, 2013) identified some hot spots and specializations in science and technology (Table 6.1). The health sector, energy, food and agriculture, ICT, environment, automobiles, and electronic and advanced materials were shown to be the most promising areas of future development in the emerging CE countries.

Facilitating institutions (R&D labs, multinational firm research centers, university-based science and innovation centers)

Overall globalization and rapid advances in new technologies, ICT in particular, have enabled novel forms of competition and access to new markets for innovative products and services. Integration with Europe and globalization in general compelled emerging CE economies to refine their products and services and engage in a nonstop process of adjustment and innovation. Innovation and research infrastructure plays an important role in this process. During the period of transition of the emerging CE countries to a market economy, the development and upgrading of research infrastructures attracted little attention and less financial support. As the countries were undergoing structural changes and budget resources were limited, research and innovation suffered social and economic marginalization. With the accession of Poland, Hungary, the Czech Republic, and Slovenia to the EU in

2004, these countries were granted access to resources from Structural Funds for the development of large-scale research and innovation infrastructure. Croatia struggled to keep pace using national resources and EU preaccession funds till its accession to the EU in 2013.

The improvements in SMEs' performance in Europe are, from a broad perspective, underpinned by the impressive number of policy measures that have been introduced by the EU and the member states since 2008. These policy developments took place under the umbrella of the Small Business Act (SBA)[8] for Europe. Adopted in 2008, SBA acknowledges the central role of SMEs in the EU economy and establishes a comprehensive SME policy framework. SBA generated a pro-SME policy momentum which helped to alleviate the effects of the 2008 global crisis that hit numerous EU countries. In 2010–12 alone, a total of nearly 2,400 policy measures to support SMEs were implemented in the EU member states, i.e. an average of 800 measures annually, and almost 90 measures per country (European Commission, 2014a).

In Europe, it is necessary to distinguish between research infrastructures and facilities in the domain of science (e.g. university research centers, private laboratories) and infrastructures for supporting business innovation (e.g. innovation centers, technology parks, technology transfer centers). While the former are used by the scientific community to conduct top-level research and achieve technological advances in their respective fields, the latter are mostly specialized institutions for innovative business support to entrepreneurs.

The innovation and business innovation infrastructure was usually developed in coevolution with the regional policy and industrial/enterprise policy conducted by the EU, which are converging on the objective of supporting clusters at the regional level. Clusters lie at the core of regional development in the EU since focus is placed on groups of firms, related economic actors, and institutions which derive productive advantages from their shared proximity and connections. In the emerging CE countries innovation infrastructures for small business development and entrepreneurship are therefore closely related to the promotion of cluster-based approaches by linking firms, people, and knowledge at a regional level.

All member states are obliged to establish regional innovation agencies to assist entrepreneurs and innovators "on the spot" with a wide range of activities and services. To this purpose, innovation agencies cooperate with their many partners: research institutions, training organizations, suppliers of innovation finance, etc. Hungary, for example, established a network of seven regional innovation agencies called RIÜNET in 2005 in order to harmonize and coordinate regional development, organize technological innovation networks, and to provide innovation services to SMEs and start-ups. The regional innovation agencies receive minimal government support and their operation is mainly funded from international programs (i.e. EU FP7[9] and Interreg[10]) (Erawatch, 2013a). The South Moravian region in the Czech Republic represents a national model of regional innovation policy with its dedicated authorities, well-functioning innovation agency, and productive dialogue with the business community. A number of other regions have achieved varying degrees of success in emulating this model, namely the Moravia-Silesia, Liberec, Zlín, and Hradec Králové regions.

Thanks to the comprehensive policy measures for enterprise development, regional development, and clusters, supportive quasi-banking financial institutions (e.g. seed capital funds, business angel networks, regional and local loan funds) are becoming increasingly common all over CE. The same policies resulted in intensive development of institutions enabling the introduction of novel products and services (e.g. technology parks, technology incubators, preincubators, academic entrepreneurship incubators, and technology transfer centers).

Institutions which operate at the intersection of science and business in order to support innovative business ideas play a key role in determining the success of the economy. They act as a bridge, facilitating communication between the worlds of business and science. The Polish Agency for Enterprise Development is a business research interface institution which supports technology parks, incubators, and transfer centers on multiple levels. Slovenia established a new agency, SPIRIT, and put it in charge of entrepreneurial support and financing of R&D activity in the business sector. Support for business sector R&D is also partially provided through the Slovenian Enterprise Fund, especially as regards bank guarantees for SMEs engaged in R&D projects and technological restructuring, and support for start-ups in an innovation environment. For the purpose of enhancing SME development and promoting investment and innovation, Croatia has established a new agency, HAMAG-BICRO. The National Innovation Office is the governmental body responsible for research, development, and technological innovation in Hungary. It also coordinates the activities of the regional innovation agencies.

Technology parks are the most common institutions linking research and business spheres. The Nickel Technology Park Poznań is the first and the largest private technology park in Poland. In recognition of its pioneering role in creating a national commercial platform for cooperation between science and business, *Nowy Przemysł* business magazine awarded it the title of "The One Who Is Changing Polish Industry." The Nickel Technology Park focuses on supporting development of the biotechnology and IT industry. The Nickel BioCentrum, which began operation in 2012, offers a meeting point for Polish and foreign companies and research teams in the biomedical industry.

The Infopark in Budapest is the oldest innovation and technology park in Central and Eastern Europe. It is primarily an innovation center for IT, telecommunication, and software development companies. The Infopark houses the head offices of multinationals such as Lufthansa Systems, Hungarian Telekom, and IT-Services Hungary alongside the head offices of young, innovative companies. The Infopark also hosts the headquarters of the European Institute for Innovation and Technology (EIT). The Czech Technology Park adjacent to the Brno University of Technology is acknowledged as the premier institution for the development of business and technology in the CE region. Situated next to the Brno University of Technology, the park facilitates access to research staff, facilities, and a skilled graduate workforce. Among other clients, the Czech Technology Park already houses offices of IBM, FEI, Motorola, Vodafone, and Silicon Figure.

The technology parks in Slovenia (e.g. Technology Park Ljubljana) and Croatia (e.g. Technology Park Zagreb) are smaller in terms of space for rent and less internationalized than the parks in the other countries. They are, however, hubs for the development of local high-tech companies. The Technology Park Ljubljana hosts around 290 companies and more than 1,500 professionals. The Technology Park Zagreb has assisted in the foundation of around 70 technological start-ups, some of which have grown to become specialized technological leaders in the region (e.g. ALTPRO, railway safety).

In contrast with innovation and business infrastructures which are mainly coordinated by national governments and supported by ESI funds, the development of large research infrastructures is more centralized at the level of the EU, which is mainly due to their pan-European character (shared access by many countries) and the significant resources needed for facilities to be established. The European Strategic Forum on Research Infrastructures (ESFRI) established in 2002 constitutes the most important coordination platform for developing pan-European infrastructures. It also aims to enable national governments to harmonize national roadmaps for research facilities development with European ones. ESFRI fulfills its function by means of the ESFRI roadmap both on the European and on the member state level. The ESFRI roadmap highlighted the importance of research infrastructures in the EU and has, consequently, had a huge impact on national policies for research infrastructures. The latest roadmap lists 48 new or significantly upgraded infrastructures to be developed in 2015–2020, mainly in the fields of the environment, biology, and energy. A conservative estimate of the total development cost of these projects amounts to nearly €20 billion, and, on average, €2 million will be required annually for their efficient operation (Technopolis, 2014).

Europe, as one of the leaders in nuclear magnetic resonance technology, is currently facing growing international competition and needs to fully exploit its scientific and technological potential.

Firm-level innovation

Not enough firms from CE are able to use the European single market as a springboard to the global economy and thus they fail to provide growth and jobs back home in Europe. Compared to the 1980s, CE counts more small companies and fewer big ones. Today's entrepreneurs are different from those of the past. They start their companies with fewer people, focus on freelancers, and hire on a project basis. Insufficient numbers of CE SMEs grow fast and enter the global marketplace. Among them, only a few become industry leaders, but they can make the growth in 2014 look very different from that of a few decades ago. Successful new global start-ups – sometimes dubbed micronationals – could contribute immensely to the economic ecosystem and below we present several cases to support this view. We will use these cases to challenge Emmanuel Carraud's statement: "Europe doesn't believe in its own assets, Europe is not a risk-taker; there are multiple barriers to capital and growth for European entrepreneurs and no culture of failure!"

Prezi

Prezi is a presentation software program created by Szabolcs Somlai-Fischer, Péter Halácsy, and Péter Árvai, two Hungarians and a Swede. The start-up was founded in 2009 and the software was developed at Zui Labs, Budapest, Hungary, with the support of Kitchen Budapest and Magyar Telekom. Prezi is a groundbreaking alternative to PowerPoint as it replaces the ordinary slide-based presentations. It is based on the concept of a "zoomable canvas": presentation imagery initially needs to be created on a single canvas. It is then possible to zoom in and out to highlight the various elements of the presentation. Prezi makes it easy for the presenter to convey their message clearly and audiences can quickly grasp the structure of the entire story and drill into the details, as necessary.

Arvai was born and raised in Sweden where he was very active on the Stockholm start-up scene, Stockholm being perceived as a pretty advanced start-up city, behind only London and Berlin. Budapest is not usually mentioned in this context but Arvai realized that the amount of talent and creativity Budapest offers is very much on a par with Stockholm. Truth be told, a great many Hungarian companies are achieving global success but we do not necessarily associate them with Hungary. What is more, Hungarians do not think of themselves as a nation of start-ups. This mistaken perception is not likely to cloud the success of Prezi. Commercially oriented markets are charging ahead in adopting Prezi. For example, the Netherlands has a great record in international trade and business and it has been among the first nations to adopt Prezi. In South Korea, which is booming thanks to a very successful import/export industry, Prezi is also very popular. Prezi is, furthermore, listed as a Career-Launching Technology Company on Wealthfront's list of rapidly growing mid-sized private technology companies. Wealthfront is the world's largest and fastest-growing automated investment service with over $1 billion in client assets and a company can only qualify for their career-launching list if it has revenues between $20 and $300 million, and is on a trajectory to grow at a rate in excess of 50 percent over at least the next three years.

Prezi chose to commercialize their product in San Francisco, Silicon Valley, USA. Prezi's founders believed that selling their product on the EU market would require too much effort. They also worried that their business plan would be copied by competitors by the time they got it off the ground in a market like the UK (Koekoek, 2012; Prezi, 2014).

Ravimed Ltd, the Republic of Poland

The Polish company Ravimed Ltd is a medical devices manufacturer located about 25 km north of Warsaw. It produces single-use medical devices, medicinal products, and individual antichemical protection sets. Ravimed Ltd is the only producer of containers for blood collection, storage, and preparation in Central Eastern Europe. It also provides a range of laboratory services, especially for testing innovative drugs.

Founded in 1991, Ravimed Ltd first started operating in the Polish market but soon expanded to the former Soviet Union, African, and Middle East countries. Presently Ravimed Ltd employs about 100 staff and its net revenues from sales and equivalent rose by *c.* €780,000 in the period 2011–13. Ravimed's initial success on the domestic market can be explained by the fact that it introduced a new product on the market (blood bags rather than bottles) and that customers wanted to support Polish companies.

Initially rather simple, Ravimed's production process for blood bottles was refined through competing with foreign companies which entered the Polish market. In the process, the owners turned down two lucrative takeover offers and chose to develop the company further. The adversities experienced by the company (e.g. the Russian financial crisis) led to product portfolio diversification and contracts for highly specialized products presently produced by only two companies in the world (e.g. autoinjectors used for self-treatment by individuals who were exposed to toxic chemical warfare agents). All technologies employed by Ravimed were developed in-house and RAVIMED Ltd currently possesses seven patents. Ravimed Ltd closely collaborates with top research centers in the region and is frequently engaged as a subcontractor in their scientific projects.

An active member of the European Parliament of Enterprises, Ravimed Ltd is devoted to the constant development and study of advanced products and medical technologies. In 2010 the company established its own R&D department and has since been employing 13 researchers simultaneously running projects of different degrees of complexity. The innovation process relies on constantly monitoring customers' needs, in-house development of potentially valuable ideas, and analysis of competitors' market and R&D activities. This intense innovative activity results in one to two new products and two to three major modifications of existing products per year, which all need to meet high quality and safety standards and get certification. The costly certification process is mostly financed from various EU project funds.

Invea-tech, Czech Republic

In 2007, several researchers from Masaryk University got together with others from the Brno University of Technology and CESNET to form Invea-tech as a university spin-off. The company was incubated by the South Moravian Innovation Centre (JIC) and still retains strong links with it. Invea-tech's headquarters are located in Brno, Czech Republic. It currently has about 35 employees and more than 10 external consultants. Invea-tech manufactures, produces, and sells network monitoring and security solutions based on NetFlow[11]/IPFIX[12], Network Behavior Analysis (NBA), and FPGA[13] acceleration. Simply put, Invea-tech develops and markets comprehensive network solutions internationally. It also provides network solutions to customers like T-Mobile, Siemens, Stanford University, Drogeria market, Konica Minolta, Allianz insurance, Hewlett Packard, and Raiffeisen Bank. Their FlowMon system was commended at the CeBIT fair in Hanover twice and in 2013 and 2014 the company was recognized among the Deloitte CE Technology Fast 50 as one of the 50 fastest-growing tech companies in Central and Eastern Europe. Over the past five years they achieved a growth of 326 percent, which earned them 46th place in the recently published Deloitte list. Invea-tech is involved in the ACE European acceleration program which helps innovative start-ups and high-growth ICT companies to find partners, clients, and financing to accelerate their move into cross-border and international markets.

Invea-tech leads in the flow monitoring and NBA market, likely to be the next-generation network security trend. Following in the footsteps of AVG, AVAST, and ESET, Invea-tech is a good representative of the new generation of cyber security companies.

Invea-tech was the first company globally to release new 100GbE (100 Gigabit Ethernet) models of FPGA-based network adapters, thus proving its global dominance in the development of high-speed monitoring probes. Using this technology, clients can monitor and analyze network traffic in the most modern and fastest networks. The company penetrates the global market via security and IT events/fair trades and through its channel partner program. Its biggest partners are Orizon (Japan), SecTec (Slovakia), VUMS DataDom, DMS, Veracomp, Bull, ComSource, ICZ (Czech Republic), Tamkaroo (Germany), Aexux, Vosko (Benelux), Passus, and Clico (Poland).

Tovarna olja GEA, Slovenska Bistrica, the Republic of Slovenia

Tovarna olja GEA from Slovenska Bistrica, a small town in the center of the Styria region, Republic of Slovenia, has been producing pumpkin seed oil from roasted and pressed pumpkin seeds since 1904. It presently employs 103 workers and markets its products in more than 20 countries. The main products of the company are vegetable oils, mayonnaise, salads, sauces, fried onion, and feeding stuff components. The company holds the majority share in the domestic market, and besides the Balkan markets, the markets in Europe and Asia present a greater and greater challenge. GEA's operations are carried out with an emphasis on high-quality and health-promoting products which entirely satisfy the need for vegetable fat and supplementary foods.

From 1904 to 1923 the production process was artisan-like but as the demand for pumpkin oil increased, the company started implementing state-of-the-art industrial solutions for refining, bottling, and increasing production levels. Finally, environmental awareness led the owners to build a new treatment plant. In 2007, GEA introduced the most modern production line for sorting seeds and the production of pumpkin seed oil. In 2010 it started bottling refined oil into PET plastic bottles in the new, entirely continuous, bottling line. The trademarks of GEA include 21 different kinds of high-quality and specialized oils.

GEA is constantly innovating and responding to market changes and final consumers' needs. Each year the company launches several new products and carries out improvements in its operations on the level of production processes and technological equipment, quality control, and purchasing, as well as on the level of sales and environmental protection.

Rimac Automobili, the Republic of Croatia

"Rimac Automobili" is an automobile manufacturing company established in 2009 in Zagreb, Republic of Croatia. Even though it all started in the founder's garage just six years ago, its number of employees tripled in 2014 and the company currently employs 60 highly skilled employees. It all started with the founder's dream that he would build an electric supercar, a unique fusion of light build and pure power. Since the required electric systems were not available at the time, the company developed the necessary parts and patented 24 innovations. In order to develop Concept One, their first electric supercar, researchers at Rimac Automobili focused on direct current electric motors and a new propulsion system: "permanent magnet synchronous motor generation." Its unique powertrain is divided into four subsystems (but each consisting of one motor, inverter, and reduction gearbox), each driving one wheel and controlled by a sophisticated engine control unit (ECU). This novel approach to vehicle dynamics is called all wheel torque vectoring (AWTV).

At Rimac Automobili, they are especially proud of the fact that all key components are designed, engineered, and produced in-house: the whole Concept One architecture is developed from scratch, around the powertrain and battery pack. The car was first introduced to the public at the 2011 Frankfurt Motor Show and it was positively reviewed at the 2012 Paris Concours d'Elegance. As at October 2014 altogether eight of these supervehicles had been sold to buyers from all over the world.

Thanks to its highly skilled and creative staff, Rimac Automobili is successfully innovating in other areas as well. For example, the company produced an electric bike, the Greyp G12, which combines the best of motorcycles and bicycles. Its top speed is 65 km/h and one can ride for 120 km without pedalling. Finally, the Greyp G12S can easily be fully recharged from a standard 220 V outlet in only 80 minutes. It runs on a state-of-the-art battery pack manufactured by Rimac Automobili.

The company raised most of its capital from three big investors who were drawn to Rimac Automobili because they admired the fact that such revolutionary technology had been developed with so few resources. The single largest investor with 10 percent of shares based on a valuation of €70 million is China Dynamics, a company active in the electric vehicle industry in China. The capital which was raised is dedicated to the development and commercialization of future Rimac sportscar models and to improving the company's production capacity.

Conclusions and recommendations

Seven years after the global financial crisis finished, the emerging economies of CE are still struggling with the consequences of the economic downturn and trying to reach precrisis levels of economic prosperity. Not only do the countries differ in their rates of recovery but their economic indicators also seem to diverge from those of the majority of the old EU member states. Poland and Hungary are forging ahead, with positive growth rates in 2013. They are followed by Slovenia and the Czech Republic, while Croatia is lagging behind due to seven consecutive years of economic recession. The main reason for the sluggish recovery is rooted in the excessive credit boom in Europe before the global financial crisis, and in economic growth being mostly based on foreign borrowing instead of innovation-based competition and exports.

Although Poland, the Czech Republic, Hungary, the Republic of Slovenia, and the Republic of Croatia share many socioeconomic and cultural similarities in their historical transition from planned to market economies, their posttransitional development is characterized by significantly uneven development in terms of economic strength and resilience to the economic crisis. The "shock therapy" economic reforms carried out in these countries in the early 1990s have paved the way to a market economy as a precondition for integration with the EU. However, the speed of transition depended on national specificities and differences which were rooted in various factors such as the pace of carrying out structural reforms (e.g. of public administration, the pension system, labor market regulations, privatization of the banking system), the size of internal markets, ability to absorb EU Structural Funds, share of private versus public sector in the economy, perception of sources of economic growth, etc. The Croatian economy is, for example, oriented towards low-tech/skill services, primarily tourism, as a source of economic growth, with the state playing an important part in economic life and a lack of structural reforms which caused persistent recession. On the other hand, Poland was oriented to industrialization and manufacturing and nurturing an economy dominated by private companies with a reduced share of the state in the economy. There are estimates that the private sector contributes 76 percent of GDP and employs 74 percent of the labor force in Poland, which is quite high for European economies. Poland is among the best performers in the region and the Polish economy has evolved into a significant constituent of the European economy. SMEs are vital to the success of both new and old European economies and are a key focus for innovation support. There are over 20 million SMEs in the European member states, 15 percent of which, or over 3 million, are located in the emerging CE countries.

Of these, 95 percent are really small, employing sometimes just one or two people or family members. With the exception of large companies mainly in the energy, transport, or trade sectors, these countries are still lacking the layer of medium to large companies that are the backbone of the old European, and notably German or French, economies.

Despite the huge progress these countries have made in the last 20 years of a market economy, the private-sector companies are still risk-averse and lack the ambition to cross national borders and become true regional companies. Entrepreneurship is hindered by too much bureaucracy, red tape, regulation, and high labor costs which impede growth and innovation. Therefore, changes are needed in both private-sector companies and public policies for fostering innovation and entrepreneurship to respond to global challenges and competition with China, India, Brazil, Korea, etc., which have demonstrated faster growth than Europe or America. The companies should pay special attention to developing management and innovation capabilities to take risky business ideas and develop new business models to respond to the needs of the internationalization of innovation and conducting businesses on a global basis. Introduction of new service models, organizational innovations, and strengthening the knowledge and technology base of companies are crucial factors in success. Closer cooperation with universities and commercializing academic research results may be a valuable innovation-generating channel and an opportunity for growth. The companies in the CE economies receive remarkable backup from EU supporting programs for SMEs, technology, and research. Programs like Horizon 2020 (e.g. collaborative projects, the SME instrument, Eurostars) and the ESI are tailored to foster innovation-based companies and promote socioeconomic development and cohesion among the "old" and "new" states to enable the prosperity of both.

Notes

1. Central Europe is sometimes divided into West-Central Europe and East-Central Europe. The countries of West-Central Europe are Germany, Austria, Liechtenstein, Switzerland, and Slovenia. These countries exist on the border of Central Europe and Western Europe, and they can be classified as either. East-Central Europe includes Slovakia, Poland, Hungary, and the Czech Republic. Because these countries exist on the border of Central Europe and Eastern Europe, they are sometimes classified as Eastern Europe instead of Central Europe. Other countries sometimes included in Central Europe are Serbia, Romania, and Croatia. According to the German tradition of geographical delineation, the countries of Northern Europe are classified as part of Central Europe.
2. Statistical data on business demography in the European Union (EU), treating aspects such as the total number of active enterprises in the business economy, their birth rates, death rates, and the survival rate. In the business demography domain, the business economy covers sections B to N, excluding activities of holding companies – K64.2 (NACE Rev.2).
3. Croatia was included in the official statistics of the EU in 2013 when it became the 28th member of the EU; therefore the statistical data before 2013 refer to EU-27, and after 2013 to EU-28.
4. GEM identifies four categories of economic sectors: extraction, production, B2B services, and B2C services.
5. GEM Poland 2012, p. 24.
6. "Smart Specialization – the Concept," Policy Briefs Nos. 5–9, pp. 25–9.
7. COM(2011) 615, http://www.ipex.eu/IPEXL-WEB/dossier/document/COM 20110615.do
8. http://ec.europa.eu/enterprise/policies/sme/small-business-act/index_en.htm
9. FP7 is the short name for the Seventh Framework Programme for Research and Technological Development. This is the EU's main instrument for funding research in Europe which was running from 2007–2013. The current programme is Horizon 2020 (http://ec.europa.eu/programmes/horizon2020/) but there are many projects funded under FP7 which are still running.
10. INTERREG EUROPE is one of the instruments for the implementation of the EU's cohesion policy, a programme which provides funds and supports policy learning, sharing of knowledge and transfer of good practices between regional and local authorities and other actors of regional relevance in the EU (http://www.interregeurope.eu/).
11. NetFlow is the industry standard for network traffic monitoring and the most widely used measurement solution today.
12. Internet Protocol Flow Information Export (IPFIX) was created by the IETF working group due to the need for a common, universal standard of export for IP flow information. The IPFIX standard defines how IP flow information is formatted and transferred from an exporter to a collector. Previously many data network operators were relying on the proprietary Cisco NetFlow standard for traffic flow information export. The IPFIX is a much more flexible successor of the NetFlow format. NetFlow is the industry standard for network traffic monitoring and the most widely used measurement solution today.

13. An FPGA is a kind of programmable hardware that in a very effective way can be used to implement algorithms and computations.

References

Archibugi, D. and Coco, A. (2005). "Is Europe Becoming the Most Dynamic Knowledge Economy in the World?" *Journal of Common Market Studies*, 43 (3): 433–59.

Bečić, E. and Dabić, M. (2012). *An Analysis of Research and Innovation in Croatia: Entrepreneurship and Innovation.* Zagreb, Croatia: University of Zagreb; Maribor, Slovenia: University of Maribor.

Bučar, M. and Stare, M. (2014). *Evolution of Innovation Policy in Slovenia since 2004: Promises and Pitfalls.* Maribor, Slovenia: Studia Historica Slovenica.

Carraud. (2014). Retrieved from http://noweurope.org/ [accessed 23 November 2014].

Deloitte. (2013). "Top 500, Central Europe." *Deloitte.* Retrieved from http://www.deloitte.com/assets/DcomSerbia/Local%20Assets/Documents/2013/Top500%20 2013%20final%20.pdf

Erawatch. (2013a). "Hungary." Retrieved from http://erawatch.jrc.ec.europa.eu/

Erawatch. (2013b). "Poland." Retrieved from http://erawatch.jrc.ec.europa.eu/

European Commission. (2013). "Research and Innovation Performance in EU Member States and Associated Countries." Retrieved from http://ec.europa.eu/

European Commission. (2014a). *Annual Report on European SMEs 2013/2014 – A Partial and Fragile Recovery, Final Report – July 2014.* Retrieved from http://ec.europa.eu/

European Commission. (2014b). "Innovation Union Scoreboard 2014." Retrieved from http://ec.europa.eu/

Eurostat. (2012). Retrieved from http://ec.europa.eu/eurostat.

Koekoek, Peter. (2012). "Innovative Companies: 'Making Europe the Best Place to Grow Highly Innovative ICT SMEs.' Report of a High-Level Panel Discussion DG Connect." Retrieved from http://ec.europa.eu/digital-agenda/sites/digital agenda/files/2._Innovative_companies_Report_Final.doc.pdf

Lundvall, B.-A. (ed.). (1992). *National Systems of Innovation. Towards a Theory of Innovation and Interactive Learning.* London, UK: Pinter Publishers.

Prezi. (2014). Retrieved from http://www.prezi.com

Radošević, S. (1996). "Restructuring of R&D Institutes in Post-Socialist Economies: Emerging Patterns and Issues." In Andrew Webster (ed.). *Building New Bases for Innovation: The Transformation of the R&D System in Post-Socialist States.* Cambridge, UK: Anglia Polytechnic University, pp. 8–30.

Švarc, J. (2011). "Does Croatian National Innovation System (NIS) Follow the Path towards Knowledge Economy?" *International Journal of Technology Transfer and Commercialisation*, 10 (2): 131–51.

Technopolis. (2014). Retrieved from http://www.technopolis-group.com/wp-content/uploads/2015/01/EPIRIA_Final-Report_2014.pdf

7
Innovation in Africa: A View from the Peaks and Hilltops of a Spiky Continent

David Wernick

Introduction

In a widely read 2005 article in the *Atlantic Monthly*, author Richard Florida argued that with respect to innovation, the world is anything but flat. Given the way that creative talent, technical expertise, and financial capital tend to cluster in a handful of hubs or "peaks" around the world – places such as New York, San Francisco, London, Berlin, and Tokyo – the world's innovation topography is best described as "spiky" (Florida, 2005). Although Africa, to borrow Florida's metaphor, has few innovation peaks of global significance and lots of valleys (and indeed, plenty of chasms), the continent does contain a growing number of hills, wherein entrepreneurs are profiting from latecomer advantages in technology to design new products, reengineer old ones, and launch bold new business models. African entrepreneurs and corporations have made particular inroads in sectors such as telecommunications and financial services, where firms like Safaricom and MTN Group are world leaders in mobile money. But they have also achieved success in manufacturing, consumer goods, and agriculture (Juma, 2011; Ware, 2013).

And while Africa's private sector is the key driver of innovation on the continent, a host of other actors, including venture capitalists, tech accelerators, private equity firms, multinational enterprises (MNEs), and nonprofit organizations, are playing a supporting role by lending their skills, know-how, and capital to unlock new business opportunities on the continent. Meanwhile, national and local policy makers are undertaking far-reaching reforms to improve the business climate and make government more efficient and transparent. This chapter takes stock of these developments, particularly within sub-Saharan Africa, while identifying key obstacles to the continent's continuing innovation-led economic expansion.

Africa's innovation landscape

For many in the West, the word "Africa" conjures up images of poverty, despotism, war, famine, and disease. On the economic front, the continent is often portrayed as a basket case, prone to graft and mismanagement and kept afloat only by foreign loans and humanitarian aid. But this picture – which was never fully accurate – is quickly becoming anachronistic. Sweeping economic and political reforms initiated during the 1990s and 2000s are transforming the business environment of the continent in a positive way, while attracting foreign direct investment and spurring economic growth (World Bank, 2014a). The improved business environment has also proven a fertile ground for innovation. Much of this innovation has involved African firms devising low-cost, homegrown solutions to local problems. This bottom-up approach to innovation has been variously termed "frugal innovation" (*The Economist*, 2010a) and "juggad" (Radjou et al., 2012). Examples of this phenomenon include the Mi-Fone line of inexpensive mobile phones launched in 2008 by a Ugandan entrepreneur and former Motorola executive (Nsehe, 2013), and the Hippo Water Roller, a lightweight yet durable, high-capacity water transportation drum created by two South African entrepreneurs that can be rolled on the ground rather than hoisted on heads or shoulders, thereby preventing spinal injuries (Holland et al., 2012).

One indicator of Africa's growing innovation prowess is the continent's steadily improving performance in the Global Innovation Index (GII), which annually ranks over 140 nations according to their technological capabilities, capital resources, and human competencies. In 2014, Rwanda, Gambia, Mozambique, Burkina Faso, and Malawi joined the ranks of "innovation learners" alongside Kenya, Uganda, and Senegal. These are economies that performed at least 10 percent higher on the index than their peers with similar GDPs. At present, nearly half of all the nations designated as innovation learners emanate from sub-Saharan Africa (Global Innovation Index, 2014).

What explains the surge of innovation in the world's poorest, least developed, and most politically unstable continent? One catalyst has been forward-looking government policy. Leaders like Rwandan President Paul Kagame and Babatunde Fashola, Governor of Lagos State, Nigeria, have championed innovation at the national and state levels, seeking ideas and inspiration from places like Singapore and Dubai (Chu, 2009). Other leaders, like Kenya's President Mwai Kibaki (2002–2013), have tapped technocrats to spearhead crucial economic development projects, including the laying of undersea cables and 4G networks, the creation of innovation hubs, and even the building of an entire new city dedicated to high-tech business and research, dubbed the "Silicon Savannah."

Rapid population growth has been a second catalyst for innovation. Between 1985 and 2014, Africa's population doubled from 500 million to 1 billion. It is projected to double again between 2014 and 2040 (UNDESA,

2013). At present, more than 200 million Africans are between the ages of 15 and 24, making Africa the youngest continent in the world (UN Habitat, 2014). The youth bulge will likely generate an economic dividend, since economies tend to grow when there is a high ratio of working-age adults to dependants (Devarajan and Fengler, 2013). And given the way that young Africans are using their technological skills to spearhead innovations in education, health care, and agriculture, there may be an innovation bonus too (UNESCO, 2014).

Urbanization has been a third catalyst for innovation. Between 1980 and 2010, the number of Africans living in cities jumped from just over one quarter to 40 percent of the population, and the number of city dwellers is expected to rise to nearly 50 percent by 2030 (UN Habitat, 2014). Already, Africa has 52 cities with at least 1 million residents – the same number as in Europe (Berman, 2013). This mass migration from rural villages to megacities like Lagos, Kinshasa, and Johannesburg has motivated enterprising Africans to come up with creative solutions to satisfy basic needs, such as paying for goods and services, navigating congested highways, finding jobs, and gaining access to health care, clean water, electricity, and education. Meanwhile, the growth of Africa's urban middle class, which now represents about 375 million people, has attracted the interest of Western consumer goods firms seeking to capitalize on the growing demand for everything from groceries to cosmetics (Ware, 2013; Deloitte, 2014).

A fourth innovation catalyst has been the mobile telecommunications revolution. At the dawn of the millennium, relatively few Africans had mobile telephone subscriptions, so making or receiving a telephone call required using the notoriously slow and unreliable government-owned, fixed-line telephone systems. Today mobile phones are ubiquitous and in some African countries more people have access to a mobile phone than to clean water, a bank account, or even electricity. The wide-scale adoption of smartphones, which can now be purchased in many African cities for under $100, will open new possibilities for web-based applications and e-commerce, allowing Africa to "leapfrog" past more developed markets (World Bank and African Development Bank, 2012).

Already mobile phones are transforming the way that business is conducted on the continent. In rural areas throughout Africa, farmers can use mobile applications like Esoko, developed by a Ghanaian start-up, or M-Farm, the brainchild of a 26-year-old Kenyan computer scientist, to check market prices for their crops while they are still in the fields, giving them enhanced negotiating leverage with local buyers (McKinsey, 2013). And in cities like Nairobi, commuters can pay their fares on private minibuses (*matatus*) without having to dig into their pockets for change thanks to Bebapay, a cashless, prepaid card system launched in 2013 by Google and Kenya's Equity Bank. Passengers simply tap their cards to their conductors' smartphones and money is digitally transferred from card to phone. The service has been so successful that Kenya's transportation authority is in the process of moving entirely to a cashless model (BBC, 2014).

A fifth innovation catalyst has been the growth of the Internet. Owing to massive investments in network infrastructure, Internet penetration across the continent grew from negligible levels in 2000 to about 16 percent in 2013, with far higher levels in urban areas (McKinsey, 2013). Already there are more Internet users in Nigeria than in either the UK or France, with the number of subscribers in Africa's largest country set to jump to 84.3 million in 2018 from 51.8 million in 2013 (eMarketer, 2014).

In addition to the role that the Internet has played as an agent of change in the telecommunications and banking sectors by giving formerly unbanked citizens access to mobile payments, it has served as a platform for innovation in other sectors such as health care, where entrepreneurs like Arthur Zang from Cameroon are designing diagnostic tools like the Cardiopad, a touch-screen medical tablet computer, to expand medical services to rural areas (Holland et al., 2012); in education, where entrepreneurs like Rapelang Rabana from South Africa are launching new mobile tools like ReKindle Learning to help students practice lessons outside the classroom (McKinsey, 2013); and in agriculture, where entrepreneurs like Su Kahumbu from Kenya have introduced mobile applications like iCow that provide dairy farmers with real-time pricing information and SMS reminders about milking schedules and immunization dates (Hussey, 2015).

And there is great potential for future Internet-driven growth, as smartphone penetration increases, bandwidth costs decline, and the size of Africa's middle class mushrooms. Indeed, by 2025, Africa could have as many as 600 million Internet users, generating $75 billion in annual e-commerce sales, and contributing as much as $300 billion a year to Africa's GDP (McKinsey, 2013).

Innovation as national policy

Africa has traditionally been the world's poster child for poor governance – a continent dominated by corrupt dictators and authoritarian "big men" who have plundered national treasuries and doled out government jobs, contracts, and favors to members of their patronage networks. At the heart of the problem are the "extractive institutions" that underpin Africa's economic and political life (Acemoglu and Robinson, 2012). Unlike the "inclusive institutions" prevalent in the West, which tend to emphasize competition, procedural fairness, and respect for property rights, extractive institutions tend to concentrate power and opportunity in the hands of a few and encourage rent-seeking activities that generate private wealth yet public misery. One manifestation of extractive institutions is the tradition of single-party political rule that has been a hallmark of African politics during the postcolonial period. Africa's penchant for political monopolies has resulted in social exclusion, strife, and a business environment marked by protectionism and corruption (Harrison et al., 2013).

During the early 1990s, many African countries put political reform front and center, casting out military leaders, convening multiparty elections, and imposing term limits on elected officials (Berman, 2013). Today, nearly all of Africa's 55 nations hold regular, multiparty elections and only four – Eritrea, Swaziland, Libya, and Somalia – lack a multiparty constitution (*The Economist*, 2012a). And while the progress towards democracy across the continent has been uneven with regular setbacks – as the 2012 military coups in Guinea-Bissau and Mali attest – the overall trend has been in the direction of political reform, with countries like Mauritius and Botswana leading the way. Indeed, those two countries rank ahead of more affluent countries such as France and Italy in the Economist Intelligence Unit's 2013 Index of Democracy, which measures the breadth and depth of democratic governance in 165 nations (The Economist Intelligence Unit, 2013).

One of the most salutary results of political reform has been the change in macroeconomic policy across the continent, with governments eschewing populism in favor of sound money, balanced budgets, and debt reduction (Devarajan and Fengler, 2013). Another consequence of political reform has been the changing of the guard in government ministries across the continent, with a new generation of globally experienced technocrats assuming leadership positions (Green and Whitehead, 2013). Not surprisingly, the influx of technocrats into positions of political power has led to the growing embrace of ICT-based solutions for government functions. A pioneer in the e-government space is South Africa. The South African Revenue Service launched its eFiling tax system in 2003, which streamlined the tax preparation process for companies and shortened the turnaround time for refunds. Today it takes South African companies on average 68 fewer hours to complete and file their tax returns than the global average (PWC, 2014).

Case study in innovative governance: Rwanda

Rwanda is an unlikely governance success story. The landlocked East African nation of 12 million was convulsed by genocidal violence in 1994 orchestrated primarily by the majority ethnic Hutus against the minority Tutsis. Between 800,000 and 1 million civilians were killed during 100 days of bloodletting – more than have been killed during any similar period of time in human history including the Holocaust (Gettleman, 2013). Meanwhile, the country's social institutions were destroyed and the economy was left in a shambles. When Tutsi rebels seized power and ended the genocide, Rwanda was teetering on the brink of chaos. With no oil, natural gas, or other natural resources at its disposal, Rwanda appeared destined to become a failed state or permanent United Nations protectorate (Ensign and Mukantabana, 2014).

Two decades later Rwanda has risen from the ashes and is widely regarded as an African economic showcase. While still amongst the poorest countries in the world, Rwanda has experienced rapid economic growth in recent years, averaging 8 percent annually between 2001 and 2013 (World Bank, 2014b). That growth, coupled with large infusions of foreign assistance from the US and Europe, has helped lift 1 million Rwandans out of poverty. Rwanda has also experienced dramatic improvements in life expectancy, literacy, and infant mortality. And thanks to an innovative national health insurance program launched in 1999, nearly all Rwandans have health care.

Rwanda has also made great strides on the gender equality front. During the genocide period, Rwanda's women suffered unspeakable crimes, including an estimated 250,000–500,000 being raped (Topping, 2014). Today, gender rights are enshrined in its constitution, and women have the legal right to inherit land, share the assets of a marriage, and obtain credit (Topping, 2014). Moreover, Rwanda has the highest percentage of female legislators in the world, with women holding nearly two-thirds of all seats in parliament and one-third of all cabinet positions, including the ministries of foreign affairs, agriculture, and health (Olopade, 2014).

Rwanda has also improved its business environment. It is perennially ranked among the least corrupt nations in Africa by antigraft watchdog Transparency International, and its capital, Kigali, is widely regarded as one of the cleanest and safest cities on the continent. Rwanda is also among the easiest places on the continent to do business. Indeed, in 2014, Rwanda ranked 46 on the World Bank's Ease of Doing Business Index – third in Africa behind only Mauritius (28) and South Africa (43), and ahead of some more developed countries like Italy. Its neighbors, by comparison, ranked 131 (Tanzania), 150 (Uganda), 152 (Burundi), and 184 (Democratic Republic of Congo) (World Bank, 2014a).

Most observers attribute Rwanda's stunning turnaround to its visionary, if controversial, leader Paul Kagame. A commander of the rebel Rwandan Patriotic Army, which seized power in 1994, Kagame has led the country since 2000 and presided over its economic renaissance with ruthless efficiency. A devotee of Singapore's Lee Kuan Yew, the twice-elected Kagame has sought to create a modern, developmental state fueled by foreign investment.

Under Kagame's Vision 2020 plan, launched early in his presidency, Rwanda has taken steps towards reducing its dependence on foreign aid and moving from subsistence agriculture to services. A key pillar of the plan has been investment in ICT. In recent years the Rwandan Information Technology Authority has installed over 1000 miles of fiber-optic cable, and in 2013 it signed a deal with Korea Telecom to build a nationwide 4G LTE broadband network that will provide broadband access to 95 percent of Rwanda's citizens by 2016 (Smith, 2013). Other ICT initiatives include the Smart Kigali program, launched in 2013, which has provided free Wi-Fi to schools, hotels, taxi stations, and public bus stations throughout the capital (Gasore and Kanyesigye, 2013), and the Smart Rwanda Master Plan, unveiled in 2014, which aims to provide better services to citizens through e-government (Melhem, 2014). Kagame himself has been dubbed the "Digital President" by the International Telecommunications Union and has an active social media presence, with more Twitter followers than any other African leader (IT News Africa, 2014).

The Rwandan government's zeal for technology-based solutions has attracted the interest of foreign multinationals like Visa, which recently selected Rwanda as the test market for MVisa, an interoperable mobile payments service (McGroarty and Sidel, 2013). Foreign universities too have been attracted by Rwanda's ICT push. Carnegie Mellon University established a campus in Kigali in 2011, offering graduate programs in information technology and electrical and computer engineering (Juma, 2011). Both of these investments will enhance Rwanda's stock of human capital and position the country as a technology hub for East Africa.

And while Kagame has come under increasing criticism in recent years for his authoritarian proclivities and interventions in the affairs of his neighbors (Gettleman, 2013), most agree that he has been a tireless and effective champion of innovative governance and economic development. As Chu (2009) observes:

> Kagame sends fact-finding missions to Asia. He pursues the Rwandan diaspora. He speaks at Google and meets American entrepreneurs. He recruits more friends. And it's beginning to look as if his personal strategy – selling people on Rwanda's story and its promise, telling them that this is a place where they can make a difference as well as profits – just might work.

Case study in innovative governance: Lagos, Nigeria

The city-state of Lagos, Nigeria, is another surprising governance success story. For years the sprawling megacity on West Africa's Gulf of Guinea was regarded as virtually ungovernable – a morass of crime, corruption, pollution, poverty, and decay. But the city of 21 million began to turn things around in the late 1990s under reform-minded Governor Bola Tinubu (1999–2007), who launched an ambitious tax collection initiative that helped wean the city-state from its dependence on financial transfers from the federal government. Lagos's fortunes took a dramatic turn for the better under Tinubu's successor and former chief of staff, Babatunde Fashola (2007–present). Fashola redoubled the government's tax-collection efforts by dispatching teams of private collectors to visit companies to explain the tax payment process and urge compliance. Tax audits, meanwhile, increased by over 300 percent between 2006 and 2011. The result has been a fourfold increase in tax revenue – a feat accomplished without raising tax rates (de Gramont, 2015).

Fashola has used the tax windfall, along with funds raised in global capital markets, to launch an ambitious $50 billion public works program that has included the construction of new highways, a commuter rail metro system, a bus rapid transport system, and West Africa's first suspension bridge (Kaplan, 2012). The new infrastructure, coupled with the installation of scores of solar-powered traffic lights at heavily congested intersections, has eased Lagos's infamous "go-slows." Fashola has also curbed blight by demolishing shanty towns and encouraging occupants to return to their rural villages, and battled street crime by deploying more police officers to high-crime areas and hiring "area boys" – unemployed youth known for preying on pedestrians and motorists – to sweep city streets (Rice, 2012). Fashola's government has further earmarked millions of dollars in funds for sanitation, affordable housing, and "beautification projects" involving the planting of trees and creation of public parks.

In addition to delivering public services to his constituents, Fashola has impressed Lagosians with his commitment to transparency and the rule of law. His administration is one of the few in Nigeria that publishes a detailed budget on the Web, and he himself is widely seen as incorruptible (*The Economist*, 2011). His reputation for integrity was burnished in 2012 when he chased down and arrested a senior military officer for driving his vehicle in one of the city's dedicated bus lanes. While television cameras rolled, Fashola chastised the officer, telling him that there would be zero tolerance for lawlessness. As one foreign journalist observed, "It was one of the first times Nigerians had ever seen a civil servant confronting a member of the security forces, whose fondness for committing crime rather than fighting it has long contributed to Lagos's legendary reputation for lawlessness" (Freeman, 2014).

Fashola has also championed technological innovation. In 2011, with the assistance of Harvard University's Kennedy School of Government, he established the Lagos Innovation Advisory Council, a 20-member body charged with advising the state government on how to promote entrepreneurship and technological dynamism (Juma, 2011). Comprising members from the private sector, government, and academia, the Council has helped launch the Lagos Innovation Hotspots Map and the Lagos Angel Network. The former, a partnership with Co-Creation Hub, a private technology incubator, provides investors and policy makers with data on innovation clusters across the city; the latter brings entrepreneurs together with potential investors to help jump-start fledging businesses (*The Economist*, 2012b). These initiatives have encouraged a wave of technology start-ups, earning Lagos the reputation of "Africa's smartphone app capital" (Moules, 2014).

Fashola's accomplishments have made him a popular figure at home. He won reelection in 2011 with 81 percent of the vote. And thanks in part to his expert handling of the Ebola crisis in 2014 (Freeman, 2014), he would be a heavy favorite to win a third mandate in 2015 if eligible to stand again.

Fashola's efforts have also garnered recognition from abroad. In 2012, Lagos was nominated as one of the world's 25 most innovative cities (Akinsanmi, 2012). Meanwhile, the city's turnaround has attracted the interest of foreign MNEs like Nissan, which has announced plans to build an assembly factory there (Walt, 2014).

Firm-level innovation

Notwithstanding the important governance reforms being implemented throughout Africa, arguably the most important action on the innovation front is taking place at the firm level, where both start-ups and established firms are launching new products and services to satisfy consumer demand and fill market voids. What follows is a survey of developments in three key sectors: financial services, retail and consumer goods, and food and beverages.

Financial services

In recent years Africa has emerged as the undisputed world leader in mobile banking. Ground zero for the revolution is Nairobi, Kenya, the birthplace of M-Pesa, a mobile payments service launched by Safaricom, the country's largest wireless operator and the Kenyan subsidiary of Vodafone Group PLC.

Prior to the advent of M-Pesa ("pesa" means money in Swahili), many Kenyans kept their savings stuffed in mattresses and transferred money to relatives by handing off cash-filled envelopes to bus drivers or couriers, and hoping for the best. As a largely unbanked society – four-fifths of Kenyans do not have bank accounts – Kenya was ripe for technological disruption. Enter M-Pesa, the brainchild of former Safaricom CEO Michael Joseph, who saw the potential of mobile money to transform Kenya's economy. Taking advantage of lax governmental rules on financial services and only token opposition from incumbent financial institutions, which did not perceive the mobile payments scheme as a threat to their core business, Joseph and his colleagues quietly rolled out the service in 2007 (Olopade, 2014).

M-Pesa caught on fast with Kenyans. Today an estimated 20 million Kenyans have M-Pesa accounts and use the service to conduct 700 million transactions worth more than $20 billion annually – a number equivalent to 45 percent of Kenya's annual GDP (Njogu, 2014). In addition to improving the lives of millions of Kenyans, mobile payments services like M-Pesa are serving as a platform for the emergence of new businesses, many of which cater to the poor. A case in point is M-Kopa LLC, a Kenyan start-up that finances the purchase of solar panels by villagers in rural Kenya living off the power grid. The villagers make a small down payment on the equipment – a solar panel and control box – and then daily payments using M-Pesa for about a year until the equipment is paid off. The solar panels, which are typically installed on the rooftops of homes or mud-walled huts, generate electricity that is considerably cheaper, safer, and cleaner than kerosene, and can be used to power reading lamps, refrigerators, and mobile phones (McGregor, 2012; Vogt, 2015). Buoyed by strong demand for its services and $20 million in annual revenue, M-Kopa has begun licensing its technology in other markets such as Ghana (Jackson, 2015).

Other mobile-payments-based businesses that have emerged to serve Kenya's poor include Kopa Kopa Inc., a firm that offers digital payment solutions for merchants, and M-Changa Ltd, a company that has created a mobile crowd-funding platform allowing individuals to tap the resources of extended families to pay for events like weddings and funerals (Vogt, 2015). Meanwhile, established companies like UAP Insurance and Britam Kenya have used M-Pesa as a platform to offer new services to their customers, including crop insurance and affordable medical insurance (*The Economist*, 2010b).

Retail and consumer goods

The digital payments revolution launched by M-Pesa in Kenya, coupled with the explosive growth of the Internet across the continent, is shaking up Africa's retail and consumer goods sector. And nowhere has the shake-up been felt more acutely than in Nigeria, where a host of start-up online retailers (e-tailers) are threatening bricks-and-mortar incumbents and jockeying for position in Africa's largest and fastest-growing consumer market.

Among the most promising of these e-tailers is Jumia.com. Founded in 2012, Jumia is an online shopping emporium that sells more than 100,000 products, including books, DVDs, consumer electronics, clothing, and appliances. Jumia's customers typically place their orders online with a PC or mobile phone (the company has developed its own Android app), or by dialing into its call center. But Jumia also offers in-person sales with teams of tablet-toting agents, which it dispatches to businesses, churches, and even private homes to educate consumers about web commerce (Kay et al., 2013). In addition to accepting credit and debit cards, Jumia allows customers to pay for purchases in monthly installments. It also accepts payment upon delivery – a key innovation given that many Nigerians continue to be wary about using credit cards for online purchases.

Thus far the privately owned company, which has raised over $150 million in venture capital, has enjoyed steady sales growth. And while the 170-million-consumer Nigerian market offers plenty of room for growth, the company has wider aspirations. It has already launched its service in South Africa, Kenya, Ivory Coast, Egypt, Morocco, Ghana, Uganda, and Cameroon, and plans to expand further throughout the continent. It is also views the Nigerian diaspora as a potentially lucrative market and has rolled out a website in the UK (jumia.co.uk), where Nigerian expatriates can purchase goods and have them delivered to family and friends back home in less than half the time it normally takes for Amazon to ship them from the UK (Pooler, 2014).

Food and beverages

Whereas Internet start-ups like Jumia are leading the innovation charge in retailing and consumer goods, large multinationals – many from South Africa – are the primary source of innovation in the continent's food and beverages sector. One of those is SABMiller. Founded in South Africa in 1895, the London-headquartered SABMiller is the world's second-largest brewer, with operations in 75 countries and a portfolio of more than 150 brands that together generate $31 billion in revenue (Berman, 2013). SABMiller began its African expansion in the 1990s, acquiring privatized breweries in Mozambique, Ghana, Uganda, Zambia, and Tanzania, and spending millions of dollars to modernize them. Today the company has operations in 15 African countries and a stake in 21 other breweries through its alliance with France's Castel. These investments have paid handsome dividends: the Africa region (not including South Africa) today accounts for 13 percent of the firm's profits – more than twice the global average (*The Economist*, 2014d).

And while SABMiller's biggest market on the continent continues to be South Africa, where its brands account for 90 percent of local consumption, its focus is increasingly on fast-growing markets to the north, like Uganda, where it recently spent $80 million to build a new brewery to double the capacity of its Nile Breweries subsidiary (Lucas and Manson, 2012), and Nigeria, where it has built breweries in the cities of Port Harcourt, Ibadan, and Onitsha (Ryan, 2014).

One of the key elements in SABMiller's African strategy is branding: it aims to give each of its beers a distinct local identity by choosing names and symbols that resonate with local populations. In Zambia its leading brand is called Mosi (the local name for Victoria Falls); in Tanzania it is Kilimanjaro. Hero, its top seller in Nigeria, carries a rising sun on its label – an icon of the Igbo tribe (*The Economist*, 2014e).

Another key element of SABMiller's strategy is pricing its beers competitively. This is important because large numbers of Africans continue to consume a variety of cheap and often dangerous concoctions known as "home brews." Getting them to switch to safer and higher-quality, factory-produced beer requires offering a product that is aspirational but affordable. It has achieved that with Hero, its Nigerian lager, which sells for about 25 percent less per bottle than Star, the market leader, and 40 percent less than Guinness, another top seller (*The Economist*, 2014a).

A third element of its strategy is developing new alcoholic beverages, like Chibuku, that appeal to low-income consumers who have traditionally opted for home brews. Made from locally grown maize and sorghum, Chibuku, also known as "Shake-Shake," since its contents tend to separate unless shaken, was developed in Zambia and has been a huge hit there, and in neighboring South Africa, where sales nearly doubled between 2008 and 2013 (Ryan, 2014). Sold in cardboard boxes, Chibuku is popular with the rural poor who value its tangy flavor and porridge-like consistency.

Cheaper to manufacture than traditional lagers – thanks in part to generous tax concessions awarded by governments anxious to promote farm-sector employment – Chibuku delivers high profit margins (Sonne et al., 2013). The beverage does, however, have a serious drawback: since it is sold in a partially fermented state, it may explode if not consumed within a week. To solve this issue, SABMiller developed "Chibuku Super," an upscale version sold in plastic bottles with a longer shelf life. The company expects the product, launched in 2013, to have strong appeal in Zimbabwe, Zambia, and Botswana, with sales ultimately outstripping those of traditional lagers by two to one (Ryan, 2014).

SABMiller has also innovated by introducing budget-priced, lager-style beers made from locally produced roots and tubers (Dontoh and Kew, 2013). Its first offering, Eagle Lager, hit Uganda's shelves in 2002. The translucent, sorghum-based ale quickly became one of East Africa's best-selling beers (*The Economist*, 2012c). As a follow up, it introduced Impala, the world's first commercially made cassava beer, in Mozambique in 2011. That too has enjoyed success, while generating new sources of income for over 2,000 smallholder cassava farmers that supply the brewer. SABMiller has since launched a cassava-based beer in Ghana, and may do so elsewhere on the continent (Sonne et al., 2013).

As for its home market of South Africa, SABMiller continues to innovate there too. It recently kicked off a program to finance renovations at 6,000 pubs across the country to attract more female patrons, who represent a fast-growing segment (Kew and Fletcher, 2014). It has also rolled out new products with feminine appeal, including fruit-flavored drinks like Brutal Fruit Mango Goji and fresh orange and crushed lemon varieties of its popular Flying Fish lager (Kew and Fletcher, 2014).

Facilitation institutions as agents of innovation

While private firms and governments may be the leading catalysts of innovation in Africa, a wide range of institutions are playing the critical role of innovation facilitators. These institutions, which range from small, nonprofit organizations to large multinational companies, are providing entrepreneurs with technical information, know-how, connections, and capital to grow their businesses. Foremost amongst these institutions is the growing number of innovation hubs and tech accelerators that cater to the continent's aspiring high-tech entrepreneurs.

Innovation hubs and tech accelerators

One of Africa's most heralded innovation hubs is Kenya's iHub (ihub. co.ke). Established in 2010 with funding from the Omidyar Network, a US-based philanthropic investment firm, and Hivos, the Dutch international development NGO, iHub bills itself as the center of Nairobi's burgeoning technology movement. Like other innovation hubs in Africa, iHub provides a communal space where entrepreneurs can collaborate, along with an in-house consulting arm that helps members develop investor-ready business plans. It also offers an incubation space for firms developing mobile applications (*The Economist*, 2012d). Boasting more than 13,000 members, iHub claims to have helped launch more than 150 Kenyan companies, among them M-Farm and iCow, the makers of mobile apps for farmers, and MedAfrica, a software start-up that has devised a program that enables users to self-diagnose medical ailments and connect with specialists (Hussey, 2015). In recognition of the key role that iHub has played in Kenya's recent development, *Fast Company* selected it in 2014 as one of the ten most innovative "companies" in Africa.

Nigeria's Co-Creation Hub (CcHub) is another highly touted innovation hub. Established in 2011 by Nigerian tech enthusiasts with support from the Omidyar Network and the UK-based nonprofit Indigo Trust, CcHub describes itself as a preincubation space where entrepreneurs can "work to catalyze creative social tech ventures" (cchubnigeria.com). In addition to hosting brainstorming sessions, focus groups, meet-ups, lectures, workshops, and hackathons, CcHub provides a mentoring service for entrepreneurs with up to 20 start-ups receiving support at any given time (Nsehe, 2011).

Like innovation hubs, tech accelerators work with start-ups to speed up their maturation. But in addition to providing a shared space, mentoring, and networking opportunities, accelerators typically provide intensive training (i.e., boot camps), seed money, and introductions to potential investors, usually in exchange for equity stakes in the fledgling companies. Among the most prominent of Africa's tech accelerators is Kenya's 88mph. Founded in Nairobi in 2011 by a Danish entrepreneur and a Kenyan businessman who met at Stanford University, 88mph has already raised more than $4.5 million from African and European investors and funded nearly three dozen companies through its three-month "start-up garage" program (Coetzee, 2015).

The latest addition to the list of accelerators and accelerator-like programs is the Tony Elumelu Foundation Entrepreneurship Program. Founded in 2014 by Nigerian billionaire Tony Elumelu, the $100 million program aims to help launch over 10,000 new businesses over the next decade. Each of the 1,000 African entrepreneurs selected for the program each year will receive $5,000 in seed capital and an additional $5,000 in "returnable capital," which they will use to establish and grow their businesses. The program, which includes 12 weeks of online mentoring and training, will culminate in a three-day forum in Nigeria where participants will meet and pitch their ideas to investors. The Foundation hopes the program will ultimately spawn 1 million new jobs and generate $10 billion in new revenues, while boosting intra-African trade (Keeler, 2014).

Venture capital and private equity firms

Another important category of innovation facilitators are foreign venture capital (VC) firms, which have been drawn to tech centers like Nairobi and Lagos in recent years in search of the next Internet juggernaut. Among the most active of these VC firms are Rocket Internet, a Berlin-based "clone" firm that seeks to replicate successful business models like that of Amazon. com in emerging markets (*The Economist*, 2014c), and Tiger Management, a New York-based investment fund. The former has raised $150 million for Nigerian e-tailer Jumia, whereas the latter has raised $100 million for South African web commerce giant Takealot, and millions more for iROKOtv, a Nigerian online music and movie distributor hailed as the "Netflix of Africa" (Moules, 2014; Coetzee, 2015). Other VC firms active on the continent include Sweden's AB Kinnevik and South Africa's Naspers, which together have raised $40 million for Nigerian e-tailer Konga.com (Rice, 2013). The capital provided by these VCs has allowed their start-up clients to expand product offerings, redesign websites, modernize warehouses, rationalize supply chains, and launch innovative marketing campaigns.

Private equity (PE) firms represent another important vehicle for innovation, raising a record $4 billion for investments in African companies in 2014 (*The Economist*, 2015). Leading PE investors in Africa include the Carlyle Group, the Blackstone Group, and Helios Partners. While much of the PE money entering Africa in recent years has been earmarked for utilities and heavy industry, a growing share is going to companies in consumer-facing industries. Emerging Capital Partners (ECP), for instance, made a major investment in the Kenyan coffee-house chain Nairobi Java House in 2012, enabling it to expand its footprint to Mombasa, Nakuru, and Kisumu, and launch a chain of frozen yogurt outlets (*The Economist*, 2015). And in 2014, the New York-based PanAfrican Investment Co. announced an undisclosed investment in Mobius Motors, a Kenyan automobile start-up with plans to build an affordable all-road vehicle for Africa's mass market (Nsehe, 2014).

Multinational enterprises

MNEs are also playing a key role in facilitating innovation on the continent through greenfield investments and acquisitions. Such investments often result in the transfer of valuable technology, know-how, and best practices to affiliates and host societies (UNCTAD, 2014). But a handful of leading MNEs are taking their commitment to the continent's technological development even further by establishing innovation hubs and research labs on the continent. IBM is one such company. The US tech giant announced plans in 2013 to build a research lab in Nairobi. The lab – its first in Africa and one of only a dozen worldwide – will work on developing applied, data-driven solutions to problems caused by rapid urbanization, including water and energy scarcity, education and health-care inadequacy, and traffic congestion (*The Economist*, 2013). Dutch electronics firm Philips also opened a research laboratory in Nairobi in 2013 to develop technology-based health-care and lighting solutions (*The Economist*, 2014b). One of its first projects is a hand-powered ultrasound fetal heart rate monitoring device for use in rural areas with irregular electricity supplies (Jack, 2014).

Nonprofit organizations

No discussion of innovation facilitators would be complete without mentioning the key role played by foreign nonprofit organizations in promoting novel solutions to the continent's social and environmental challenges. Foremost among these organizations are the Bill and Melinda Gates Foundation and the Clinton Global Initiative, which fund a variety of important projects in the areas of health, education, sanitation, and sustainable enterprise. But smaller and less well-known institutions are also making a mark. One such organization is KickStart International, a San Francisco-based nonprofit that sells low-tech irrigation pumps to smallholder farmers throughout the continent. The devices, which can irrigate up to two acres a day, are expensive by local standards – costing between $75 and $160 each – but are well worth the investment, as they allow farmers to switch from seasonal to year-round production, thereby boosting returns (Lesle, 2013).

Conclusion

Africa is clearly a continent on the move. The mood of pessimism that has long prevailed is giving way to one of hope and optimism. Perhaps the most encouraging aspect of Africa's recent economic rise, as Kelly (2014) observes, is that it is a largely homegrown phenomenon, being driven primarily by local companies developing local content (e.g., software and mobile apps) for local consumers. This is in stark contrast to the continent's traditional growth model, which has been driven by foreign extractive companies and agribusinesses, using foreign know-how for the benefit of foreign consumers. Taking Africa's growth to the next level, however, will require tackling thorny obstacles that continue to stymie innovation and thwart would-be entrepreneurs, such as poverty, poor infrastructure, weak governance, fragile security, and the shortage of human capital, electricity, and investment capital. Solving these challenges will require creative thinking, ingenuity, political will, and plenty of *kazi ngumu* (Swahili for "hard work"). Africa's new generation of business and government leaders clearly have their work cut out. But if they can continue to push forward with political and economic reforms, harnessing the power of information technology and the skills and resources of foreign partners including the Africa diaspora, some of the continent's nascent innovation hills may yet rise to become peaks.

References

Acemoglu, D. and Robinson, J. (2012). *Why Nations Fail: The Origins of Power, Prosperity, and Poverty.* New York: Crown Publishers.

Akinsanmi, G. (2012). "Lagos Emerges Africa's Most Innovative City." *This Day Live.* Retrieved from http://www.thisdaylive.com/articles/lagos-emerges-africa-s-most-innovative-city/128001/

BBC. (2014). "Kenya's Matatu Bus System to Go Cashless." *BBC News.* Retrieved from http://www.bbc.com/news/business-28106004

Berman, J. (2013). *Success in Africa: CEO Insights from a Continent on the Rise.* Brookline, MA: Bibliomotion.

Chu, J. (2009). "Rwanda Rising: A New Model of Economic Development." *Fast Company.* Retrieved from http://www.fastcompany.com/1208900/rwanda-rising-new-model-economic-development

Coetzee, J. (2015). "8 Exciting African Startup Programmes and Accelerators to Watch in 2015." *Ventureburn.* Retrieved from http://ventureburn.com/2015/01/8-exciting-african-entrepreneurship-programmes-watch-2015/?device=desktop

de Gramont, D. (2015). "Governing Lagos: Unlocking the Politics of Reform." *Carnegie Endowment for International Peace.* Retrieved from http://carnegieendowment.org/2015/01/12/governing-lagos-unlocking-politics-of-reform/hz99

Deloitte. (2014). "The Deloitte Consumer Review: Africa: A 21st Century View." Retrieved from http://www2.deloitte.com/uk/en/pages/consumer-business/articles/africa-a-21st-century-view.html

Devarajan, S. and Fengler, W. (2013). "Africa's Economic Boom." *Foreign Affairs.* Retrieved from http://www.foreignaffairs.com/articles/139109/shantayanan-devarajan-and-wolfgang-fengler/africas-economic-boom

Dontoh, E. and Kew, J. (2013). "SABMiller Sells Cassava Beer to Woo African Drinkers." *BloombergView.* Retrieved from http://www.bloomberg.com/bw/articles/2013-05-23/sabmiller-sells-cassava-beer-to-woo-african-drinkers

eMarketer. (2014). "Internet to Hit 3 Billion Users in 2015." Retrieved from http://www.emarketer.com/Article/Internet-Hit-3-Billion-Users-2015/1011602

Ensign, M. M. and Mukantabana, M. (2014). "Rwanda 20 Years Later: A Model for Progress and Reconciliation." *Christian Science Monitor.* Retrieved from http://www.csmonitor.com/Commentary/Opinion/2014/0407/Rwanda-20-years-later-A-model-for-progress-and-reconciliation

Florida, R. (2005). "The World Is Spiky." *The Atlantic.* Retrieved from http://www.theatlantic.com/past/docs/images/issues/200510/world-is-spiky.pdf

Freeman, C. (2014). "Meet the Man Who Tamed Nigeria's Most Lawless City." Retrieved from http://www.telegraph.co.uk/news/worldnews/africaandindianocean/nigeria/11184759/Meet-the-man-who-tamed-Nigerias-most-lawless-city.html

Gasore, B. and Kanyesigye, F. (2013). "Rwanda: 'Smart Kigali' Brings Free Internet to City." *The New Times (Kigali).* Retrieved from http://allafrica.com/stories/201309231105.html

Gettleman, J. (2013). "The Global Elite's Favorite Strongman." *The New York Times.* Retrieved from http://www.nytimes.com/2013/09/08/magazine/paul-kagame-rwanda.html

Global Innovation Index. (2014). *The Global Innovation Index 2014: The Human Factor in Innovation.* Retrieved from https://www.globalinnovationindex.org/content.aspx?page=GII-Home

Green, A. and Whitehead, E. (2013). "Africa's Next Generation." *This Is Africa.* Retrieved from http://www.thisisafricaonline.com/Policy/Africa-s-next-generation?ct=true

Harrison, A., Yifu Lin, J., and Xu, C. (2013). "Explaining Africa's (Dis)advantage." *The National Bureau of Economic Research*, paper 18683. Retrieved from http://www.nber.org/papers/w18683.pdf

Holland, M., Tucker, I., Mark, M., Kelly, A., and Honigsbaum, O. (2012). "Africa Innovations: 15 Ideas Helping to Transform a Continent." *The Guardian*. Retrieved from http://www.theguardian.com/world/2012/aug/26/africa-innovations-transform-continent

Hussey, M. (2015). "Silicon Savannah – How Start-ups in Africa Are Taking on Some of Humanity's Biggest Challenges." *The Huffington Post*. Retrieved from http://www.huffingtonpost.co.uk/matthew-hussey/african-startups-take-on-challenges_b_6416676.html

IT News Africa. (2014). "Rwanda's President Tweeting up a Storm." Retrieved from http://www.itnewsafrica.com/2014/10/rwandas-president-tweeting-up-a-storm/

Jack, A. (2014). "Low Technology and Innovative Funding Carry Africa's Hopes." *Financial Times*. Retrieved from http://www.ft.com/intl/cms/s/0/ca2e7f66-6f3a-11e4-8d86-00144feabdc0.html?siteedition=intl#axzz3Qcq7PRqQ

Jackson, T. (2015). "Africa's New Breed of 'Solar-Preneurs.'" *BBC News*. Retrieved from http://www.bbc.com/news/business-30805419

Juma, C. (2011). "Africa's New Engine." *International Monetary Fund*. Retrieved from http://www.imf.org/external/pubs/ft/fandd/2011/12/juma.htm

Kaplan, S. (2012). "City Development States: Why Lagos Works Better than Nigeria." Retrieved from http://www.policyinnovations.org/ideas/innovations/data/000212/:pf_printable?sourceDoc=000065

Kay, C., Spillane, C., and Kew, J. (2013). "Trying to Build the Next Amazon in Nigeria." *Bloomberg*. Retrieved from http://www.bloomberg.com/bw/articles/2013-11-21/jumia-africas-amazon-dot-com-takes-cash-and-delivers-by-motorbike

Keeler, D. (2014). "Elumelu Foundation Launches $100m Program to Boost African Business." *Frontier Markets*. Retrieved from http://blogs.wsj.com/frontiers/2014/12/01/elumelu-foundation-launches-100m-program-to-boost-african-business/

Kelly, T. (2014). "Tech hubs across Africa: Which will be the legacy-makers?" The World Bank Informational and Communications for Development (IC4D) Blog. Retrieved from http://blogs.worldbank.org/ic4d/tech-hubs-across-africa-which-will-be-legacy-makers

Kew, J. and Fletcher, C. (2014). "SABMiller Cleans up South Africa's Bars to Attract Women." *Bloomberg*. Retrieved from http://www.bloomberg.com/bw/articles/2014-05-29/sabmiller-cleans-up-south-africas-bars-to-attract-women

Lesle, T. (2013). "The Simple Water Pump that's Changing Lives across the World." *WIRED*. Retrieved from http://www.wired.com/2013/12/2112kickstart/

Lucas, L. and Manson, K. (2012). "Brewers See New Horizons in Africa." *Financial Times*. Retrieved from http://www.ft.com/intl/cms/s/0/7b7e1c84-5343-11e1-aafd-00144feabdc0.html?siteedition=intl#axzz3Qcq7PRqQ

McGregor, S. (2012). "Kenya's M-KOPA Gives Phone-Loans to Put Solar Power in Reach." *Bloomberg*. Retrieved from http://www.bloomberg.com/news/2012-10-04/kenya-s-m-kopa-offers-cheaper-solar-power-to-off-grid-villages.html

McGroarty, P. and Sidel, R. (2013). "Visa Plants a Seed for Growth Abroad." *Wall Street Journal*. Retrieved from http://www.wsj.com/articles/SB10001424127887323687604578466671317052766

McKinsey. (2013). "Lions Go Digital: The Internet's Transformative Potential in Africa." Retrieved from http://www.mckinsey.com/insights/high_tech_telecoms_internet/lions_go_digital_the_internets_transformative_potential_in_africa

Melhem, S. (2014). "Smart Africa Returns with a Focus on Rwanda." *Information and Communications for Development*. Retrieved from http://blogs.worldbank.org/ic4d/smart-africa-returns-focus-rwanda

Moules, J. (2014). "Apps Mean Jobs for Young Nigerians." *Financial Times*. Retrieved from http://www.ft.com/intl/cms/s/0/77df79a8-c716-11e3-889e-00144feabdc0. html#axzz3PJbocDnu

Njogu, L. (2014). "Top 10 List of Most Destructive Technologies in Kenya." *PIVOT East*. Retrieved from http://www.pivoteast.com/top-10-list-destructive-technologies-kenya-last-5-years/

Nsehe, M. (2011). "EBay Billionaire Omidyar Gives Nigerian Tech Incubator $200,000." *Forbes*. Retrieved from http://www.forbes.com/sites/mfonobongnsehe/2011/07/20/ebay-billionaire-omidyar-gives-nigerian-tech-incubator-200000/

Nsehe, M. (2013). "Meet the Entrepreneur Working to Challenge Nokia, Blackberry and Samsung in Africa." *Forbes*. Retrieved from http://www.forbes.com/sites/mfonobongnsehe/2013/04/30/meet-the-entrepreneur-working-to-challenge-nokia-blackberry-and-samsung-in-africa/

Nsehe, M. (2014). "American Billionaire Ronald Lauder Is Funding Africa's Cheapest Car." *Forbes*. Retrieved from http://www.forbes.com/sites/mfonobongnsehe/2014/05/12/american-billionaire-ronald-lauder-is-funding-africas-cheapest-car/

Olopade, D. (2014). *The Bright Continent: Breaking Rules and Making Change in Modern Africa*. New York: Houghton Mifflin Harcourt.

Pooler, M. (2014). "Internet Retailer Tapping into Diaspora Dollars." *Financial Times*. Retrieved from http://blogs.ft.com/beyond-brics/2014/06/20/internet-retailer-tapping-into-diaspora-dollars/

PWC. (2014). *Paying Taxes 2015: The Global Picture*. Retrieved from http://www.pwc.com/gx/en/paying-taxes/assets/pwc-paying-taxes-2014.pdf

Radjou, N., Prabhu, J., Ahuja, S., and Roberts, K. (2012). *Jugaad Innovation: Think Frugal, Be Flexible, Generate Breakthrough Growth*. San Francisco, CA: Jossey-Bass.

Rice, X. (2012). "Africa: Lessons from Lagos." *Financial Times*. Retrieved from http://www.ft.com/intl/cms/s/0/8b24d40a-c064-11e1-982d 00144feabdc0.html?siteedition=intl#axzz3Qcq7PRqQ

Rice, X. (2013). "Internet Sales Flourish in Nigeria." *Financial Times*. Retrieved from http://www.ft.com/intl/cms/s/0/3f455b7e-b1bb-11e2-9315-00144feabdc0. html#axzz3PlSJq45m

Ryan, F. (2014). "A New Frontier: Africa's 'Explosive' Homebrew Market." *Financial Times*. Retrieved from http://blogs.ft.com/beyond-brics/2014/05/23/a-new-frontier-africas-explosive-homebrew-market/

Smith, D. (2013). "Rwanda Strikes 4G Internet Deal with South Korean Telecoms Firm." *The Guardian*. Retrieved from http://www.theguardian.com/world/2013/jun/11/rwanda-4g-internet-south-korea

Sonne, P., Maylie, D., and Hinshaw, D. (2013). "With West Flat, Big Brewers Peddle Cheap Beer in Africa." *Wall Street Journal*. Retrieved from http://www.wsj.com/articles/SB10001424127887324034804578348533702226420

The Economist. (2010a). "First Break All the Rules." *The Economist*. Retrieved from http://www.economist.com/node/15879359

The Economist. (2010b). "Security for Shillings." *The Economist*. Retrieved from http://www.economist.com/node/15663856

The Economist. (2011). "A Rare Good Man." *The Economist*. Retrieved from http://www.economist.com/node/18652563

The Economist. (2012a). "A Glass Half-Full." *The Economist*. Retrieved from http://www.economist.com/node/21551494

The Economist. (2012b). "Angels in Lagos." *The Economist*. Retrieved from http://www.economist.com/blogs/baobab/2012/10/nigerias-entrepreneurs

The Economist. (2012c). "From Lumps to Lager." *The Economist*. Retrieved from http://www.economist.com/node/21551092

The Economist. (2012d). "Upwardly Mobile." *The Economist.* Retrieved from http://www.economist.com/node/21560912

The Economist. (2013). "The Next Frontier." Retrieved from http://www.economist.com/news/business/21571889-technology-companies-have-their-eye-africa-ibm-leading-way-next-frontier

The Economist. (2014a). "Africa's Testing Ground." *The Economist.* Retrieved from http://www.economist.com/news/business/21613341-make-it-big-africa-business-must-succeed-nigeria-continents-largest-market-no

The Economist. (2014b). "On the Rise." *The Economist.* Retrieved from http://www.economist.com/news/middle-east-and-africa/21611112-scientific-research-africa-gathering-momentum-rise

The Economist. (2014c). "Rocket Machine." *The Economist.* Retrieved from http://www.economist.com/news/special-report/21593586-how-build-companies-kit-rocket-machine

The Economist. (2014d). "Slim SIMS." *The Economist.* Retrieved from http://www.economist.com/blogs/baobab/2014/09/disrupting-mobile-banking-kenya

The Economist. (2014e). "The Beer Frontier." *The Economist.* Retrieved from http://www.economist.com/news/business/21602999-long-established-african-firm-went-global-only-find-fastest-growing-market-was-its

The Economist. (2015). "Unblocking the Pipes." *The Economist.* Retrieved from http://www.economist.com/news/leaders/21640349-africa-needs-lot-capital-private-equity-offers-lessons-how-get-it-there-unblocking

The Economist Intelligence Unit. (2013). "Democracy Index 2013: Democracy in Limbo." Retrieved from http://www.ihsnews.net/wp-content/uploads/2014/06/Democracy_Index_2013_WEB-2.pdf

Topping, A. (2014). "Rwanda's women make strides towards equality 20 years after the genocide." The Guardian. Retrieved from http://www.theguardian.com/global-development/2014/apr/07/rwanda-women-empowered-impoverished

United Nations Conference on Trade and Development (UNCTAD). (2014). *World Investment Report 2014.* UNCTAD. Retrieved from http://unctad.org/en/publicationslibrary/wir2014_en.pdf

United Nations Department of Economic and Social Affairs (UNDESA). (2013). *World Population Prospects: The 2012 Revision.* UNDESA. Retrieved from http://esa.un.org/wpp/

UNESCO. (2014). "Africa's Minds: Build a Better Future." Retrieved from http://unesdoc.unesco.org/images/0022/002278/227858e.pdf

UN Habitat. (2014). "The State of African Cities 2014." Retrieved from http://unhabitat.org/the-state-of-african-cities-2014/

Vogt, H. (2015). "Making Change: Mobile Pay in Africa." *Wall Street Journal.* Retrieved from http://www.wsj.com/articles/making-change-mobile-pay-in-africa-1420156199

Walt, V. (2014). "Lagos, Nigeria: Africa's Big Apple." *Fortune.* Retrieved from http://fortune.com/2014/06/12/lagos-nigeria-big-apple/

Ware, G. (2013). "The Rise of Africa's B-brands." *The Africa Report.* Retrieved from http://www.theafricareport.com/North-Africa/the-rise-of-africas-b-brands.html

World Bank. (2014a). *Doing Business 2015: Going Beyond Efficiency.* Washington, DC. Retrieved from http://www.doingbusiness.org/~/media/GIAWB/Doing%20Business/Documents/Annual-Reports/English/DB15-Chapters/DB15-Report-Overview.pdf

World Bank. (2014b). *Rwanda Overview.* Retrieved from http://www.worldbank.org/en/country/rwanda/overview

World Bank and African Development Bank. (2012). *The Transformational Use of Information and Communication Technologies in Africa.* Retrieved from http://www.infodev.org/infodev-files/resource/InfodevDocuments_1162.pdf

8

Reverse Innovation in Emerging Markets

Vijay Govindarajan and Ravi Ramamurti

Introduction[1]

The locus of innovation in the global economy appears to be changing because of the rise of emerging economies, such as China and India, and the "flattening" of the world. Poor emerging markets no longer just borrow innovations from developed countries; from time to time they also contribute innovations to the rest of the world, including developed countries.[2] We refer to this as reverse innovation, that is, the case where an innovation is adopted first in a poor country before being adopted in rich ones. Instances of reverse innovation still appear to be rare, and it is hard to tell if this will change materially in the future. Therefore, this chapter is about a nascent phenomenon whose future potential is uncertain. Nevertheless, reverse innovation is a promising area for research by international business and strategy scholars because it provides the opportunity to enrich and extend mainstream theories in a number of areas. At the same time it raises the level of awareness necessary for governments trying to foster innovation, the role facilitating institutions might play (NGOs, R&D labs around the world), and finally the potential benefits that multinational companies might gain to sustain a competitive advantage.

Regarding the innovation drivers introduced in this book: national innovation policy, facilitators of innovation, and firm-level innovation (product, process, and business model), the chapter provides guidelines for the last of these. Issues related to national innovation policies and the facilitators of innovation are blended through the perspective of local firms and developed-country multinational enterprises (DMNEs). The emphasis is on the nature of innovation in emerging markets, the reasons why such innovations might diffuse to other parts of the world, including the role of local firms and DMNEs in this process, and the strategic and organizational challenges faced by DMNEs that pursue reverse innovation. In each of these areas, the chapter offers illustrative hypotheses and questions for further research. However, these are only tentative ideas based on assumptions and arguments that need to be evaluated and rigorously tested.

The chapter also raises critical questions to increase the knowledge of innovation creation while offering information of practical relevance for managers and policy makers if reverse innovation becomes more widespread in the future.

What is reverse innovation?

Innovation has always been recognized as one of the pillars of multinational firms (Bartlett and Ghoshal, 1989). In the mainstream international business (IB) literature, innovation is often equated with technological innovation (Caves, 2007) and is assumed to originate in the developed countries where the world's leading multinational corporations (MNCs) are located (see Figure 8.1). In the original product cycle hypothesis, for instance, Vernon (1966) saw the United States after the end of World War II as the technology leader and the source of product innovations that targeted high-income consumers and process innovations that substituted capital for labor. He argued that US multinationals spread both types of innovation to Europe and Japan and, eventually, developing countries. By the 1970s, Europe and Japan had closed their technology and income gaps with the US, and Vernon (1979) argued that innovations would now diffuse horizontally among developed countries and downward to developing countries.

Subsequent research has recognized that DMNEs strive to capture knowledge and innovation in many locations around the world, not just their home countries (Kuemmerle, 1997; Cantwell and Mudambi, 2005), but as late as 2004, DMNEs had 90 percent of their foreign R&D affiliates in the Triad regions and only 10 percent in developing countries (United Nations, 2005). The possibility of important innovations occurring in developing countries and then trickling up to developed countries is only now receiving recognition (e.g., Immelt et al., 2009; Ramamurti, 2009a).

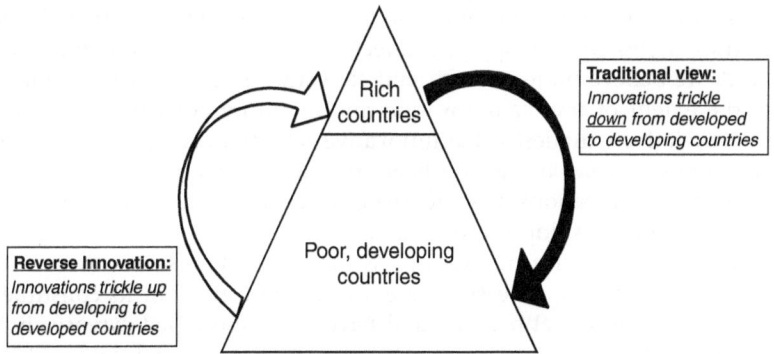

Figure 8.1 Reverse innovation
Source: Authors.

Well-known examples of innovations in emerging markets include the Tata Nano, Grameen Bank (microfinance), GE's ultrasound, Embraer's regional jets, BYD's electric cars, Bharti Airtel's ultra cheap wireless telephony, and Nokia's inexpensive cell phones. A few of these innovations have already made the journey from poor to rich countries. But innovation in poor countries seems to be deeper and wider than these examples indicate, and more of these innovations may diffuse internationally in the future. *The Economist's* special issue on innovation in emerging markets, subtitled "The world turned upside down," noted that "the emerging world, long a source of cheap labor, now rivals the rich countries for business innovation" (*The Economist*, 2010, p. 1). After citing examples from many industries, it concluded that these innovations will change "not just emerging markets but the rest of the world as well" (*The Economist*, 2010, p. 16).

Innovations occurring in emerging economies tend not to involve technological breakthroughs of the kind that drive innovation in developed countries. They do, however, involve novel and innovative combinations of existing knowledge and technologies to solve pressing local problems and the use of new processes and business models. The firms leading these innovation efforts could be local firms or foreign MNEs with operations in emerging markets. Also, innovations first adopted in a developing country may draw on talent, technology, and ideas from many parts of the world, as is to be expected in today's integrated world. General Electric's ultralow-cost ultrasound and ECG machines, for instance, were led by development teams in China and India, initially for use in these countries, but with inputs from local and foreign subsidiaries. The same was true of the development of the Tata Nano car. Our focus in this chapter is on where an innovation is first *adopted*, not where it is *developed*. Likewise, the adjective "reverse" refers specifically to the flow of innovations from poor to rich countries, which is opposite to the predominant direction of technology flow in modern times.[3] It is not meant to connote any other kind of value judgment.

Compared to the flow of innovations and technology from rich to poor countries, flows in the opposite direction probably are still minuscule. Indeed, we believe more research and case studies are needed to establish the true extent of reverse innovation and its future potential. Our claims here are relatively modest – that innovative activity in emerging economies is much greater than it has ever been, that it is growing rapidly, and that the resulting innovations may increasingly find their way to the rest of the world, including developed economies.

In its totality, reverse innovation, as described here, seems to be without historical parallel, at least in recent times, but many of its constituent ideas do have historical parallels and have been researched extensively by strategy and IB scholars. The latter include studies of how location affects innovation (Porter, 1990; Nelson, 1993), why firms innovate outside the home country (Kuemmerle, 1997), how and why innovations diffuse across countries (Vernon, 1966, 1979), how firms adapt products designed for one country for sale in others (Yip, 1989), and how MNEs manage the tension between global standardization and national responsiveness (Prahalad and Doz, 1987; Bartlett and Ghoshal, 1989). Also familiar is the notion of firms in industrially backward countries playing catch-up with First World firms, typically starting with low-end products and then working their way up to high-end products, e.g., Japanese or Korean firms in the postwar period (Tsurumi, 1976; Amsden, 1992). For these reasons, extant theory can help us understand many aspects of reverse innovation. For example, the notion that innovations occur in response to opportunities and constraints in a firm's immediate business environment seems to be just as true in poor countries as in rich ones; it is just that the opportunities and constraints in poor countries are systematically different from those in rich ones. It would be a mistake, however, to conclude that because some facets of reverse innovation have a familiar ring or are well-understood, there is nothing new about it in whole or in part and that there is no need to rethink some mainstream theories.

DMNEs may have innovated outside the home country in the past, but they seldom did so in really poor developing countries, and they were seldom joined in this effort by entrepreneurial local firms, as is occurring today.[4] Similarly, DMNEs may always have adapted products designed for one Triad market to suit other Triad markets, but that is not the same as adapting Triad-based products for the mass market of poor countries. True, Japan and Korea also had poor consumers when they began to catch up after World War II, but the income gap between them and the US was much smaller than what exists today between China or India and the US. For instance, in 1985, after more than three decades of economic recovery, Japan's per capita income in purchasing power parity (PPP) terms was 81 percent of the US level (Balassa and Noland, 1988). In 2009, fully 30 years after China's economy opened up, its per capita income in PPP terms was only 15 percent of the US level; India's was only 7 percent of the US level.[5] The scale and size of the Chinese and Indian markets, each potentially as big as the US, but much poorer, are also without parallel in recent history (Wilson and Purushothaman, 2003). Likewise, Japanese and Korean companies did export cheaper and better cars to the US, disrupting the American auto industry, but that is not the same as producing ultralow-cost products, such as a $2,500 car or a $50 computer. Further, Western multinationals were unable to enter and compete in postwar Japan and Korea due to tariff, nontariff, and FDI (foreign direct investment) barriers, but this has been less true of post-liberalization China or India. As a result, DMNEs have been able to participate in the innovation process of emerging economies in ways they could not in Japan or Korea. And, finally, the world itself has changed profoundly in the last 30–40 years, becoming much "flatter" than before (Friedman, 2005), with "gateways to internationalization" that did not exist earlier (Williamson and Zeng, 2009). In summary, reverse innovation is a phenomenon with familiar features *as well as* new ones, and it is occurring in a changed world compared to even the 1960s and 1970s.

Reverse innovation in emerging markets

Reverse innovation poses interesting puzzles for understanding where innovations occur, why, and how they diffuse internationally. One question is why poor countries might be centers of innovations and what kinds of innovation they might spawn – regardless of whether these innovations trickle up to rich countries. A second question is why innovation is flowering in emerging markets today but did not in earlier decades. And a third question is when and why innovations might move in the counterintuitive direction from poor to rich countries. After all, if the large "distance" (Ghemawat, 2001) between poor and rich countries is what makes local innovation necessary in emerging markets, should not the same reason make it harder to transfer the resulting innovations to rich countries?

What kinds of innovations will emerging markets spawn?

One important line of research is to understand the type of innovations emerging markets are likely to spawn, drawing on the literature on national innovation systems (Porter, 1990; Nelson, 1993). The most important dimension on which emerging economies appear to differ from developed ones is the per capita income of average consumers, except perhaps at the very top of the economic pyramid. Therefore, the mass market in poor countries requires products with dramatically improved price–performance features, creating opportunities for "affordability innovation."

Success in these markets also requires optimizing products to suit local conditions, such as making them sturdy to compensate for the harsh conditions under which they are typically used, making them portable so equipment can be taken easily to far-flung customers whose mobility is limited, or making them easy to operate and maintain because users are technically unsophisticated. Because of institutional voids of various kinds, innovations may also be necessary in the way products are sold, distributed, and financed (Khanna and Palepu, 2005). All this adds up to a fundamental rethinking of the business model necessary for success in emerging markets (Ramamurti and Singh, 2009b).

In other words, large differences in per capita income between poor and rich countries and the conditions under which products are used seem to provide the basis for innovations in poor countries. Ghemawat (2001) captures some of these differences in his multidimensional measure of cultural, administrative, geographic, and economic distance between countries. However, much more research is needed to understand how such "distance" creates pressure for local innovation in emerging markets or how it inhibits their diffusion to developed countries.

Reverse innovation can also build on the literature on lead users and their role in diffusing innovations. Von Hippel (1986) views lead users as early adopters of cutting-edge innovations who have demanding standards and are not very price sensitive. Firms work with them to launch new products that later can be targeted at mainstream consumers. In this framework, the left tail of the distribution consists of laggards – individuals who are extremely value conscious, happy with "good enough" quality, and costly to reach and serve, a profile that resembles the mass-market consumer in emerging economies. In other words, just as lead users can be leveraged to diffuse an innovation to mainstream users, reverse innovation can be used to turn laggards into mainstream users. Future research can examine such issues as the conditions under which laggards, rather than lead users, will be a more advantageous source of new product ideas and whether some industries are more likely than others to benefit from reverse innovation.

Why now and not earlier?

Another interesting question is why innovation may be flourishing in emerging economies now and did not in prior periods. This would require historical analysis of the national and international environments at different times to discern key driving forces. On this question, our tentative guess is that the demand for and supply of local innovation have only just come together in emerging markets. On the demand side, we referred earlier to accelerating growth in emerging economies and the simultaneous slowing down of developed economies. In the coming decades, as much as two-thirds of world GDP growth is likely to occur in these markets. The needs of middle-class consumers in emerging markets are, therefore, of strategic significance to firms everywhere. For local innovation to actually occur, however, demand for such innovation must be matched by supply, and it is on this front that conditions have changed dramatically in the last two decades. Economic liberalization in many emerging markets, coupled with a "flatter" world brought on by technological change (Friedman, 2005), has spawned local firms in emerging economies that can draw on local and global resources to innovate for local markets (Williamson and Zeng, 2009; Ramamurti, 2009a). An innovation like Bharti Airtel's telecom revolution could not have happened in preliberalization India or without the access to international capital, suppliers, and technology made possible by a flatter world (Martinez-Jerez et al., 2006). At the same time, DMNEs have realized that it is not enough to serve just the premium segment in emerging markets with products that are close variants of those developed for rich countries. Instead, they have had to develop products with drastically lower costs and special features, which were not always achievable simply by tinkering with existing designs (Prahalad, 2005).

In theory, DMNEs have long had access to the technology and skills needed to create something like the Tata Nano but, in practice, there was little incentive for them to do so. Without a local firm taking the lead, the latent demand of Indians for such products probably would have remained unrealized. However, once local firms developed such products, DMNEs were drawn into the game out of fear those firms eventually would exploit their innovations globally and challenge DMNEs everywhere (Ramamurti, 2009a). This has added momentum to the supply side of local innovation in emerging markets.

Why might innovations "trickle up" from poor to rich countries?

Reverse innovation entails at least three stages. The first is adoption of an innovation in an emerging market, such as China or India. The second is the transferring of this innovation to other emerging markets. The third and final step is transferring it selectively to developed-country markets. Both local firms and DMNEs can engage in all three stages, but their strengths and weaknesses are likely to be quite different, as will be discussed. Because the local firms that take such innovations abroad are, by definition, multi-nationals, we refer to them as "emerging-market multinational enterprises" (EMNEs).

The diffusion of innovations from one emerging market to another with similar circumstances is not particularly surprising (Stage 2) because both may share customers who are price sensitive or need products that are port-able, rugged, and easy to use. But the trickling up of innovations from poor to rich countries (Stage 3) is perplexing. If "distance" between countries (Ghemawat, 2001) makes it necessary to innovate specifically for emerging markets, should not the same distance factors make it harder for the result-ing innovations to migrate to rich countries? Furthermore, the product cycle theory provides a persuasive argument for why innovations will dif-fuse from developed to less developed countries as the latter become rich, but since rich countries seldom become poor, why would innovations ever travel in the opposite direction?

We propose five possible reasons why innovations might trickle up from poor to rich countries (see Table 8.1). First, there are poor people in rich countries for whom "good enough" products at ultralow prices may have great appeal. This is the mirror image of the argument that products devel-oped for rich countries may have an immediate market in poor countries because there are rich people in poor countries. Thus, innovations in low-cost housing, medical care, or banking (e.g., microfinance) may appeal to the inner-city or rural poor in developed countries. Admittedly, this may be a thin slice of the market and may not be considered attractive enough by some DMNEs, but it could be a nice avenue for growth in otherwise satu-rated businesses, and it could become more important if incomes stagnate in the rich countries.

Table 8.1 Why innovations may trickle up from poor to rich countries

No	Trickle-up mechanisms	Examples
1	Innovations developed in EMs may have a ready market among poor people in rich countries.	Microfinance invented in Bangladesh, works also for the inner-city poor in rich countries like the US.
2	Dramatic cost and price reductions of 70 to 90 percent achieved to succeed in EMs can help expand demand in rich countries.	Dropping prices to 15 percent of original levels expanded demand for GE's ultrasound machines in the US.
3	New features incorporated for EMs, such as sturdiness, portability, or ease of use, may create new market segments in rich countries.	Making ECG machines portable and more compact created new market segments for GE in the US.
4	Technology of 'good enough' products developed for EMs may improve over time to satisfy high-end applications in rich countries.	Portable ultrasounds developed by GE for China were later useable for high-end radiology and obstetrics applications.
5	EMs may leapfrog to latest technologies, especially if they have large internal demand, are unencumbered by legacy technologies, and face fewer regulatory obstacles.	EMs have advanced capabilities in industries such as wireless banking, nonconventional energy, and electric cars, which have relevance and value in developed-country markets.

Source: Authors.
Note: EM = emerging market.

Second, products that have been redesigned to be ultralow-cost for emerging markets may expand overall market demand in developed countries. Because the price reductions necessary to succeed in emerging markets are dramatic (of the order of 70–90 percent of relative prices in developed countries), demand may be expanded significantly because of the price elasticity effect when new versions are introduced in developed countries at these ultralow price points. It is hard to imagine that the demand for many products would not expand dramatically if prices were cut to 10 or 20 percent of what they used to be. There are numerous historical examples of large expansion in demand because of falling costs and prices, fueled by the so-called "experience curve effect" – e.g., televisions, videocassette recorders, copiers, personal computers, and so on (Hall and Howell, 1985). Because the cost reductions demanded by emerging markets are an order of magnitude greater than those resulting from the experience curve effect, the potential expansion of demand in developed countries may be that much greater as well.

Third, as shown in Table 8.1, optimizing products for emerging markets requires adding new functionalities, such as portability or ease of use, that may also create new applications in developed countries. For instance, making a product sturdier or easier to use may expand its appeal to at-home users in rich countries. Thus, redesigning existing products from the ground up for emerging markets may help generate a second round of growth in those products in developed markets.

Fourth, research on disruptive innovations (Christensen and Bower, 1996) argues that over time, the technology of low-cost products can be improved until they satisfy the more demanding mainstream customer. Thanks to advances in technology, for instance, GE's portable ultrasounds, originally developed in China, were able by 2009 to perform radiology and obstetric functions that once required high-end conventional machines (Immelt et al., 2009).

Finally, emerging markets can also be at the forefront of global innovations when they leapfrog over legacy technologies to adopt frontier technologies. Examples include the leap in many poor countries from no telephone or obsolete electromechanical exchanges to the latest in wireless telephony. That revolution, in turn, is causing poor countries to leapfrog past brick-and-mortar banking to wireless banking for the masses. A similar dynamic may be underway in the adoption of nonconventional energy sources, such as wind and solar, led by emerging-market multinationals like Suzlon Energy Limited and Goldwind Science and Technology Co., Ltd (which are among the "global first movers" in their industries) (Ramamurti and Singh, 2009b; Williamson and Zeng, 2009). In all these examples, the large potential market in poor countries, unencumbered by sunk investments in old technologies or complex regulatory systems, may allow these countries to move faster than rich ones. But the solutions they develop are likely to have value and relevance for rich countries.

For all these reasons, it is plausible that innovations will trickle up from poor to rich countries. Reverse innovation thus provides an opportunity to update Vernon's (1966, 1979) product cycle theory to explain such things as ultralow-cost innovations in emerging markets and the trickling up of innovations. It also raises interesting questions about the circumstances under which it is more likely to happen. For instance, is it more likely in some industries than others, or between some country pairs than others? These empirical questions will benefit from carefully conducted case studies and large sample studies if and when sufficient examples become available.

Reverse innovation at the firm level

As presented before, reverse innovation has a significant impact at the firm level. And it happens at all levels: product, process, and the business model. Clearly there are significant differences between the perspectives of EMNEs and DMNEs, which is something we will discuss in the next section.

Prior research has argued that the global strategy of DMNEs can be viewed as creatively resolving the tension between economies of global scale and the advantages of local adaptation (Prahalad and Doz, 1987; Bartlett and Ghoshal, 1989; Ghemawat, 2007). Since the overall approach combines the best of globalization and localization, it can be labeled "glocalization." Although reverse innovation may appear to be a special case of a familiar problem (i.e., how to balance national responsiveness and global standardization), the amount of national responsiveness necessary to compete in the mass markets of poor countries may be considerably greater than that required to adapt products from one Triad market to another.

GE's struggle and subsequent success in penetrating the market for medical diagnostic imaging equipment in China (see the case study) suggest a fruitful line of future inquiry: how can DMNEs pursue glocalization and reverse innovation simultaneously? Glocalization centralizes strategy formulation in *global product divisions* headquartered in the rich countries, and subsidiaries in emerging markets play implementer roles, with mainly sales and distribution functions (i.e., *process adjustments*). Glocalization has been and is a useful and necessary approach, but DMNEs might struggle with reverse innovation precisely because they are so good at glocalization. The centralized, product-focused structures that have made multinationals so successful at glocalization could create challenges that might make it difficult for the DMNE to seize opportunities in emerging markets. Therefore, some adjustments to their *business model* are required.

There are basically three challenges (or traps) that need to be explicitly faced by DMNEs when exploring the potential advantages of reverse innovation: the familiarity trap, the competency trap, and the complacency trap.

Familiarity trap

Organizations favor the familiar (Christensen and Bower, 1996; Ahuja and Lampert, 2001). The global product division headquartered in a rich country understands customer needs in the rich world. It is very difficult for the same unit to identify opportunities for *business model innovations in emerging markets*. Emerging markets, after all, are two to three levels below the CEO and do not get the mind share, voice, or visibility of the bigger, more developed markets. For the *global product division*, an emerging market like India is just one country and, as such, represents only a very small fraction of its total revenues – hardly enough to justify senior leaders from the global center making many visits to India or investing a lot of energy to understand the Indian market's unique features (i.e., with the resulting adjustments in *process*). At the same time, the local India organization does not feel empowered to develop strategy. Despite its implementer role, if the local organization proposes novel ideas for growth, the global product division is likely to rank order such ideas against proposals from all countries and, given their unfamiliarity, the India opportunities stand little chance of funding. For a global product division, emerging markets represent a zone of discomfort, not a zone of opportunity.

Competency trap

The more an organization excels in certain competencies, the more it is likely to exploit those competencies instead of building new ones (March, 1991). Global product divisions typically have core competencies in marketing and R&D that support premium products. Emerging markets need value products – ultralow-cost, functional, good-enough-quality products that demand fundamentally different competencies. Worse still, value products have the potential to make obsolete the multinational's current competencies; they could cannibalize the sale of premium products and/or threaten the company's brand reputation for product leadership.

Complacency trap

As the global product division succeeds with premium products, it more likely will view success as a validation of the past. This results in organizational inertia (Tripsas and Gavetti, 2000). Value products and premium products are fundamentally different and inconsistent business models; the former focus on low margin and high volume, while the latter focus on high margin and low volume. Value products involve investments today with uncertain future revenue, whereas premium products are well-tested business models with a short-term payoff. Given the quarterly financial pressures, the global product division based in the DMNE's home country would rather invest its scarce resources in premium products – its "known" success formula.

The GE case study suggests a new organizational arrangement, known within the company as "local growth teams" (LGTs), as a way to counter these traps (Immelt et al., 2009). Building on the body of research on organizational design, future research can examine three hypotheses pertaining to LGTs. First, LGTs are likely to be full business units with their own profit and loss responsibility and dedicated local resources in all areas, including product development, sourcing, and marketing. They have to be given the power and authority to design new offerings from a blank page instead of being expected to simply adapt global products (Govindarajan and Trimble, 2005b). Second, because innovation involves experimentation and unknown outcomes the local teams may have to focus on testing assumptions (Levinthal and March, 1993) and be judged on performance criteria other than rigid, short-term financial measures (Govindarajan and Trimble, 2010). Finally, the local teams are likely to be connected to and leverage the company's global resource base (Ambos et al., 2010). After all, global technology and global resources are what give DMNEs a significant edge over local competitors (Gupta and Govindarajan, 2000; Govindarajan and Trimble, 2005a).

As the three traps illustrate, there is a significant impact at the firm level in the management of *products* (global or local), adjustment of the different *processes*, and most importantly in the adjustment of the *business models* to be implemented. Future research on organizational design that can help DMNEs excel simultaneously in glocalization and reverse innovation can extend and enrich our understanding of ambidextrous organizations (O'Reilly and Tushman, 2004). The next section explores the competitive advantages when comparing firms classified as EMNEs and DMNEs.

Competitive advantages of emerging-market firms

Some observers have argued that EMNEs rely too much on home country advantages for their international competitiveness (Rugman, 2009) and that they possess only "ordinary resources" (Madhok, 2010) or few intangible ownership advantages, such as brands and cutting-edge technology that can sustain long-run performance. Others disagree, pointing to a variety of ownership advantages possessed by EMNEs (Guillen and Garcia-Canal, 2009; Lessard and Lucea, 2009; Williamson and Zeng, 2009). EMNEs are seen as having a deep understanding of the needs of local customers, strengths in ultralow-cost design and manufacturing, and strong distribution in their home markets – all of which translate into a significant competitive advantage in many "mid-tech" industries, not only in the local market but also in other emerging markets (Ramamurti, 2009a, b). Occasionally, as noted earlier, EMNEs may even be at the forefront of an emerging industry, acting as "global first movers" (Ramamurti and Singh, 2009b). EMNEs pursuing either strategy are likely to trigger Stage 1 of the reverse innovation process and then participate in the international diffusion of those innovations. In the future, more DMNEs may trigger Stage 1 themselves to preempt local firms from getting a head start with reverse innovation.

As they innovate at home or expand abroad, EMNEs are likely to encounter competition from DMNEs, which raises the interesting question of how their relative competitive advantage is likely to play out during the three stages of reverse innovation.

How do EMNEs and DMNEs compete in the diffusion of reverse innovations?

Table 8.2 compares the competitive advantages of EMNEs and DMNEs in each stage of reverse innovation. Once again, these are only educated guesses informed by anecdotal evidence, and they require more rigorous development and testing in the future. In Stage 1, both EMNEs and DMNEs have significant strengths, but in different areas. EMNEs are likely to have a deeper understanding of local customers and their preferences, as well as stronger local brands and distribution through which local innovations can be exploited. The technological capabilities of EMNEs are apt to be narrower and shallower than those of DMNEs, but they are likely to have the capacity to optimally adapt foreign products and technologies for local markets. In so doing, they are also likely to drastically reduce capital and operating costs,

Table 8.2 EMNEs vs DMNEs in the three stages of reverse innovation

Stage	Main challenges for DMNEs	Relative advantages of DMNES	Relative advantages of emerging-market MNEs (EMNEs)
1 Winning in key emerging markets	• Zero-based innovation for a foreign market • Giving the emerging market subsidiary access to firm's global technology	• Technology reservoirs within the firm • Familiarity with several emerging markets • Deep pockets	• Customer intimacy • Flair for low-cost solutions • Clean slate approach because of fewer prior investments • Strong commitment to local market • Access to local resources and capabilities • Patient capital
2 Winning in other emerging markets	• Managing transfer to other EMs	• Preexisting distribution and brand recognition EMs	• Product pricing and features better suited to emerging markets
3 Winning in developed country markets	• Positioning the reverse innovation vs. existing offerings • Managing the risk of cannibalization	• Strong presence, customer intimacy, brand recognition, and distribution in rich country markets	• Unconstrained by prior investments or risk of cannibalization • Prospect of rising margins in moving upscale

Source: Authors.

resulting in ultralow-cost products and services (Ramamurti, 2009a). EMNEs are less constrained by prior investments or market positions that need to be protected while pursuing innovations for the home market. As local players, EMNEs are likely to be well-connected to important local actors, including local and state governments. With ownership typically in the hands of a family business, they are likely to take a long-term view and be patient in winning in the home market.

Unlike DMNEs, EMNEs live or die by their home markets and will, therefore, persevere and adapt as needed to succeed. They usually are quicker to make strategic decisions and nimbler than the local subsidiaries of DMNEs that require approval from remote headquarters (Amsden, 2009). DMNEs will also be held back by the need to protect past investments and market positions that could be disrupted by radical innovations for emerging markets. However, DMNEs have a deeper reservoir of technological capabilities, deeper pockets to fund innovation, and the prospect of a bigger market in which innovations can be exploited because of the positions they already hold in markets around the world. Overall, we suspect the scales are tipped in favor of local firms engaging in innovations for emerging markets. Without prodding from a local firm, DMNEs may not deviate from their existing business models; it is the actual or imminent threat of disruption by local firms that prompts DMNEs to pursue reverse innovation.

In Stage 2, however, the EMNE and the DMNE may be equally matched in their ability to transfer innovations from the country of origin to other emerging markets. The former is likely to have products and services better suited for emerging markets and feel more comfortable operating in the difficult environment that prevails in emerging markets (Cuervo-Cazurra and Genc, 2008). However, the DMNE is likely to have a stronger preexisting market position in emerging markets (brands, distribution, customer relationships, etc.) that it can leverage to exploit the innovation.

In Stage 3, the DMNE is likely to have a definitive edge over the EMNE in transferring the innovation to developed countries. *Now the shoe is on the other foot.* The DMNE is more likely than the EMNE to better understand customers in its traditional (i.e., developed-country) markets and it is likely to be strongly entrenched in these markets. Transferring the innovation will not require creating a new subsidiary or making expensive acquisitions, as might be the case for the EMNE. But DMNEs will have to contend with strong internal resistance to transferring radical innovations from emerging markets because such innovations have the potential to cannibalize existing sales and profits in traditional markets and might adversely affect the firm's brands and image. It is not clear what the optimal strategy is for a DMNE to manage this process of "self-disruption," and this is one of many areas in which future research is needed. However, even DMNEs that are not enthusiastic about Stage 3 of the reverse innovation process may find themselves having to engage in the first two stages just to succeed in high-growth emerging markets.

In summary, EMNEs seem to have the edge over DMNEs in the first stage of reverse innovation, but the advantage seems to shift progressively toward DMNEs in the later stages.

Conclusions

Faster growth in emerging economies, combined with a "flattening" of the world economy, is promoting innovation in emerging markets. This innovation is beginning to find its way from traditional poor countries to other poor countries and sometimes to rich ones as well. Clearly firms can take advantage of this phenomenon. The impact at the firm level at each of its levels – product, process, and business model – cannot be underestimated, especially when presenting important competitive advantages. This trend is led by local firms (EMNEs) and reinforced by foreign firms (DMNEs). Emerging economies seem likely to produce innovations that dramatically lower costs and add features that are especially valuable in the local context, such as portability and ease of use. From time to time, they will also produce innovations in new businesses, such as wireless communication or mobile banking, by leapfrogging past legacy technologies to frontier technologies. Some of these innovations have the potential to "trickle up" from poor to rich countries, challenging the position of DMNEs in their traditional markets. Some DMNEs are trying to preempt this by embracing reverse innovation themselves and allowing subsidiaries in large emerging markets to pursue local-for-local or even local-for-global innovation. However, implementing such a strategy requires DMNEs to rethink their traditional organizational arrangements.

From a theoretical standpoint, reverse innovation poses many interesting questions for mainstream models of innovation creation particularly in emerging markets. Clearly it presents interesting challenges for managers and policy makers as it becomes more widely accepted as a source of competitive advantage.

Notes

1. This chapter is an adaptation of a previously published article by the authors in the *Global Strategy Journal* (2011).
2. We use "emerging markets," "developing countries," and "poor countries" interchangeably and likewise "industrialized countries," "developed countries," and "rich countries" interchangeably. The latter include the Triad markets of North America (the US and Canada), Western Europe, and Japan, while the former include the rest.
3. To be sure, in the pre-Industrial Revolution era, important innovations did flow in the other direction, e.g. Chinese inventions such as gunpowder, silk, or paper found their way to the West.
4. One admittedly imperfect indicator of growing innovation capability in poor countries is the rapid growth of patents filed by firms from these countries. *Fortune*

reported, for instance, that in 2011 China was likely to file more patents than the US or Japan (Tseng, 2010). Although the quality of these patents is suspect, the growth in numbers is indicative of China's growing importance as a source of innovations.

5. Japan figures from Balassa and Noland (1988). In 2009, per capita income in PPP terms in China was only $6,778 and in India only $3,015, compared to $45,934 in the US (see IMF, 2010).

References

Ahuja, G. and Lampert, C. M. (2001). "Entrepreneurship in the Large Corporation: A Longitudinal Study of How Established Firms Create Breakthrough Innovations." *Strategic Management Journal*, 22 (6–7): 521–43.

Ambos, B., Asakawa, K., and Ambos, T. (2010). "A Dynamic Perspective on Subsidiary Autonomy." Paper presented at the *Global Strategy Journal* launch conference, Oak Brook, IL.

Amsden, A. (1992). *Asia's Next Giant: South Korea and Late Industrialization.* New York: Oxford University Press.

Amsden, A. (2009). "Does Firm Ownership Matter? POEs vs. FOEs in the Developing World." In Ramamurti, R. and Singh, J. V. (eds) *Emerging Multinationals in Emerging Markets.* Cambridge, UK: Cambridge University Press, 64–77.

Balassa, B. and Noland, M. (1988). *Japan in the World Economy.* Washington, DC: Institute for International Economics.

Bartlett, C. A. and Ghoshal, S. (1989). *Managing across Borders: The Transnational Solution.* Boston, MA: Harvard Business School Press.

Cantwell, J. and Mudambi, R. (2005). "MNE Competence-Creating Subsidiary Mandates." *Strategic Management Journal,* 26: 1109–28.

Caves, R. (2007). *Economic Analysis and Multinational Enterprise.* Cambridge, UK: Cambridge University Press.

Christensen, C. M. and Bower, J. L. (1996). "Customer Power, Strategic Investment, and the Failure of Leading Firms." *Strategic Management Journal,* 17 (3): 197–218.

Cuervo-Cazurra, A. and Genc, M. (2008). "Transforming Disadvantages into Advantages: Developing-Country MNEs in the Least Developed Countries." *Journal of International Business Studies,* 39 (6): 957–79.

Friedman, T. (2005). *The World Is Flat: A Brief History of the Twenty-first Century.* New York: Farrar, Straus, and Giroux.

Ghemawat, P. (2001). "Distance Still Matters." *Harvard Business Review,* 79 (8): 137–47.

Ghemawat, P. (2007). *Redefining Global Strategy.* Boston, MA: Harvard Business School Press.

Govindarajan, V. and Trimble, C. (2005a). "Building Breakthrough Businesses within Established Organizations." *Harvard Business Review,* 83 (5): 58–68.

Govindarajan, V. and Trimble, C. (2005b). *Ten Rules for Strategic Innovators: From Idea to Execution.* Boston, MA: Harvard Business School Press.

Govindarajan, V. and Trimble, C. (2010). *The Other Side of Innovation: Solving the Execution Challenge.* Boston, MA: Harvard Business School Press.

Guillen, M. F. and Garcia-Canal, E. (2009). "The American Model of the Multinational Firm and the 'New Multinationals' from Emerging Markets." *Academy of Management Perspectives,* 23 (2): 23–35.

Gupta, A. and Govindarajan, V. (2000). "Knowledge Flows within Multinational Organizations." *Strategic Management Journal,* 21 (4): 473–96.

Hall, G. and Howell, S. (1985). "The Experience Curve from the Economist's Perspective." *Strategic Management Journal*, 6 (3): 197–212.

Immelt, J., Govindarajan, V., and Trimble, C. (2009). "How GE Is Disrupting Itself." *Harvard Business Review*, 87 (10): 56–65.

International Monetary Fund (IMF). (2010). "World Economic Outlook Database." *IMF*. Retrieved from http://www.imf.org/external/pubs/ft/weo/2010/02/weodata/index.aspx

Khanna, T. and Palepu, K. (2005). "Strategies that Fit Emerging Markets." *Harvard Business Review*, 83 (6): 63–76.

Kuemmerle, W. (1997). "Building Effective R&D Capabilities Abroad." *Harvard Business Review*, 75 (2): 61–70.

Lessard, D. R. and Lucea, R. (2009). "Mexican Multinationals: Insights from CEMEX." In Ramamurti, R. and Singh, J. V. (eds) *Emerging Multinationals in Emerging Markets*. Cambridge, UK: Cambridge University Press, 280–311.

Levinthal, D. A. and March, J. G. (1993). "The Myopia of Learning." *Strategic Management Journal*, 14 (4): 95–112.

Madhok, A. (2010). "Acquisitions as Entrepreneurship: Inter-Nationalization and Emerging Market Multinationals." Paper presented at *Global Strategy Journal* launch conference, Oak Brook, IL.

March, J. G. (1991). "Exploration and Exploitation in Organizational Learning." *Organization Science*, 2 (1): 71–87.

Martinez-Jerez, F. A., Narayanan, V. G. and Jurgens, M. (2006). *Strategic Outsourcing at Bharti Airtel*. Boston, MA: Harvard Business School.

Nelson, R. R. (ed.). (1993). *National Innovation Systems*. Oxford: Oxford University Press.

O'Reilly, C. A. and Tushman, M. L. (2004). "The Ambidextrous Organization." *Harvard Business Review*, 4 (4): 74–81.

Porter, M. E. (1990). *The Competitive Advantage of Nations*. New York: Free Press.

Prahalad, C. K. (2005). *The Fortune at the Bottom of the Pyramid*. Philadelphia, PA: Wharton School Publishing.

Prahalad, C. K. and Doz, Y. L. (1987). *The Multinational Mission: Balancing Local Demands and Global Vision*. New York: Free Press.

Ramamurti, R. (2009a). "What Have We Learned about Emerging-Market MNEs?" In Ramamurti, R. and Singh, J. V. (eds) *Emerging Multinationals in Emerging Markets*. Cambridge, UK: Cambridge University Press, 399–426.

Ramamurti, R. (2009b). "Impact of the Crisis on New FDI Players." Paper presented at the Columbia University conference on FDI, the Global Crisis, and Sustainable Recovery. Columbia University, New York.

Ramamurti, R. and Singh, J. V. (eds). (2009a). *Emerging Multinationals in Emerging Markets*. Cambridge, UK: Cambridge University Press.

Ramamurti, R. and Singh, J. V. (2009b). "Indian Multinationals: Generic Internationalization Strategies." In Ramamurti, R. and Singh, J. V. (eds) *Emerging Multinationals in Emerging Markets*. Cambridge, UK: Cambridge University Press, 110–66.

Rugman, A. (2009). "Theoretical Aspects of MNEs from Emerging Markets." In Ramamurti, R. and Singh, J. V. (eds) *Emerging Multinationals in Emerging Markets*. Cambridge, UK: Cambridge University Press, 42–63.

The Economist. (2010). "New Masters of Management: Special Report on Innovation in Emerging Markets." *The Economist*. Retrieved from http://www.economist.com/node/15894358

Tripsas, M. and Gavetti, G. (2000). "Capabilities, Cognition, and Inertia: Evidence from Digital Imaging." *Strategic Management Journal*, 21 (10–11): 1147–61.

Tseng, N. H. (2010). "Behind China's Surge in Patents." *Fortune*. Retrieved from http://money.cnn.com/2010/10/14/news/international/china_patents_innovation. fortune/

Tsurumi, Y. (1976). *The Japanese Are Coming: A Multinational Interaction of Firms and Politics.* Cambridge, MA: Ballinger Publishing.

United Nations. (2005). *Globalization of R&D and Developing Countries: Proceedings of an Expert Meeting.* Geneva, Switzerland: United Nations.

Vernon, R. (1966). "International Investment and International Trade in the Product Cycle." *Quarterly Journal of Economics*, 80 (2): 190–207.

Vernon, R. (1979). "The Product Cycle Hypothesis in a New International Environment." *Oxford Bulletin of Economics and Statistics*, 41 (4): 255–67.

Von Hippel, E. (1986). "Lead Users: A Source of Novel Product Concepts." *Management Science*, 32 (7): 791–805.

Williamson, P. and Zeng, M. (2009). "Chinese Multinationals: Emerging through New Gateways." In Ramamurti, R. and Singh, J. V. (eds) *Emerging Multinationals in Emerging Markets.* Cambridge, UK: Cambridge University Press, 81–109.

Wilson, D. and Purushothaman, R. (2003). *Dreaming with the BRICs: A Path to 2050.* Global Economics Paper No. 99. New York: Goldman Sachs.

Yip, G. (1989). "Global Strategy ... in a World of Nations?" *Sloan Management Review*, 31 (1): 29–41.

9
Social Enterprise and Innovation in Emerging Markets

Leslie R. Crutchfield and Kyle Peterson

Across the poorest villages of rural India, people can count on the abundance of two things: rice and darkness. In the state of Bihar, India's poorest, around 80 percent of the households are not electrified. Villagers instead rely on kerosene lanterns and diesel generators for light and power, both of which are expensive and destructive to health and the environment. And as the world's second largest rice producer, India is estimated to have harvested more than 100 million tons in the 2012–2013 crop year. Along with the rice comes a significant amount of discarded husk, some of which can be used for fodder but most of which is burnt in fields. Three Indian entrepreneurs from Bihar and an American colleague added together these two factors to create Husk Power Systems (HPS), an off-grid electricity company that converts rice husks into power that is reliable, eco-friendly, and affordable for families that can spend only $2 a month on power (Bornstein, 2011).

The concept quickly took off: launched in 2007, HPS by 2013 had built 74 mini-power plants, providing electricity to more than 200,000 people in hundreds of villages across Bihar. By 2013, HPS had raised more than $2.8 million in financing from impact investors such as Acumen, Bamboo Finance, and Shell Foundation, and was quickly hatching expansion plans: HPS had developed a franchise model and would soon launch Husk University to rapidly train the expanding personnel base, with the bold goal of providing electricity to 10 million people by installing 3,000 plants in India by 2017. Plans were also underway to expand to Nepal and Southeast Asia, and eventually, Kenya and Nigeria.

Social enterprise and innovation in emerging markets: bold aspirations, hard realities

The founders of HPS had arrived at nirvana for the global social enterprise set – they had hatched an innovative product with the potential to tangibly improve millions of lives, which was affordable to base-of-the-pyramid consumers, and made use of an abundant local natural resource without causing environmental damage. HPS is part of the growing movement to introduce social innovations in emerging economies that hitch market forces to the challenge of solving problems that plague very poor populations. Pioneering companies like HPS are fueled by a fast-growing impact investing community – an asset class valued at $46 billion, and an array of facilitating institutions that provide technical assistance, business mentorship, and networking opportunities tailored to drive both societal and business value. Global private institutions like Ashoka, Endeavor, Skoll Foundation, and Schwab Foundation for Social Entrepreneurship have turned the international spotlight toward players like Pandey, ushering them into once-exclusive domains of the world economic elite. Today no World Economic Forum is complete without its itinerant coterie of social entrepreneurs.

It is too soon to tell whether HPS will successfully scale its model across and beyond India. Yet today, the company stands as an archetypal example of a social enterprise – a business that simultaneously generates profits and societal impact. It is an inspiring symbol for the growing field of social-purpose businesses that hope to tackle critical problems in emerging markets – issues which government aid and private philanthropy alone have so far struggled to resolve.

Husk Power Systems is also an icon for the challenges that every social enterprise confronts. For all of HPS's success in its first six years, the company had reached by 2013 less than 1 percent of the market; by then, 100,000 Indian villages still lacked grid power. Admittedly, the billions contributed in national, bilateral, and multinational government aid, and private philanthropy, had also not successfully closed the electricity gap facing India's rural poor. But it was not clear that a small and growing business like HPS could bridge the gap.

HPS faces the same limitations of any small and growing business (SGB) operating in very poor markets. While many innovations show great promise, most market-based solutions are operating at relatively low levels of scale: Monitor's 2011 analysis of 439 similar social enterprises in Africa showed that only 13 percent of them had begun to scale significantly. Succeeding at scale is exceedingly difficult in developing countries; entrepreneurs confront many more barriers than they would in more developed contexts. The economic, political, cultural, and physical environments in which social enterprises operate are dauntingly complex. The poorest emerging markets often present insurmountable hurdles that even the most adaptive entrepreneurs struggle to overcome. As the authors of *Emerging Markets, Emerging Models* (Karamchandani et al., 2009) note, "A great product idea married to a noble mission ... is rarely enough to make meaningful progress in the face of massive social challenges like improving the lives and livelihoods of billions worldwide living in impoverished conditions."

Systemic barriers must be removed if enterprises like HPS are to stand a chance of flourishing.

Beyond social enterprise: shared value, NGOs earning income, and industry facilitators

While their Silicon Valley-type origins make upstarts like HPS compelling stories, social enterprises comprise just one set of a range of actors aspiring to use market forces to drive social impact in emerging markets. Large-scale, mature global companies are awakening to their potential to contribute to solving social and environmental problems by creating shared value (Porter and Kramer, 2006, 2011; see also Emerson, 2003). These corporations are moving beyond traditional philanthropic approaches and social responsibility programs, and putting problem-solving for the poor at the center of their business models. Shared value is created as companies develop new products and services, reorganize supply and value chains, or innovate in other ways that generate profits while addressing critical societal problems.

Take Vodafone, the UK-based telecommunications company which, through its local Kenyan operator, Safaricom, rolled out M-Pesa, a mobile money transfer system. ("Pesa" is the Swahili word for cash; "M" stands for mobile.) The venture allowed unbanked households to easily transfer money and pay bills, and eventually to access a full range of formal financial services. The original idea took root as Vodafone executives investigated how their company could contribute to advancing Millennium Development Goals. The project began as an internal R&D experiment with support from a DFID grant; within five years, M-PESA had handled $10 billion in transactions globally, attracting millions of users who collectively processed more payments than Western Union did across its entire global network.

Although it has been more than a decade since C. K. Prahalad suggested that a fortune awaits at the bottom of the economic pyramid, large-scale companies are just now starting to take action. Big businesses can contribute large-scale distribution networks, financial and marketing know-how, and ready capital to rapidly scale social innovations. But to date, most commercial concerns have lacked a real understanding of how to solve social and environmental problems. And almost all have not successfully operated a business in the austere, underresourced environments that characterize most emerging markets.

Here another set of key actors on the global social innovation stage come into play: nongovernmental organizations (NGOs), which bring deep and proven commitment to solving social and environmental problems, authentic local relationships, and the know-how to operate in emerging market communities – advantages which businesses (whether big or small) often lack. Historically funded almost exclusively by government aid and private philanthropy, today many NGOs are exploring hybrid business models that combine earned income with traditional charitable support. This shift requires NGOs to meet the needs of customers in more disciplined ways, as they become directly accountable to the populations they aim to serve, rather than to third-party funders based primarily in the West. This market mind-set has not always come naturally to NGOs. Often the most effective hybrid NGO approaches involve partnerships with for-profit companies; they combine their complementary strengths to together invent new methods of tackling societal issues with market-based solutions.[1]

Finally, another type of actor is emerging that works across all sectors to play the role of "industry facilitator." The facilitator's goal is to increase the pace and scale of success for firms operating in a given industry or field (such as agriculture or health). They investigate the entire ecosystem of actors and institutions that can potentially influence the success of a market-based solution – looking within firms, across value chains, and at governmental laws and policies, among other contextual factors – and identify and address underlying systemic barriers, which are often insurmountable by any single enterprise.

The promise of market-based solutions

This chapter explores social enterprise and social innovation in emerging markets as currently practiced across multiple forms. Emphasis is given here to the increasing levels of social enterprise activity starting in the 1990s, fueled by a rising field of impact investors and amplified by supporting organizations, incubators, accelerators, and other capacity-building institutions. But the scope of the chapter is intentionally broad and diverse, with firm-level case studies ranging from emerging social enterprises like HPS, to multinational corporations creating shared value like Vodafone, and NGOs pursuing earned income solutions (see Figure 9.1 and Table 9.1). The chapter also includes an investigation of the relationship between government and social enterprise, and how policies can help or hinder the success of market-based solutions.

Sidebar context and definitions

While the practice of social enterprise and innovation is not new to emerging markets, the terminology is. The term "social enterprise" and the activities associated with it gained popularity in the West, first in the US and later in Europe, starting in the late 1980s. The term came into vogue alongside social entrepreneurship, an idea popularly attributed to Bill Drayton, who founded Ashoka in 1980 as a global network of social entrepreneurs. eBay cofounder Jeff Skoll helped expand interest in social entrepreneurship through the Skoll Foundation. In Europe, the Schwab Foundation for Social Entrepreneurship was established in 1998. Definitions of the term "social

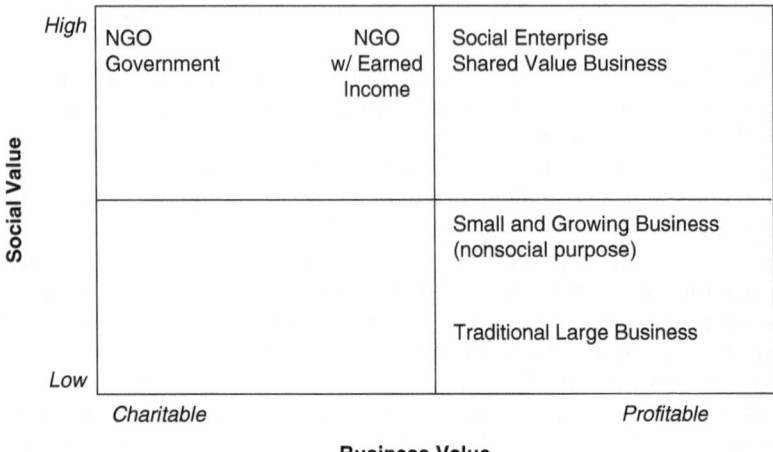

Figure 9.1   Focus on market-based solutions
Source: Authors.

Table 9.1   Featured market-based solutions

Type	Enterprise	Sample Country/ Region
Social Enterprise	Husk Power Systems	India, Africa
	Compoalto	Latin America
	WaterHealth International	India, Africa, others
Shared Value	Nestle - Milk districts	Bangladesh, India
	Mars - cocoa bean farmers	West Africa
	Vodaphone/Safaricom-M-Pesa	East Africa
NGO with	BRAC	Bangladesh
Earned Income	MercyCorps –TISA	Central America
	Friends International	East Asia, Central America
Industry	MIM Housing	India
Facilitators	Ashoka Hybrid Value Chain (HVC)	Mexico, India

Source: Authors.

enterprise" vary. A social enterprise is defined here as an organization that aims to address social and environmental problems in a financially sustainable manner (Etchart and Comolli, 2013). Social enterprises operate on a spectrum that spans from traditional charitable models to purely commercial models (Dees, 1998). (See Figure 9.2.)

The potential legal forms for a social enterprise can vary. It can be structured as a commercial business; whether the business ultimately generates a profit (or return to shareholders) depends on many factors. Alternately, a social enterprise can also be structured as a nonprofit organization that

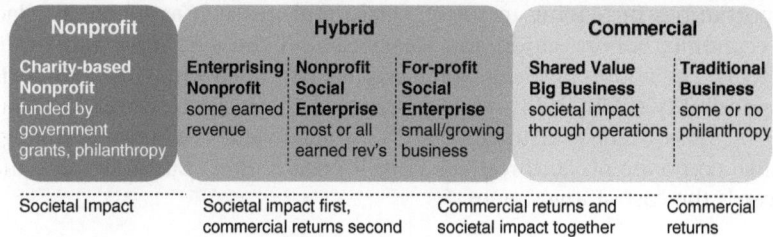

Nonprofit	Hybrid			Commercial	
Charity-based Nonprofit funded by government grants, philanthropy	**Enterprising Nonprofit** some earned revenue	**Nonprofit Social Enterprise** most or all earned rev's	**For-profit Social Enterprise** small/growing business	**Shared Value Big Business** societal impact through operations	**Traditional Business** some or no philanthropy

Societal Impact	Societal impact first, commercial returns second	Commercial returns and societal impact together	Commercial returns

Figure 9.2 The social enterprise spectrum
Source: Authors.
Some content for this figure was adapted from these sources:
Dees, J.G. "Enterprising Nonprofits," *Harvard Business Review* (Jan–Feb 1998); Cambridge, MA: Harvard Business School Press. Retrieved from https://hbr.org/1998/01/enterprising-nonprofits/ar/1
"The European Venture Philanthropy Association: An Introduction." London: EVPA. Retrieved from http://evpa.eu.com/wp-content/uploads/2010/08/EVPA-IntroductionMarch-2012.pdf

generates a significant portion of its revenue through earned income, but profits must be retained by the organization for future activities and cannot be distributed to owners or investors. In recent years, new legal forms have been developed in the US and overseas to create formal designations specifically for social enterprises, such as B corps.[2] In most emerging markets, no separate legal form exists for social enterprises; by default they operate under traditional nonprofit or commercial legal rules. One emerging market exception is Eastern Europe, where relatively sophisticated legal forms for social enterprise exist (Etchart and Comolli, 2013).

In emerging markets, social enterprises are closely associated with SGBs, or small and medium-sized businesses (SMEs). It can be argued that any SGB operating in an emerging market can improve society; SGBs provide jobs, generate profits, pay taxes, and spur development of local and national economies. However, an SGB that does not aspire to *deliberately* generate positive social or environmental impact (beyond what any business can contribute) would not generally be considered a social enterprise. Acumen and other emerging market impact investors have developed specific criteria for identifying the social or environmental "additionality" that a social-purpose business must aim to create, which separates social enterprises from mainstream SGBs.

At the other end of the spectrum, a greater number of large businesses operating in emerging markets are aiming to create social and environmental impact. Porter and Kramer use the term "shared value" to describe the efforts of big, established companies to create measurable financial value while simultaneously addressing social problems that intersect with their business model (Porter and Kramer, 2011). Emerson (2003) introduced the concept of "blended value." Shared value is described by Porter and Kramer as an approach that leads large companies to discover new business opportunities in helping to solve social problems such as inequality, lack of economic opportunities, and weaknesses in the education and health systems. The notion of shared value is often associated with corporate social responsibility (CSR) and its early pioneers in the West. However, the practice of shared value is distinct from CSR. Shared value is a business strategy; unlike corporate philanthropy or CSR, it creates direct business value that exceeds the cost of investment.

Social-purpose businesses in practice: models that work

Any business that seeks to operate in an emerging market – whether social-purpose or purely for-profit – faces extraordinary challenges. Social enterprises, like mainstream SGBs, have difficulty accessing finance, attracting and retaining talent, achieving economies of scale, and creating recognizable and trustworthy brands. But social enterprises that aim to serve the poorest-of-the-poor segments face another set of challenges: selling to customer bases with the most severely limited resources; distribution channels that are minimal or nonexistent; and a nearly total lack of market data. Social enterprises often rely on suppliers who suffer high volatility in production or have weakened loyalty due to cash flow needs. Further, social enterprises must operate in socioeconomic–political contexts that mount additional barriers to business success, such as: minimal or nonexistent access to reliable electricity; suboptimal transportation systems; inhospitable and inefficient regulatory environments; and weakened states with lower overall institutional quality (lack of rule of law or property rights) as well as degraded regulatory quality. These myriad challenges are present at some level in every base-of-the-pyramid market; they require adaptation of business models to suit these unique conditions.

Based on in-depth research of hundreds of social-purpose enterprises operating in India, sub-Saharan Africa, and other emerging market regions, Monitor Group offers a useful framework for capturing the range of business models that appear most effective in these austere contexts (Karamchandani et al., 2009).[3] For its report, *Emerging Markets, Emerging Models,* Monitor Group studied HPS and hundreds of other market-based enterprises like it, based primarily in India, exploring how these businesses had successfully tailored their products, services, and business models to survive in emerging market conditions. The report explores four workable business models that focused on serving the poor as customers, and three additional business models focused on engaging low-income suppliers or producers. The models include the following.

Customer-based business models[4]

1. *Pay-per-use:* Consumers pay lower costs for each use of a product, service, or group-owned facility.

2. *No frills:* Pared-down services and products that meet basic needs while generating positive cash flow and products through high volume, high asset utilization, and service specialization.
3. *Para-skilling:* Combines "no-frills" services while reengineering complex services and processes into a set of disaggregated standardized tasks that can be undertaken by workers without specialized qualifications.
4. *Shared channels:* Distribution networks that reach into remote markets by piggybacking products and services through existing customer supply chains.

Supplier- or producer-based business models

1. *Contract production:* "Tip to tail" supply chain organization; players at the top are usually large companies; they provide critical inputs, specifications, training, and credit to suppliers; suppliers provide assured quantities at fair and guaranteed prices; typically involve small-scale farmers or producers in rural supply chains.
2. *Deep procurement:* Arrangements that bypass "middle-men" and reach into the base of the pyramid, enabling direct purchases from large networks of low-income producers and farmers in rural markets.
3. *Demand-led training:* Enterprises pay a third party to recruit, train, and place employees for job openings, akin to a "temp agency."

The HPS model is an example of an enterprise that successfully combined a "no-frills" technological solution with a "pay-per-use" pricing model that enabled success in previously underserved markets in India. The HPS factories are pared-down versions of more technologically advanced outfits. And the owners prioritize recruiting rural employees so that factories can be locally operated and managed. One issue that HPS faced early on was that its customers were using more electricity than they were paying for, whether inadvertently or as theft. So the company developed a stripped-down prepayment smart-card reader for home installation. That way, customers would use only the allowance paid for in advance. Another challenge that HPS confronted was recruiting and training employees with the necessary technological and management know-how to satisfactorily operate the facilities. Since indigenous laborers often lacked the education and training to run even the pared-down, no-frills facilities that HPS had invented, the idea for Husk University was developed to address this critical input. Each of these adaptations has enabled HPS to flourish in hundreds of rural Indian villages.

WaterHealth International provides another innovative example of "no-frills" and "pay-per-use" approaches. This profitable large-scale business provides clean water at a fraction of a cent per liter to over 5 million people in Africa, India, Bangladesh, and the Philippines through decentralized purification stations. A 1992 cholera epidemic in West Bengal inspired Ashok Gadgi, an Indian-born physicist, to develop a technology designed to address the issue of waterborne diseases.[5] Several years later, he founded WaterHealth International (WHI) in California with a mission to provide scalable, safe, and affordable water solutions to populations lacking access to clean water. WHI develops, installs, and operates water purification and disinfection in rural and periurban areas. WHI houses its water treatment systems in small-scale, decentralized facilities called WaterHealth Centers in local communities. At an initial investment of less than $10 per person, WHI can provide more than a decade of healthy drinking water to communities in need.

Other examples of workable business models in emerging markets include the following.

Para-skilling model

Aravind Eye Care System, or Aravind, is a world-class leader in applying the para-skilling approach to the challenge of giving sight to ultralow-income people who are blind. An end-to-end eye care provider founded in 1976 in India, Aravind grew to become one of the world's largest facilities for eye care, having treated by 2013 more than 32 million patients and performed 4 million eye surgeries. Two health-care bottlenecks inspired the Aravind approach: the extremely high cost of a surgeon's time, and the dearth of medical professionals available to treat the millions of blind people in India who would experience dramatically improved lives and productivity if they could see. The Aravind business model centered on reengineering the surgical eye care process. Instead of a doctor seeing a patient at each step, the doctor attended only the preliminary exam, final diagnosis, and surgery. The others were handled by para-skilled paramedics, trained in a range of clinical tasks from outpatient care to ward management. As a result, Aravind doctors have been highly productive: Aravind performed 2,400 surgeries per doctor per year, compared to 300 in standard Indian clinics, with comparable success rates.

The para-skilling model has been applied in other emerging market countries and across a range of industries. In Bogotá, Colombia, three medical professionals launched Campoalto, an educational institute originally designed for nursing assistants, which they soon expanded to include business administration, auto mechanics, cooking, and clothing manufacturing, among other industries. Campoalto's founders had earned their medical degrees in the US but when they returned to Colombia to begin practice, they could not effectively treat patients because of a shortage of qualified nursing assistants. So they launched Campoalto as an affordable, high-quality training program offering real-work competencies for Bogotá's bottom-of-the-pyramid workers. The program grew to become Colombia's largest vocational health training school. Since 1997, the social enterprise has expanded to include 16 different training programs which have collectively graduated more than 20,000 alumni. And in a compelling example of reverse innovation, with support from Endeavor, Campoalto had established plans in 2014 to open operations in Miami, Florida, where demand for para-skilled workers was very high, especially in medical fields. Campoalto training was uniquely suited for this southern US market because the programs were developed in Spanish and better-positioned to meet the needs of the vast and growing market of primarily Spanish-speaking potential trainees than existing US vocational tech programs.[6]

Shared channels model

By folding products and services into existing customer supply chains, social enterprises are able to take advantage of informal distribution networks and reach remote markets. For instance, the international NGO, Mercy Corps, established in Guatemala TISA – a network of financially sustainable health stores – by building on preexisting distribution channels comprised largely of *botequines* (informal pharmacies). Rural Guatemalan communities traditionally relied on *botequines* to obtain medicine and other health-care items, but these informal suppliers struggled to meet customer needs due to unreliable product availability and lack of financial sustainability. They often stocked neither enough medicine nor the right products. As a result, malnutrition and mortality were exacerbated in rural areas by the limited availability of common medicines. For instance, if an infant spiked a fever in a remote village, hours or even days of travel could be necessary to find an open pharmacy that stocked a simple fever-reducing medication.

Mercy Corps saw an opportunity to improve health outcomes in rural Guatemala by transforming the informal *botequines* into a network of financially viable social enterprises that could provide reliable access to medication. With grant support from a private foundation, Mercy Corps began in 2009 to provide financing and business training to existing operators, and ultimately helped create or transform 50 microbusinesses into financially self-sustaining enterprises. The stores had multiple levels of impact: they served impoverished people living in remote rural areas, and also provided financial income for the store owners – creating a shared value proposition for both local business owners and vulnerable Guatemalans. The stores were supported initially by Mercy Corps and later operated by Farmacias de la Comunidad (FedC), a Guatemalan pharmaceutical company, providing health products and medicine to nearly 88,000 people.[7]

Deep procurement, contract production, and demand-led training

In this second category of workable business models identified by Monitor Group, the approaches include supplier- or producer-based models that provide ways to connect base-of-the-pyramid producers and suppliers with global customer markets. These approaches are particularly suited to larger

Table 9.2   Business models to connect base-of-the-pyramid producers and suppliers with global customer markets

Model	Example
Deep procurement	Global companies support local suppliers, such as Nestlé Milk Districts, which work with rural dairy farmers to ensure a reliable supply of fresh milk
Contract production	Large companies negotiate buy-back deals with small-scale farmers or producers in rural supply chains
Demand-led training	A "temp agency" model in which companies pay a third party to identify, train, and place employees for job openings at the edges of the formal and informal sectors

Source: Authors.

companies creating shared value, as a multinational corporation can leverage its global customer base and connect it to emerging market actors. Table 9.2 contains examples of these approaches.

Whether the business is large or small and growing, it must adapt its model to succeed in emerging market contexts. Commercial enterprises that flourish in middle- and upper-income markets almost never translate effectively into emerging market contexts. Large companies like GE and Novartis had to overhaul their products, paring them down using "no-frills" technologies and "pay-per-use" pricing strategies in order to meet the demand for their products at prices that emerging market customers could afford. As the CEO of GE Healthcare-South Asia said, "We realized that the biggest impediment was that we were selling what we were making [rather than] making what the customers here needed" (Borgonovi et al., 2011).

Further, smaller social enterprises often offer products in "push" categories, which makes success even harder as they must simultaneously convince customers of the importance or value of their product, such as condoms or bed nets. Whereas "pull" products are easier to introduce because customers are already hoping to find them at a price they can afford, such as cell phones and electricity.

Even with the right technology and a no-frills model, social-purpose businesses face many other hurdles. They must find capital to support and expand operations, a unique challenge because most emerging market start-ups require lower levels of financing than traditional Western investors are used to providing. Enterprises must also build management and operating capacity, using adaptive methods to accommodate locally based workers who often lack the appropriate education and technical skills to be immediately productive. Further, if the enterprise is operated as a hybrid NGO, these challenges can be even higher, because NGOs often lack business know-how in marketing, managing, and finance that are more closely

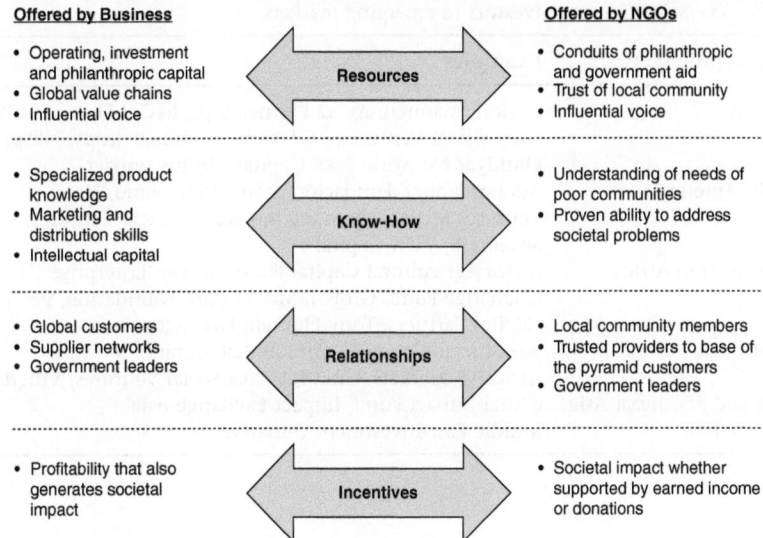

Offered by Business		Offered by NGOs

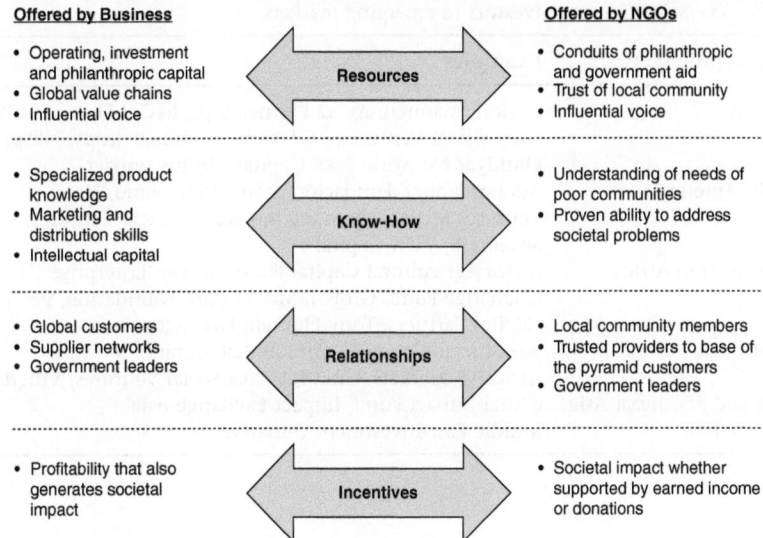

Offered by Business

- Operating, investment and philanthropic capital
- Global value chains
- Influential voice

Resources

Offered by NGOs

- Conduits of philanthropic and government aid
- Trust of local community
- Influential voice

- Specialized product knowledge
- Marketing and distribution skills
- Intellectual capital

Know-How

- Understanding of needs of poor communities
- Proven ability to address societal problems

- Global customers
- Supplier networks
- Government leaders

Relationships

- Local community members
- Trusted providers to base of the pyramid customers
- Government leaders

- Profitability that also generates societal impact

Incentives

- Societal impact whether supported by earned income or donations

Figure 9.3   Businesses and NGOs offer distinct assets to social innovation
Source: Authors.

associated with (though not ubiquitous in) the private sector. (See Figure 9.3 for an illustration of the relative strengths offered by businesses versus NGOs.) Thus, in emerging markets, the role of supporting organizations and impact investors has become increasingly important, as these institutions can help connect enterprise leaders with the know-how, networks, and capacity-building support that they may need in order to flourish.

Supportive ecosystems: impact investors, grant makers, capacity builders, and other facilitating institutions

The recent acceleration in the number of social enterprises launching and attempting to scale in emerging markets is partially attributable to the fast-growing global network of organizations dedicated to supporting them, whether with equity investment capital, traditional philanthropic grants, or business know-how, mentoring, and technical training.

Take the field of impact investment: during the past two decades, impact investing has evolved from a practice that was almost exclusively the domain of government development aid actors, into a distinct investing asset class comprised of a wide diversity of philanthropic, commercial, and public investors with total assets under management estimated at $10.6 billion by 2013. Philanthropic investors like the Rockefeller Foundation and Acumen, and commercial investors like Accion, have joined traditional government

Table 9.3   Select impact investors in emerging markets

Geographic scope	Examples
Global	Accion, Acumen, Agora Partnerships, E+Co, Elevar Equity, Gray Ghost Ventures, LGT Venture Philanthropy, NESsT, Omidyar Network, Root Capital, Unitus Impact
Latin America	Adobe Capital, Fundacion Avina, Ignia Fund, New Ventures Mexico, Pomona Impact, Potencia Ventures, Sistema B, VOX Capital
Sub-Saharan Africa	Africa Agricultural Capital Fund, African Enterprise Challenge Fund, GroFin, Mo Ibrahim Foundation, Pearl Capital Partners, Tony Elumelu Foundation
South Asia	Aavishkaar, Grameen Capital, Lok Capital, Monitor Inclusive Markets, UnLtd, Upaya Social Ventures, Villgro
East and Southeast Asia	China Impact Fund, Impact Exchange Asia
Middle East	Middle East Investment Initiative

Source: Authors.

Table 9.4   Select grant makers, capacity-builders, and facilitating organizations

Geographic scope	Examples
Global	ANDE, Ashoka, Avina, Draper Richards Foundation, echoing green, Endeavor Global, FSG Shared Value Initiative, Monitor Inclusive Markets, NESsT, Rockefeller Foundation, Schwab Foundation for Social Entrepreneurs, Skoll Foundation

Source: The Growing Impact Investing Movement.

players such as the World Bank, DFID, and OPIC (the US' Overseas Private Investment Corporation) in global development, bringing with them the disciplines of Silicon Valley-style investing to emerging market deals.[8]

Meanwhile, nonprofit capacity-builders like Endeavor in Latin America strive to help social enterprises (and traditional SGBs) scale by sharing business know-how and training on the ground, and global networks like the Aspen Network of Development Entrepreneurs (ANDE) and NESsT bring visibility, synergy, and sometimes capital, to the field. Business incubators and accelerators housed within universities also contribute, particularly in regions like Latin America where universities are expanding their missions to contribute directly to community development. Tables 9.3 and 9.4 depict a range of impact investors, philanthropic funders, and facilitating organizations operating in emerging markets.

Impact investing has become a significant global movement. The Global Impact Investing Network (GIIN) was formed in 2008 at the impetus of the Rockefeller Foundation and several other early impact investment proponents; GIIN has grown to include nearly 60 member institutions representing the range of actors in the community. At the same time, many systemic obstacles exist that continue to prevent social enterprises from scaling impact.

The mismatch between capital supply and deal flow

Even in the most advanced emerging markets – those with relatively mature impact investment fields and endemic cultures accepting of commercial enterprise as a vehicle for social impact – the pipeline of investable companies is still underdeveloped. Part of the misalignment occurs at the intersection of supply and demand for impact capital, as investors and entrepreneurs do not always share expectations around deal structures and returns. It is exceedingly difficult to strike a productive balance between realizing a firm's potential for achieving financial sustainability and also creating significant social or environmental impact: consider that by 2010, Acumen had vetted more than 5,000 social-purpose companies operating in emerging market regions of India and nations in Africa, but invested in just 65 of them during a decade of active due diligence.

While investing in promising social enterprises has emerged as a popular approach to sustainable development in poor markets, capacity-builders and funders also support other types of market-based solutions. Proponents of market-based social innovations have awakened to the reality that even the most promising new social enterprise confronts myriad obstacles as it attempts to scale; many of these hurdles are systemic and can be insurmountable by any single firm.

As demonstrated in previous segments of this chapter, the financial, political, and cultural contexts in which emerging market entrepreneurs operate are cumbersome, complex, and usually not conducive to traditional scaling approaches. In response, some actors have begun to play the role of "industry facilitator," helping to create or bolster markets that serve the needs of ultra-poor populations by aligning interests of influential players. In its 2014 report, "Beyond the Pioneer," Monitor Inclusive Markets (MIM) leaders examine the places where potential barriers can exist to scaling market-based social innovations, grouping them into four segments: within the firm; across the industry value chain; among public goods relevant to the industry; and through governmental laws, policies, and actions.

Successful examples of industry facilitation include the Sun Quality Health model for improved reproductive health services in Myanmar, which was facilitated by Population Services International. This systemic approach has delivered more than 2 million consultations annually through 1,200 franchised doctors across the country. MIM has played a catalytic role in developing affordable housing markets for low-income segments in India; as part of its housing initiative, MIM is facilitating several pilots designed to create low-income housing for 10–30 million families in urban India over the next decade. And Ashoka hybrid value chain (HVC) initiatives have contributed to transforming housing markets for the poor in Mexico and India, among other industry advancements.

Government, social enterprise, and social innovation

The relationship between governments and social enterprise varies among emerging market countries. Government directly influences social enterprise through vehicles such as legal and regulatory policies and tax systems, as well as through financial investment, procurement of goods and services, and other means. The impact of specific government policies at national, state, and local levels can promote, hinder, or be neutral for social innovators of all types. As social enterprises have begun to proliferate, and as certain successful social entrepreneurs have gained increasing international attention spurred by support from private funders like Ashoka, Schwab, and Skoll, proponents of social enterprise have advocated for governments to adopt new policies designed specifically to recognize and benefit social-purpose businesses and nonprofits. Some Western nations have passed specific policies. But few emerging market countries have yet done so. Emerging market governments are constrained by budgets and limited technical capabilities, among other factors. Policy makers are becoming more aware of, and interested in, aiding the growth of social enterprises, as these can be vehicles for the delivery of important public services such as water or energy, and potentially fill gaps where government falls short.

Monitor Group offers a useful analysis of the role of government policy in emerging markets in India and sub-Saharan Africa, based on case studies of hundreds of social enterprises (or market-based solutions) targeting the $2-a-day segment in various sectors such as agriculture, health, sanitation, and others. Government influence can be divided into two categories: "Rules of the game" such as regulation and tax policies, and "Infrastructure and support" such as helping address risk and volatility issues in industries vulnerable to major climate changes (agriculture), or using its purchasing and procurement power to provide anchor demand for products and services provided by social enterprises (see Table 9.5).

Depending on the way in which governments enact policies and implement supportive programs, social enterprise can be aided or thwarted, or the effects can be mixed or neutral. An example of a government intervention that has been introduced for the purpose of benefiting social enterprise is the priority sector lending (PSL) requirement in India. First enacted in 1969 during the nationalization of India's banking sector, PSL regulations were intended to promote banking and financial services to underserved and underfinanced populations in specific regions of India. More than 90 percent of India's small businesses were part of the informal economy, where

Table 9.5 How government influences social enterprise and innovation

Rules	Infrastructure and support
– Tax incentives (exemptions, credits)	– Public investments
– Corporate legal forms ("B Corps")	• challenge grants
– Patent regime/policy (licensing)	• social impact bonds
– Industrial policy (preferred procurement, R&D funds)	• loan guarantee funds
	• direct investments of pension funds
– Regulation	– Funding capacity building and technical assistance
	– Certification
	– Subsidies

Source: Authors, based on Karamchandani et al. (2009).

capital was accessible but very expensive. For the first several decades, the PSL requirements achieved limited success. Banks discovered ways to avoid lending to what were considered high-risk borrowers. In 2009, members of a national financial reform commission proposed Priority Sector Lending Certificates (PSLCs) as a way to balance the tension between profitability for banks and the inclusion of India's poorest participants in the financial systems. Lenders like microfinance institutions could issue PSLCs, and state-owned banks could purchase them to fulfill regulatory requirements. More than $276 billion in capital was directed in 2009 to Indian borrowers in underserved or target sectors under the new rules. In its analysis of the effectiveness of select government programs, Monitor Group concludes that mandated approaches such as PSL overall have achieved mixed success. While PSL has led to increases in access to basic financial services for many previously excluded actors, evidence suggests the depth of inclusion is insufficient, as PSL advances often do not reach the intended targets.

Beyond regulation, government can facilitate social enterprise through a range of initiatives, such as by providing well-run extension programs in agriculture or health care, so that workers update their skills and social enterprises can stimulate more demand. Governments can also accelerate campaigns to build and improve physical infrastructure, especially transportation and upgrading agricultural extension services. And they have the capability to publish more information about low-income population segments, locations, and education levels to help social entrepreneurs and large businesses target more effectively.

Conversely, social enterprises have also been impeded by public policies and programs. This is particularly problematic for social enterprises operating in industries that compete with public provisions of services, or when subsidies are provided to certain actors that result in skewed incentives or crowding out competitors. For instance, in certain sectors such as health, education, and water, some governments in sub-Saharan Africa (SSA) provide free or subsidized services, which pose a high barrier to social enterprises. Tariffs on clean water kiosks can vary greatly from country to country, which makes it difficult for a social enterprise to scale up across the SSA region as different business models will need to be employed. In the education sector, the Ghanaian government permits private operators to offer service beyond what the state provides, so social enterprises are flourishing and can offer innovation, affordability, and access to segments of the poor that are not well-serviced or where state services are oversubscribed. Meanwhile, the Senegalese government does not allow private health-care clinics to compete with state facilities. In Kenya, regulations prohibit private firms from operating microgrids, potentially denying access to millions of unelectrified villages. And even when government allows competition from privately held social enterprises (whether for-profit or nonprofit), other policies can prohibit the diffusion of their models.

Policy innovation and social enterprise

The Schwab Foundation, among others, offers practical advice for policy makers that seek to create a more hospitable environment for social enterprise. These include, first, developing regulatory frameworks that allow for both public and private provision of goods and services, such as schools or health clinics; in this way nongovernmental actors (whether nonprofit or commercial) can fill gaps where government falls short. Other recommendations include the use of purchasing power for anchor demand, among others. And given the proliferation of business models such as the inexpensive "no-frills" and "pay-per-use" approaches explored above, policy makers should take into consideration the impact of regulations and policies that potentially favor large companies over small and microenterprises, whether advertently or inadvertently.

In terms of specific policies that recognize social enterprise as a unique legal form, most emerging markets governments do not offer them. Etchart and Comolli's analysis of the regulations and legislation in five Latin American countries revealed "no regulatory or legal framework recognizing social enterprises" (2013, p. 6). This poses unique challenges for social enterprises; as noted on a Schwab Foundation report, "Corporate legal forms do not recognize dual purpose business models ... and tax systems rarely distinguish between companies that benefit society and the environment, and those that damage it." As a result, social enterprises are subject to the same tax schemes as mainstream SGBs. And social enterprises operate under existing legislation that dictates rules for nonprofits either as associations, corporations, or foundations, or in the case of for-profits, as limited liability companies. Some advocates call for national governments to establish specific legislation recognizing social enterprise, or "fourth sector" organizations, as a means of removing barriers.

Unleashing the power of enterprise for social good

Changing governmental policies is one step toward unleashing the potential of free enterprise to address social and environmental problems in emerging markets. And while policy and regulatory reform can help advance social innovation, a multitude of other factors present in emerging markets create significant barriers to success at scale. The limitations of operating in austere economic environments force adaptations of business models. The amount of, and timing for, capital infusions needed by social-purpose businesses are different than what Western investors are accustomed to. As a result, a mismatch exists between the levels of capital that entrepreneurs can absorb at very early stages, and what investors expect to place. If they want to realize the full financial and societal returns promised by social enterprise, investors will need to adapt their expectations, and find ways to infuse the right levels of capital at the appropriate times. And entrepreneurs on the ground will need to continue to adapt their approaches to the conditions inherent to emerging markets, innovating "no-frills" and "pay-per-use"-type models so that they can appropriately serve the demands of very poor customers. Yet even with business models optimally suited to emerging market environments, many entrepreneurs will still need the help of industry facilitators to help overcome the systemic economic, political, and physical infrastructure barriers they confront. Proponents of market-based social innovations can also look increasingly to large companies to help scale promising approaches, as established businesses awaken to their potential to create shared value. And similarly, NGOs that have traditionally relied on philanthropy and aid can adopt market-based solutions to advance their missions in more financially sustainable ways.

For most of modern world history, Western societies have provided, and emerging market countries have accepted, government philanthropy and aid to address pressing problems. While many populations would undeniably be much worse off without it, traditional aid has fallen short of its ambitious goals. Social enterprises and large businesses, working in tandem with NGOs embracing earned-income models, and philanthropists and other actors playing industry facilitation roles, are now tackling societal challenges in emerging markets. Social innovators are building on the inroads laid primarily by philanthropists and governments during the first half of the last century, and are supported by a fast-growing cadre of impact investors eager to bring market forces to bear on social and environmental issues. More than a decade after Prahalad opened the world's eyes to the fortune at the bottom of the pyramid, it may be that it is not so much a financial boon, but rather a wealth of social and environmental impact that will be mined by the next generation of emerging market entrepreneurs.

Notes

1. "Market-based solutions" is a term used by Monitor Group to describe a range of social-purpose business models which generate earned income rather than rely on philanthropic donations or government subsidies.
2. B Corps are certified by the nonprofit organization, B Lab, to meet rigorous standards of social and environmental performance, accountability, and transparency. Retrieved from http://www.bcorporation.net/what-are-b-corps/the-non-profit-behind-b-corps
3. For its 2009 report, *Emerging Markets, Emerging Models: Market-Based Solutions to the Challenges of Global Poverty,* Monitor Group conducted case study research on more than 270 market-based solutions, primarily based in India, and also operating in Africa and other emerging market regions. The study encompassed social-purpose businesses operating under several different legal forms; more than two-thirds of the sample were operating either as SGBs (for-profit) or as NGOs (nonprofit); less than 20 percent were large corporations operating at significant scale (pp. 132–3).
4. This list of business models developed by Monitor Group is not exhaustive; others exist. Often, businesses will incorporate more than one of these approaches into their models.
5. WaterHealth data adapted from the FSG report, "Creating Shared Value in India: How Indian Corporations are Contributing to Inclusive Growth While Strengthening Their Competitive Advantage," October 2011, pp. 9–10.
6. Interview with David Wachtel, SVP Marketing, Communications and Partnerships, Endeavor, September 2, 2014.
7. The Mercy Corp TISA case example data are based on an unpublished FSG shared value case study, "Mercy Corps Guatemala: Improving Health through TISA, a Network of Health Stores," written in 2013, and also the Mercy Corps publication, *Sustainable Community Health Stores: The Agency's First Microfranchise Business Model Successfully Transitions to the Private Sector.* Portland, OR: Mercy Corps. Retrieved from http://www.mercycorps.org/sites/default/files/GuatemalaHealthStoresCaseStudy16May2013.pdf/
8. Acumen Fund is a nonprofit that accepts charitable donations and uses them to make long-term debt or equity investments in early-stage companies that serve low-income customers in emerging markets; it was established by the Rockefeller Foundation and other philanthropies committed to impact investing in emerging markets.

Further Reading

Bugg-Levine, A. and Emerson, J. (2011). *Impact Investing: Transforming How We Make Money While Making a Difference.* Hoboken: Jossey-Bass.

Kerlin, J. A. (2009) *Social Enterprise. Civil Society: Historical and Contemporary Perspectives* [Kindle Edition]. Medford: Tufts University.

Koh, H., Hegde, N., and Karamchandani, A. (2014) Beyond the Pioneer: Getting Inclusive Industries to Scale. Deloitte Touche Tohmatsu India Private Limited. Available from http://www.beyondthepioneer.org/wp-content/themes/monitor/Beyond-the-Pioneer-Report.pdf

Koh, H., Karamchandani, A., and Katz, R. (2012) From Blueprint to Scale: The Case for Philanthropy in Impact Investing. A Report published by Monitor Institute in collaboration with Acumen Fund. Available from http://monitorinstitute.com/downloads/what-we-think/blueprint-to-scale/Blueprint_to_Scale.pdf, p. 8.

References

Bornstein, D. (2011). "A Light in India." *The New York Times*. Retrieved from http://opinionator.blogs.nytimes.com/2011/01/10/a-light-in-india/?_php=true&_type=blogs&_r=0

Borgonovi, V., Meier, S. and Vaidyanathan, L. (2011). "Creating Shared Value in India: How Indian Corporations Are Contributing to Inclusive Growth While Strengthening Their Competitive Advantage." Published by FSG. Retrieved from http://www.fsg.org/Portals/0/Uploads/Documents/PDF/India_CSV.pdf. p. 11.

Bugg-Levine and A. Emerson (2011); Karamchandani, A., Kubzansky, M., and Frandano, P. (2009); and Boschee, J. (2001) *The Social Enterprise Sourcebook*. Duluth, MN: Northland Institute.

Dees, J. G. (1996). *Social Enterprise Spectrum: Philanthropy to Commerce."* HBS No. 396343. Boston, MA: Harvard Business School Publishing.

Dees, J. G. (1998). "Enterprising Nonprofits." *Harvard Business Review* (Jan–Feb 1998). Retrieved from https://hbr.org/1998/01/enterprising-nonprofits/ar/1

Emerson, J. (2003). "The Blended Value Proposition: Integrating Social and Financial Returns." *California Management Review*, 45 (4): 35–51.

Etchart, N. and Comolli, L. (2013). *Social Enterprise in Emerging Market Countries: No Free Ride*. Basingstoke, UK: Palgrave Macmillan.

Karamchandani, A., Kubzansky, M., and Frandano, P. (2009). *Emerging Markets, Emerging Models: Market-Based Solutions to the Challenges of Global Poverty*. Monitor Group. Retrieved from http://www.beyondthepioneer.org/wp-content/uploads/2014/04/emergingmarkets_full.pdf

Porter, M. E. and Kramer, M. R. (2006). "Strategy and Society: The Link between Competitive Advantage and Corporate Social Responsibility." *Harvard Business Review*, 84 (12): 78–92.

Porter, M. E. and Kramer, M. R. (2011). "Creating Shared Value." *Harvard Business Review*, 89 (1–2): 62–77.

10

Social Media and Innovation

Michael Shoag and Tory Colvin

Introduction and overview

It is 6:30 am in the small town of Morogoro, Tanzania. Saibaba (meaning "spiritual master"), one of 82 mine detection trainees, has just finished his breakfast when he and his fellow soldiers are loaded onto a truck and driven to the nearby "landmine detection training field" at SOKINE University of Agriculture. When he arrives, Saibaba is outfitted with a harness and clipped onto a series of ropes, which will guide him through his individual training plot. On a good day, he will cover around 200 square meters, and he will be rewarded for every mine he discovers.

Saibaba, in fact, is an African pouched rat – a "HeroRAT." Detection of landmines is difficult, dangerous, and expensive. Yet rats are light enough that they can successfully locate these deadly mines without detonating them, long before they can harm innocent lives. HeroRATs offer a cost-effective means to get communities back on their land as quickly as possible. In 2013, their detection skills helped over 900,000 people to return to their villages and fields in order to work and play without fear. Saibaba's end-game? Being deployed as an elite, four-legged member of the "mineaction team," responsible for detecting *real* landmines in nearby Mozambique.

HeroRATs are trained and deployed by APOPO, a nonprofit that focuses on identifying innovative solutions to landmine clearance in post-conflict areas, as well as detecting TB (tuberculosis) in poverty-stricken communities. In the past, APOPO would need to approach a long list of potential benefactors one by one, or by writing a host of grant applications that would have taken months or years to get approved. Yet this incredible program, which works in six countries across Southern Africa and Southeast Asia, raises over $250,000 worth of public donations annually, from over 4,000 online donors across the globe. It gained wide exposure through the Global Giving platform, a charity fund-raising website that gives social entrepreneurs and nonprofits throughout the world a chance to raise money. Last year Global Giving was able to raise over $25,000 for APOPO through 500 donors, and they support APOPO with regular training, most importantly concerning the use of social media channels to crowd-fund its work from people throughout the world.

When companies want to test out new ideas, they often consider engaging the United States because it is such a highly developed market. But engagement levels in developed countries are typically low compared to emerging markets in less developed regions. Social media is a perfect, low-cost way to try something new. And because "a little" typically goes much farther in developing regions, the impact of a successful idea is often more profound (Carlman and Pursey, 2014).

Purpose of this chapter

This chapter illustrates social media's impact on innovation across Asia, Latin America, Central Europe, and the Middle East. We explore ways in which social media has shaped innovation in emerging markets, and we also highlight less obvious ways in which these platforms have accelerated progress across the globe.

For the purposes of our research, we have narrowed our focus to platforms like Facebook, Twitter, and Baidu (China), video-sharing platforms like YouTube, and socially driven data visualization tools. The examples we have chosen span critical global topics like natural disasters, international public health crises, and public diplomacy. They illustrate the dual role social media plays in promoting good (predicting pandemics), while fueling upheaval (using social media to instigate rebellions); in incurring passion (money raised globally for niche projects), while accelerating hate (beheadings publicized for millions to see). The chapter concludes with an analysis of the future implications social media is likely to have on society.

Social media as an information marketplace

Social media is defined as any form of electronic communication that allows people to share information, ideas, personal messages, and other content. Its goal is to foster a more open and connected world by giving people the power to create and share information instantly, and without barriers. The overwhelming surge in social media is driven by the simple fact that information is power, and social media platforms have created a perfect marketplace whereby information can be produced and consumed.

Today, anyone can become an *information producer*. Governments throughout the world no longer have complete control over the flow of information, and there are few to no production costs associated with creating and disseminating information across these channels. Likewise, *information consumers* are now able to gain immediate access to a much broader range of facts and opinions. Complex stories are distilled and discussed in a more simplified way. One-way communication via mainstream media has been replaced by "feedback loops," where robust, two-way dialogue allows opinions, examples, and personal stories to generate a stronger pulse across global topics.

Its impact across the globe

The international impact of these channels is overwhelming. According to a recent study by the Pew Research Center's Global Attitudes Project: "While the internet still has a limited reach in the emerging and developing world, once people do gain access to the internet, they quickly begin to integrate it into their lives" (Pew Research Center, 2014).

Facebook currently has 829 million daily active users, with approximately 82 percent of its accounts outside of the United States and Canada. Twitter has 271 million monthly active users, with 77 percent of its accounts outside of the United States, and currently it supports over 35 different languages. These statistics are impressive as they are, but they become staggering when compared to the fact that only 40 percent of the world's population is online. As the digital divide continues to close and more people gain access to these types of online tools, social media will play an even more critical role in shaping the global landscape, both economically and politically. As we will see in the next section, some countries are much better at leveraging the power of social media than others.

I. Innovation as national policy

Governments throughout the world operate in a much more transparent environment than they did just a few decades ago. In today's globally connected world, a significant portion of social, economic, and political activity now happens online. As people share information, purchase products, and connect with others on social media, they leave a digital footprint that is a gold mine of information.

The private sector has become extremely savvy at combing through this data to understand consumer behavior and tailor their marketing messages in an attempt to influence purchasing decisions. Governments are now following suit. Almost every government agency in the United States has a Facebook and Twitter page, and more often than not, agencies will have a YouTube presence and an Instagram account too. This trend has extended internationally, including to many developing countries. Those in power recognize the value of these tools. Hillary Clinton, for example, described social media as the pulse of the planet. India's Narendra Modi uses social media to understand the needs of the populace and adapt policies to meet those needs. These politicians understand that social media is one of the most reliable checks on where the public's mood is on a particular topic and can be used to influence that mood and guide policy decisions.

This section looks at how social media has helped promote innovation within national policy, and in particular its role in emerging markets and in fueling democracy building across the globe.

Democracy building

Democracy building paves the way for new markets to emerge by controlling volatility and allowing economies to scale. Social media offers tools for reshaping national policy and promoting democracy that ordinary citizens may not have historically had. These channels foster collaboration, social space, a two-way dialogue of policy issues, and allow people to communicate broadly about critical topics.

Antidemocratic regimes understand the power that social media has over political arguments. Many try to stymie prodemocratic, social participation by limiting access to social media. We have seen this in regimes across the globe. China often limits access to specific keywords and websites. Other countries, such as Iran, turn off access to specific social media sites during times of protest. And of course the extreme example is North Korea, which not only forbids social media, but allows no web access for ordinary citizens.

The reshaping of public diplomacy is a unique example of how social media has transformed the national policy landscape across the globe. In this section, we compare traditional public diplomacy, elections, and revolutions to their modern-day counterparts in the digital age.

Reshaping public diplomacy and mobilizing public opinion

Social media allows people to disseminate information – classified or not – to millions of people. It takes away intermediaries' ability to control what types of information are allowable and not allowable. And it increases transparency around information, giving broader reach to critical content that may not have been so easily accessed before the onset of the digital revolution.

When comparing public diplomacy pre-social media to its characteristics in the digital age, three themes emerge:

1. Traditional diplomacy has been inherently slow, with process and red tape often hampering efforts to make progress on an issue. Social media accelerates movement by allowing people to instantly blast out information to hundreds, sometimes thousands, of their followers. These channels quickly bring critical topics to the forefront of people's minds, forcing movement and response by political leaders.
2. Traditional diplomacy relies heavily on a top-down flow of information. Social media has enriched the information environment by allowing exponentially more people to weigh in on critical topics. In doing so, it has become more difficult for nations to control the flow of information.
3. Traditional diplomacy centered on leadership and a pyramid-based leadership structure, whereas social media is much more disparate. This

flattened structure allows for more extreme attitudes to rise, amplifying voices at the edges of a given issue. Given that, it also creates a situation where compromise can be harder to achieve.

Hashtag diplomacy. Twitter campaigns have become an increasingly common way for political leaders, intergovernmental organizations, and diplomats to bring attention to a specific topic. This brand of "hashtag diplomacy" happens when a specific hashtag like "#stopjosephkony" spotlights a particular social or political issue and drives public discourse around that issue. With over 271 million active users on Twitter as of spring 2014, tweets carrying a specific hashtag are able to reach an incredibly broad array of people. They bring deeper awareness to stories that take place in less media-heavy regions of the world (Joseph Kony's "Lord's Resistance Army" largely operated in the underdeveloped areas of northern Uganda, the Democratic Republic of Congo, and South Sudan, for example) and that may not have otherwise reached the global stage.

In many ways, governments take advantage of the way social media has changed the foreign relations landscape (Kojo Nnamdi Show, 2014a). It is a great way for them to create more transparency around their work. They can comb social media posts and pull out content to buttress their own positions. And they are now better equipped to listen to the people they serve. Long-standing barriers between the public and elected officials have been deconstructed, giving rise to an environment in which ordinary people have much greater accessibility to those in power. The asynchronous nature of this medium means government officials do not have to be available at a specific time in order to hear from the people they serve (e.g., at a public hearing or speech). They can respond on their own time, from their computer, tablet, or mobile phone.

Future implications of social media on public diplomacy. The increased transparency that social media has fostered has created unique challenges for governments moving forward. Historically, heads of state have been able to discuss critical topics behind closed doors. But now, more than ever, the world is watching, which means they can no longer control the flow of information. The traditional top-down approach has given way to two-way diplomacy. Anyone can post their thoughts, opinions, and beliefs, and depending on how much traction a particular topic gets, governments may be forced to address the issue or even make diplomatic changes to respond more formally.

A force for good and evil

Social media amplifies everything, regardless of whether it is used to promote good or to instill harm. It levels the playing field in many regards, enabling ordinary citizens to weigh in on critical topics, and giving them undeniable power where they once had none.

Often, though, it is the opinions at the far edges of the argument that are loudest, which creates an environment where extremist attitudes are amplified, and proactive compromise becomes harder to achieve. This section explores ways in which social media has been used as a powerful force for both good and evil.

Waging war and terror

It should come as no surprise that social media has become an extremely popular platform for waging war. Social media is used to recruit new fighters, spur the populace to action, help prove facts on the ground, and mostly to empower a whole new breed of propaganda.

Social media has become a ubiquitous part of fighting, terrorism, and war. During the fighting in Syria, in Ukraine, and between Israel and Hamas in Gaza, all sides use social media to fight in the war of public opinion. It seemed that with every bomb there was a new video or posting to Twitter, YouTube, Instagram, Vine, and other social sites. When Israeli bombs hit four young boys on a beach, it predictably sparked outrage and defensive posts on popular social media channels.

After the shooting down of a Malaysian airliner with 298 passengers aboard over Ukraine, a rebel leader asserted responsibility. Igor Strelkov posted to a Russian networking site "We just downed an An-26 near Torez." A video of the plane falling accompanied the post. The post was quickly deleted when it became clear that it was a civilian aircraft that had been hit.

These stories point out many aspects of how social media has changed the ability to wage propaganda during wartime including: (1) information can be distributed in near real time – even if, as in this case, or with the embarrassing accidental "reply all," it is not in the best interest of the sender; (2) once on the Web, information is nearly impossible to pull back; and (3) cameras are in the pockets and purses of over a billion people and they can easily take a video and post it to social media with just a few finger taps.

There is also an ever-growing crop of journalists and bloggers who are happy to pontificate on the success, or failure, of social media campaigns (and with this chapter we add ourselves to this list). These produce headlines such as:

- "Israel is Winning the Social Media War in China" (LegalInsurrection. com)
- "Who's Winning the Eukraine Social Media War" (LewRockwell.com)
- "@ISIS is #Winning: Why Is a Barbaric Medieval Caliphate So Much Better at Social Media than Washington?" (*Foreign Policy Magazine*)

Case study: ISIS and social media

As with every war since time immemorial, propaganda is a staple. Social media does not change the basic tenets of propaganda, but dramatically changes who can create propaganda, how it is disseminated, whom it reaches, and how it can be countered.

The Islamic State has been particularly effective at using social media to spread their message of death and fear. They have used a combination of well-produced photographs and videos, and a sophisticated understanding of how Twitter works to trick the platform into spreading their message to a far wider audience than it would normally reach.

ISIS has been very concerned with their brand from the very start. Even their name has changed several times to promote their changing brand. While name changes are always tricky for brands to pull off, ISIS has seen it as a necessity as their ambitions and reach have grown. Wikipedia documents no fewer than nine names for the group.

While at one time they referred to themselves as the Islamic State of Iraq, it was later expanded to the Islamic State of Syria and Iraq (ISIS). It later grew to the Islamic State of Iraq and the Levant (ISIL). The "Levant" refers to a geographic region in the Middle East. Finally, once their ambitions grew beyond the Middle East, they shed the limitations of their name to arrive upon "Islamic State." This has led to some social media and communications challenges for the group.

The Islamic State learned quickly how to game the system to grow their social media reach beyond those who were looking for them. They have been particularly good at using Twitter.

In Twitter, as with many other social networks, users may tag their posts with a hashtag to show its relevance. One would expect the Islamic State to use hashtags such as #ISIS, #ISIL, and #beheading. They have instead co-opted hashtags such as #WorldCup2014, #WC2014, #NapaEarthquake, and other trending hashtags. As a result, when someone on Twitter searched for #WC2014 during the World Cup they may have seen an ISIS recruitment video that they almost certainly were not looking for.

By co-opting the most popular hashtags (and there are easy-to-find lists of rising and popular hashtags), the Islamic State can ensure their message is viewed by many who would not have otherwise seen it.

The Islamic State also created a mobile app for Android phones called "The Dawn of Glad Tidings." Ostensibly the app allows users to follow ISIS and their messages. In reality, it also enabled ISIS to post messages to each app user's Twitter accounts without their knowledge. At times tens of thousands of app users would have the same message posted to their Twitter accounts simultaneously. This greatly amplified their propaganda by enabling it to be viewed by millions more Twitter users than through traditional distribution channels. Google soon removed the app from their app store (Berger, 2014).

Social media sites including Twitter and YouTube have worked hard to take ISIS's messages off their networks when they violate the company's terms of use. YouTube's terms of use provide easy cover for removing IS videos including prohibitions on graphic or gratuitous violence, inciting violence, intent to shock, and hate speech. The Islamic State and its supporters, however, have shown resilience at creating new accounts and finding other platforms to stream their graphic and often fear-inducing videos. By one estimate over 60,000 pro-Islamic State accounts have been created on Twitter. Such proliferation has meant that Twitter cannot pull down sites fast enough to stop the messages from getting out.

The US State Department has also created its own Twitter account to counter the Islamic State. The @ThinkAgain_DOS Twitter handle includes links to articles, images, and videos – many taken from pictures and videos first posted by the Islamic State. The "Think Again Turn Away" campaign is targeted toward those considering joining the Islamic State. It shows the horrors of terrorism and explains why it is a bad choice. Some of their Twitter posts include: "US troops are punished for misconduct, #ISIS fighters are rewarded #thinkagainturnaway" and "#ISIS has no shame, using children's images to promote its cult of death #thinkagainturnaway."

But even the videos produced by the US State Department to counter the Islamic State tend to reinforce their message of fear and brutality. One State Department video, "Welcome to the 'Islamic State' Land (ISIS/ISIL)" uses grisly videos and images of beheaded corpses, blown-up mosques, crucified bodies, and whippings to try to dissuade potential followers. It is overlaid with statements such as "Learn useful new skills," "Blowing up mosques," and "Travel is inexpensive because you won't need a return ticket."

None of the Islamic State social media tactics is new, yet their savvy demonstrates how social media can be used for harm instead of good. It shows how the medium is optimized for those willing to break the rules (e.g. using irrelevant hashtags to promote a message, and posting to Twitter accounts without permission). It reinforces how difficult it is for government agencies to be successful on social media. It also shows how social media companies can be overwhelmed by a flood of messages that violate their policies (Warren, 2014).

II. Facilitating institutions

The private sector is seen as the "engine of growth," whereas universities are widely recognized as a vast repository of information and knowledge creation. Historically, these institutions have worked independently of one another, but the rise of a global knowledge economy has increased the need for strategic partnerships between the two. When effective, university–industry collaboration can help solve very big problems. This section explores how collaboration between universities and privately held companies has generated innovations related to social media that have helped to address recent global challenges.

Detecting illnesses and saving lives

Social media has led to innovative ways to monitor disease progression, and fill gaps in traditional epidemiology. Take the case of "HealthMap," an online mapping tool developed by a group of researchers, epidemiologists, and software engineers at Boston Children's Hospital, a "teaching hospital" affiliated with Harvard Medical School.

HealthMap relies on an algorithm that analyzes social media posts, news reports, medical workers' social networks, and government websites. It aggregates content from information freely available on other websites (e.g., ProMed Mail, WHO (World Health Organization), FAO (Food and Agriculture Organization of the United Nations)), many of which are generated by privately owned companies or universities.

HealthMap takes that information, distills it using an algorithm, and tracks the data outputs on an interactive map. The map itself uses the Google Maps API (application program interface), a desktop and mobile web-mapping service application and technology provided by Google.

Challenges with traditional epidemiology

Academically driven institutions, like Boston Children's Hospital, typically innovate in response to a long-standing challenge with traditional practices. In this case, traditional epidemiology techniques have historically been expensive, time-consuming, and heavily reliant on in-person data collection and analysis.

On-the-ground disease surveillance can be difficult, with few medical resources and labs available in remote, underdeveloped areas. Likewise, large international organizations like the WHO often face budgetary constraints, which can result in delayed access to resources across affected areas. According to Randal Shoepp, chief of the Applied Diagnostics Branch of the US Army Medical Research Institute of Infectious Diseases, "Aid workers have to bring everything with them. When they arrive, they have to set up entire laboratories, train medical staff, and supervise them as they continue their investigation" (Kojo Nnamdi Show, 2014b). Health officials often spend weeks interviewing victims, gathering test results and data, and completing their investigation and analyses. Most of this work is done in private, and it can take weeks or even months for final reports to be released.

Using socially driven data to predict Ebola

HealthMap was able to predict the Ebola outbreak nine days before the WHO issued its first statement on the epidemic. One of the first indications that an Ebola outbreak was starting was word of a "mystery hemorrhagic fever" spreading in Guinea, which had already killed eight people in that country (Schlanger, 2014). Ebola had struck numerous times before in the Democratic Republic of Congo, South Sudan, and Uganda, but never as far west as Guinea and Liberia (CDC, 2014a). As a result, local health officials did not see the warning signs as residents increasingly developed diarrhea, vomiting, and muscle pain.

Health officials initially started testing for more common causes of sickness in Western Africa, like malaria and Lassa fever, which meant they were not enforcing "barrier nursing techniques" to prevent the spread of Ebola. Barrier nursing requires the use of head-to-toe protective clothing (masks, gloves, gowns, and goggles), using infection-control measures (such as complete equipment sterilization and routine use of disinfectant), and isolating patients with Ebola from contact with unprotected persons (CDC, 2014b). And even as early indications of the possibility of Ebola arose, those discussions were restricted to private conversations between health-care workers and the family members of victims.

By March 19, 23 people in Guinea had died of the mystery illness, and HealthMap.org issued an alert by posting its first dot on that country's map. The tool also provided links to local news stories about reported cases in the region. HealthMap continued to plot data points in the days that followed, as scores more people in Guinea died, and the virus spread across borders to neighboring Sierra Leone. Finally on March 23, nine days after initial symptoms were reported in Guinea, the WHO confirmed the Ebola outbreak (Public Health Watch, 2014).

HealthMap is just one example of how technology has transformed the disease-hunting process. Another effort underway is the PREDICT project, led by the One Health Institute at the University of California, Davis. PREDICT works in concert with the US Agency for International Development (USAID)'s Emerging Threats Program, and "builds on the understanding that humans, wildlife and the environment are inextricably linked" (Kerlin, 2012).

The goal of this public–private partnership is to create a global early warning system for emerging diseases that move between wildlife and people. It does so by combining risk modeling, computerized data collection, and wildlife field sampling to identify specific situations and locations that are most likely to prompt the next pandemic (UC Davis, 2012). PREDICT utilizes the same HealthMap tool to visualize the global, relative risk of a new infectious disease emerging across Latin America, Southeast Asia, the Congo Basin, and other "hot spot" regions.

Economic impacts of Ebola.   The economic implications of the current Ebola outbreak in Liberia are significant, with costs associated with fighting the disease expected to reach $32 billion by the end of 2015. There is a very real possibility that countries will start attempting to close their borders (as best they can) to prevent people in Western Africa from entering and spreading the illness further.

Money to fight Ebola is not coming in nearly as quickly as it did for natural disasters like the Haiti earthquake or the Japanese tsunami. These crises offered a wealth of photo opportunities for journalists and ordinary citizens to capture the horror of the event and increase the emotional impact they had on viewers. Pictures coming from the Ebola outbreak, on the other hand, usually show the same scene – medical workers clad in biohazard suits wheeling out body bags of disease-stricken victims. Even though the stories associated with this disease are awful, the Ebola crisis simply has not pulled at the heartstrings of people across the globe.

Private donations are only trickling in as a result. Four major US aid organizations recently surveyed had received a combined total of $19.5 million as of October 2014 (approximately six months after the outbreak began), much of which came from nonprofit foundations as opposed to individual donors. In contrast, major charities raised nearly double that amount within one week of the typhoon Haiyan (which struck the Philippines), and US relief organizations raised $1.3 billion within six months of the 2010 earthquake in Haiti (Cohen, 2014).

Furthermore, people living in heavily impacted parts of Liberia and other African nations can no longer work because many are in preventative quarantines. The reduction in work leads to even more money shortages in already low-income regions. According to Margaret Chan, head of the WHO: "I have never seen a health event threaten the very survival of societies and governments in already very poor countries," she said. "I have never seen an infectious disease contribute so strongly to potential state failure" (Chan, 2014).

Chan warned that "rumors and panic spreading faster than the virus" could create even more economic impact, with the World Bank recently estimating that 90 percent of the cost of the outbreak would arise from "irrational attempts of the public to avoid infection."

The future of technology in disease hunting

As technology continues to transform traditional epidemiology, many scientists are latching on to the importance of using social networks to track new outbreaks. According to Dr Taha Kass-Hout, deputy director for information science at the CDC (the US' Centers for Disease Control and Prevention), "Given that the next SARS probably can travel at the speed of an airliner from continent to continent in a matter of hours [Ebola has already been reported in Texas as a result of someone flying to the United States from Liberia], it just makes perfect sense to adapt the speed and flexibility of social networking to disease surveillance" (Kass-Hout, 2011).

International health agencies will no doubt continue to play a critical role in disease hunting, but these online tools can supplement their work by incorporating real-time, socially driven data into their studies (Garrity, 2011). The challenge that likely will plague scientists in the immediate future is how best to comb through the ever-present noise on social media channels, which is often ripe with misinformation and fear.

III. Firm-level innovation

Private sector organizations are constantly seeking out opportunities to expand their reach, and the innovations they pursue are often in direct response to real-world challenges that have not yet been addressed. This section explores the effects that globalization has on firm-level productivity and innovation, and addresses common challenges that these companies face throughout these endeavors.

Western companies and the attraction of going abroad

Western entrepreneurs have become increasingly enthusiastic about launching start-ups in developing countries, and while there are many reasons why companies would decide to forge this path rather than stay stateside, most entrepreneurs who go abroad cite competitive advantages resulting from lower staff wages as the main benefit over following the path of similar firms operating from within the United States (TNW, 2014). These companies launch in developing countries like Southeast Asia, but in many cases the clients they serve continue to be stateside. According to Jan Jones, founder of "Oozou," a web and mobile app development studio based in Bangkok, Thailand: "Our Western customers get the same quality that they would get Stateside, but at considerably less cost" (Jones, 2014). Others, like Chilean start-up "Easy Vino" – a company that helps restaurants develop wine lists and provides an app to customers to select wines with their meal – gain so much success outside of the United States that their owners then try to bring these ideas to Silicon Valley (Borison, 2014).

And even other companies have recognized that the consumer base in developing and lower-income countries promotes ripe conditions for the sale of specific products or services. Asian consumers, for example, tend to be much more social, mobile, and gaming-centric, which means that products geared towards that user base will fare better than they would in another region with different target users.

The benefits of starting tech-centered businesses in developing countries are not without some unique challenges, with cultural differences, language barriers, corruption, and currency exchanges as some of the most notable. Time zone differences add to an already unique business communications environment, where asynchronous conversations with a globally disparate clientele are commonplace, and early morning or late evening calls via Skype are the only real way to connect with customers on opposite sides of the world in real time.

Public/private sector roles in promoting technical innovation in Southeast Asia

One Southeast Asian country in which technology start-ups are quickly gaining traction is Vietnam. Under a one-party political system, Vietnam has historically discouraged the kind of free speech and open communications promoted by online communities and services. While the country is only in the early stages of becoming a true tech hub like Silicon Valley or Singapore, Vietnamese regulatory policies around IT services are becoming increasingly progressive. In 2010 the prime minister, Nguyen Tan Dung, launched a campaign that was centered on transforming Vietnam into an advanced

IT country by 2019. The Vietnamese science and technology ministry has committed money in infrastructure and high-tech industrial parks, and the World Bank awarded a $100 million loan to be used towards tech-related development within the country. These improvements will provide a significant boost to the emerging start-up industry in Vietnam, and it will enable more companies to take advantage of high-tech subsidies (M.I. Hanoi, 2014).

Private sector-led start-ups are also gaining momentum in Vietnam. One example is the site "Chomp," which is an online platform that allows marketers to interact with target audiences through hashtags. Described by the company as "ActionTags," marketers can create several types of campaigns, including photo contests, prize money contests, giveaways, and voting challenges. The campaigns can be promoted across multiple social media channels like Facebook, Twitter, and Instagram, and marketers can track everything related to the campaign via a centralized dashboard.

Vietnamese start-ups face cultural nuances that more established technical industries are perhaps more experienced at navigating. Language barriers are always a top concern, and these companies must choose whether to create their websites in Vietnamese only – which limits readership to the 90 million people in the country, English – which may exclude many Vietnamese users, French, or in both/multiple languages. Some have stuck strictly to Vietnamese. "Lozi.vn," for example, is a "Yelp meets Pinterest"-style food site in Vietnam, where users can take photos of food, "pin" photos of food that other users have posted, and share this information via social media channels like Facebook. That site is only offered in Vietnamese, whereas other companies like Chomp offer English and Vietnamese language options to help bridge cultural divides among its users.

Using technology to conquer developing world challenges in Africa

For many countries throughout the world, going mobile is an afterthought – a second step that companies take to augment the services and products they already offer on desktop websites. Countries in Africa, however, are not just mobile first – they are *mobile only*, with mobile devices acting as the "digital equivalent of a railroad" and allowing these people to connect with others, even if they are in the most remote parts of the world. As of 2014, "Only 16 percent of the continent's one billion people are online, but that share is rising [...]. More than 720 million Africans have mobile phones, some 167 million already use the Internet, and 52 million are on Facebook," a recent McKinsey report stated (Manyika, 2013).

A team of software developers, engineers, and technologists in Nairobi, Kenya, recently developed a self-powered, mobile WiFi device called "BRCK." BRCK's goal is to provide Internet connectivity in a go-anywhere device that can provide electricity and Internet access where urban and rural areas have none. The device looks just like a brick. It is able to connect to multiple networks, it has a huge battery that can keep 20 devices connected for eight hours, and it is robust enough to handle power failures and poor connectivity speeds. According to BRCK's founders, "We realized that the way the entire world is connecting to the web is changing. We no longer only get online via desktops in our office with an ethernet connection, we have multiple devices, and mobile connectivity is crucial" (http://www.brck.com/). Throughout Africa's poorest countries, many people still rely heavily on feature phones (i.e., non-"smartphone"), and as a result they have a hard time accessing some of the largest Western social media sites like Facebook and Twitter. Smart entrepreneurs have figured out how to tap the technology mismatch and help bridge the digital divide that exists in developing countries. In Zimbabwe, for instance, Internet connectivity is expensive, and, for many, only available in Internet cafes or offices. "ForgetMeNot" is an African "Optimiser Platform," which uses eTXT technology to connect mobile phones on any mobile network with a number of social media and Internet services. An "eTXT" is a message that can be sent and received as SMS, as an email, or even as a chat message. It allows users in rural areas to have instant two-way chats on their mobile phones with people who have full access to social media channels, etc., without having to download anything. And since nearly all phones have SMS capabilities at a very low cost, people who could not otherwise afford to use these services now can.

Limited access to lines of credit, online banking, online shopping, shipping, and delivery are additional obstacles deeply embedded in the way of life for many people living in Africa, and a company called "Jumia.com" is navigating these challenges by striving to become the Amazon.com of Africa. Jumia is the biggest online shopping mall in Africa and offers *cash-on-delivery* products to the cities of Lagos and Abuja in Nigeria. People living there can order online or via mobile phone, and the products are then driven by motorcycle to homes or businesses, where cash is paid upon delivery.

Best places to launch a start-up in South America

Latin America's high-tech industry has blossomed in recent years, and even though there is no single Silicon Valley within Latin America at the moment, several tech hubs have emerged, largely in the southern portions of the continent. Argentina's tech community is probably the most mature, with Buenos Aires and Palermo Valley counting start-ups like Globant (an IT and software development company operating in Argentina, Colombia, Uruguay, the UK, Brazil, and the US) and Mercado Libre (an online marketplace dedicated to e-commerce and online auctions) as hometown successes.

As parts of Europe have reduced their R&D budgets towards technology investments, countries like Chile, Argentina, and Brazil have increased tech-related funding, and as a result have begun to attract high-profile companies to their burgeoning tech hubs. The Brazilian government has injected significant funding into the tech industry to compete with more established tech sectors throughout Europe (Essinger, 2013). It is estimated that

Brazil invested approximately R$6.8 billion through its Ministry of Science and Technology in 2012, and even though the rest of the country experienced major budget cuts, science and technology investments were not affected (Group of Eight Australia, 2014). And some incredibly successful start-ups have popped up in Brazil in recent years, including:

- *Easy Taxi*, which is the "Uber" of Brazil. It is funded by Rocket Internet and has approximately 30,000 drivers in its system.
- *Dafiti*, which is Brazil's version of Amazon. It was launched in 2011 and has already raised more than $249 million.
- *XMarket*, which is the "Craigslist meets Amazon" of Brazil. This company is slated to be an online e-commerce site for real estate, cars, and second-hand items. It has $850,000 in seed funding and launched in the fourth quarter of 2014.

Countries like Chile and Peru are often referred to as Latin America's "tech accelerators" or "incubators" for start-ups. An accelerator takes single-digit portions of equity in externally developed ideas in return for small amounts of capital and mentorship. They are typically consolidated into a three- or four-month program, after which that start-up "graduates." Incubators, on the other hand, incorporate external management teams to manage an idea that was developed internally within a company. These ideas typically require more time to conceptualize, and as a result incubators take a larger amount of equity as compared to accelerators (Desmarais, 2012).

Abandonment and acquisitions

While Chile and Peru offer a friendly incubator/accelerator environment for fostering innovation in technology, lots of companies launch in areas like this but end up abandoning the continent after graduating from these programs. Likewise, start-ups that become incredibly successful are often acquired by larger companies. Before working on XMarket in Brazil, for example, the start-up's CEO Lonny Szneiberg founded Investing.com, which was acquired. The Indian-born company "Bookpad" is another example of acquisition, in which Yahoo acquired the company for an estimated $15 million. This follows acquisitions of other Indian start-ups, including by Facebook, which acquired Little Eye Labs (estimated at $10–15 million), and by Google, which took over Impermium ($9 million).

An interesting aspect of some of these international acquisitions is that sometimes the teams stay in place, whereas other times the whole company moves shop. When Yahoo! acquired Bookpad, for example, the team joined the Yahoo team in Bangalore. With Little Eye Labs, on the other hand, *The Telegraph* reported that "The entire Little Eye Labs team will move to Facebook's headquarters in Menlo Park, California" (Shu, 2014). Quite an upheaval for a three-year old company also based in Bangalore, India.

Altruism, or smart business moves?

The private sector will continue to look for opportunities to expand their own reach and impact by reinvesting in their companies to develop products and services that solve critical global challenges. Google's "Project Loon" (http://www.google.com/loon/), for example, is an attempt to help close the gap of nearly two-thirds of the world's population that does not yet have Internet access. Project Loon is a network of balloons traveling on the edge of space, designed to connect people in remote areas, help fill coverage gaps, and bring people back online after natural or other disasters. It began as a pilot test in 2013 in New Zealand and continued with experiments in California and Brazil. Project Loon partners with telecommunications companies to share the cellular spectrum, and end users can connect to the balloon's network directly from their smartphones and other devices.

Google is a $367.6 Billion company (as of May 2015), with a stock price that hovers around $776 per share (as of May 2015). It dominates the search landscape, with nearly 65 percent of the total share across all popular search engines (Search Engine Land, 2013).

While Project Loon is looking to help close a critical gap in Internet access worldwide, it also increases the company's name recognition and impact every time a user posts any reference to the project on a website or on a social media feed. It also highlights the fact that Google remains at the cutting edge of all things technology, outshining traditional media sources and government agencies alike.

Conclusion

Social media will continue to play a critical role in shaping the global landscape, both economically and politically. Its novelty has largely worn off, leaving the world with a common understanding that social media offers something truly valuable – and occasionally dangerous – for the world. It distills complex topics into shareable information that can be discussed in simplified ways. It creates feedback loops where two-way dialogue and personal stories humanize issues and generate more engagement. And it offers widespread reach that augments traditional media sources, with television, print, and social media reinforcing each other and creating even broader distribution channels.

A common thread woven throughout each of the examples we have shared is that social media often advances transparency and accountability. It can help to simplify complex topics, making them easier to grasp. It can be used to augment real-world activities – to recruit volunteers, raise money, or identify disease outbreaks. It can even be used to collect enough funds from a thousand people in a distant land to ensure that Saibaba, the rat, can continue to do his job as a minesweeper.

Yet for every benefit social media brings to global innovation come questions and potential challenges for its role in the future. Perhaps one of the most obvious questions is whether social media creates *too* much information.

Millions of people can now instantaneously share content and personal opinions about global topics. This creates a genuine risk of misinformation when that content is not properly vetted and verified. Others would point to the fact that social media dumbs down critical arguments, because there is only so much sophistication a person can bring to a topic in 140 characters or fewer. This medium is asynchronous, which allows people to publish their thoughts before they edit them for the world to see. Yet it also allows people to hide anonymously behind their writing. Social media amplifies opinions at the far edges of an argument, which creates an environment where extremist attitudes tend to be loudest, and proactive compromise becomes harder to achieve. And although social media played a starring role in the Arab Spring, some have questioned whether the still-evolving results produced anything positive in the long term for the citizens of those countries.

Despite these questions, the conversations that take place across social media as a whole are often more detailed and comprehensive than mainstream media can offer. They are ripe with global opinions, personal stories of success and failure, and passion that have not yet been diluted by intermediaries.

Whether to stop protests, create revolutions, uncover corruption, prepare communities for disaster, or raise money to support HeroRATs in their pursuit of conquering minefields in Mozambique, social media will only become a more influential factor in the coming years.

References

Berger, J. M. (2014). "How ISIS Games Twitter." *The Atlantic*. Atlantic Media Company, June 16. (URL: http://www.theatlantic.com/international/archive/2014/06/isis-iraq-twitter-social-media-strategy/372856/).

Borison, Rebecca. (2014). "The 9 Hottest Up-and-Coming Startups in South America." *Business Insider*, July 15. (URL: http://www.businessinsider.com/the-9-hottest-startups-in-south-america-2014-7?op=1#ixzz3N9d2qWZQ).

Carlman, Alison (Global Giving) and Pursey, James (APOPO). (2014). "APOPO Hero Rats Interview." E-mail interview, September.

CDC (Centers for Disease Control and Prevention). (2014a). "Outbreaks Chronology: Ebola Virus Disease." (URL: http://www.cdc.gov/vhf/ebola/outbreaks/history/chronology.html).

CDC. (2014b). "Prevention." (URL: http://www.cdc.gov/vhf/ebola/prevention/).

Chan, Dr Margaret (2014). "WHO Director-General's speech to the Regional Committee for the Western Pacific." Keynote address to the Regional Committee for the Western Pacific, Sixty-fifth session Manila, Philippines, October 13, 2014. (URL: http://www.who.int/dg/speeches/2014/regional-committee-western-pacific/en/).

Cohen, Elizabeth. (2014). "Ebola: Five Ways the CDC Got It Wrong." *CNN*. Cable News Network, October 14. (URL: http://www.cnn.com/2014/10/13/health/ebola-cdc/).

Desmarais, Christina. (2012). "Accelerator vs. Incubator: What's the Difference?" *Inc.*, February 7. (URL: http://www.inc.com/christina-desmarais/difference-between-startup-accelerator-and-incubator.html).

Essinger, James. (2013). "Latin America Creates Rival to Silicon Valley." *World Finance*, April 23. (URL: http://www.worldfinance.com/inward-investment/americas/latin-america-creates-rival-to-silicon-valley).

Garrity, Bronwyn. (2011). "Social Media Join Toolkit for Hunters of Disease." *The New York Times*, June 13. (URL: http://www.nytimes.com/2011/06/14/health/research/14social.html?pagewanted=all&_r=0).

Group of Eight Australia. (2014). "Policy Note Government Research Funding in 2014 in Selected Countries." April. (URL: https://go8.edu.au/sites/default/files/docs/publications/policy_note_-_government_research_funding_in_2014_in_selected_countries_final.pdf).

Jones, Jan. Founder of Oozu. "4 Western founders discuss what it's like to run a startup in Asia." TNW News (2014) (URL: http://thenextweb.com/asia/2014/08/08/4-western-founders-on-what-its-like-to-run-a-startup-in-asia/2/).

Kass-Hout, Dr. Taha (2011). "Social Media Join Toolkit for Hunters of Disease." The New York Times. By BRONWYN GARRITY, JUNE 13, 2011 (URL: http://www.nytimes.com/2011/06/14/health/research/14social.html?_r=0).

Kerlin, Kat (2012). "UC Davis PREDICT program a model for global pandemic prevention." UCDavis, November 29, 2012 (URL: http://news.ucdavis.edu/search/news_detail.lasso?id=10431).

Kojo Nnamdi Show. (2014a). "Hashtag Diplomacy." NPR, July 22. (URL: http://thekojonnamdishow.org/shows/2014-07-22/hashtag-diplomacy/transcript).

Kojo Nnamdi Show. (2014b). "Virus Hunters: Ebola and Tracking Viral Threats." NPR, August 5. (URL: http://thekojonnamdishow.org/shows/2014-08-05/virus-hunters-ebola-and-tracking-viral-threats/transcript).

Manyika, James, Cabral, Armando, Moodley, Lohini, Moraje, Suraj, Yeboah-Amankwah, Safroadu, Chui, Michael, and Jerry Anthonyrajah. (2013). "Lions Go Digital: The Internet's Transformative Potential in Africa." McKinsey Global Institute, November. (URL: http://www.mckinsey.com/insights/high_tech_telecoms_internet/lions_go_digital_the_ internets_transformative_potential_in_africa).

M.I. Hanoi. (2014). "Tech Start-ups in Vietnam: Bird Feeders." *The Economist*, February 20. (URL: http://www.economist.com/blogs/banyan/2014/02/tech-start-ups-vietnam).

Pew Research Centers Global Attitudes Project. (2014). "Emerging Nations Embrace Internet, Mobile Technology." Web. October, 11. (URL: http://www.pewglobal.org/2014/02/13/emerging-nations-embrace-internet-mobile-technology/).

Public Health Watch. (2014). "How a Computer Algorithm Predicted West Africa's Ebola Outbreak Before It Was Announced." *Publichealthwatch*, August 10. (URL: http://publichealthwatch.wordpress.com/2014/08/10/how-a-computer-algorithm-predicted-west-africas-ebola-outbreak-before-it-was-announced/).

Schlanger, Zoe. (2014). "An Algorithm Spotted the Ebola Outbreak Nine Days Before WHO Announced It." *Newsweek*, August 11. (URL: http://www.newsweek.com/algorithm-spotted-ebola-outbreak-9-days-who-announced-it-263875).

Search Engine Land. (2013). "Google Still World's Most Popular Search Engine by Far, but Share of Unique Searchers Dips Slightly." *Search Engine Land*, February 11. (URL: http://searchengineland.com/google-worlds-most-popular-search-engine-148089).

Shu, Catherine (2014). "Indian Startup Little Eye Labs Confirms Its Acquisition By Facebook, Deal Worth $10-$15M." TechCrunch, Jan 7, 2014 (URL: http://techcrunch.com/2014/01/07/little-eye-labs-acquisition/).

TNW. (2014). "4 Western Founders Discuss What It's Like to Run a Startup in Asia." *TNW*. (URL: http://thenextweb.com/asia/2014/08/08/4-western-founders-on-what-its-like-to-run-a-startup-in-asia/2/).

UC Davis. (2012). "UC Davis PREDICT Program a Model for Global Pandemic Prevention." *UC Davis News & Information*, November 29. (URL: http://news.ucdavis.edu/search/news_detail.lasso?id=10431).

Warren, Christina. (2014). "Kicked Off Twitter, Islamic State Flocks to Diaspora." *Mashable*, August 22. (URL: http://mashable.com/2014/08/22/diaspora-islamic -state/).

11
Innovation in Financial Services

Krzysztof Rybinski

Introduction

Two respected magazines, *Fast Company* and *Business Insider*, published lists[1] of the top ten globally most innovative companies in financial services in 2014. They did it almost simultaneously, so both titles had access to the same information universe when preparing these lists. And yet, they are completely different. No single innovator company in financial services appears on both lists. It suggests that this perception – what makes a great, possibly disruptive innovation in the financial services industry – is highly subjective. But those two lists have common features. Almost all the companies have been founded in the United States, one in Israel, one partly in Estonia, and two companies specializing in mobile payments are located in developing countries: M-Pesa in Kenya and Paga in Nigeria. It appears that – at least according to these two magazines – the United States holds almost a monopoly for innovations in financial services.

This chapter will argue that it is a very biased picture and a lot more innovation is happening in the developing world. However, access to such information is more difficult, as these innovators often operate in countries where English is not the first spoken language and they are not gossiped about in tech-savvy cafeterias in Silicon Valley. For example, Boor et al. (2014) studied the origin and types of innovations in financial services offered via mobile phones. They used a complete list of mobile financial services as reported by the GSM Association and collected detailed information about the innovation processes in these services. They found that 85 percent of all innovations originated in developing countries. Moreover, they also showed that at least 50 percent of all mobile financial services were pioneered by users, 45 percent by producers, and 5 percent by both. Finally, three-quarters of these innovations that originated in non-OECD countries had already diffused to OECD countries. Financial services offered via mobile devices are now driving the entire financial innovation universe, so one can rightly conclude that developing countries play a much more important role in this process than shown by rankings put together by people focused on US or Israeli innovation ecosystems. It is also worth noting that user-driven innovation plays a dominant role in mobile financial services. Therefore the innovation process is not developed by foreign investors, usually a large telecom company operating in a developing country, but by users of mobile services, and then it is adopted by mobile operators themselves, foreign and domestic. Petalcorin (2011) gave a very good and early example of such an innovation process. The Philippines span over more than 7000 islands, and access to financial services was poor. So mobile phone users came up with the following scheme. One person would buy a scratch card and send the activation code by SMS to another person, who would use the code to upload the credit onto his or her own phone. This was a simple and innovative way to transfer airtime between two different phone numbers. This method became so popular that Smart, the largest telecom company in the Philippines, launched *PasaLoad*, a dedicated service that allowed electronic airtime transfer between customers. A similar innovation has also been developed in Uganda (Chipchase, 2009).

But in order to understand innovation in financial services one has to spend less time in Silicon Valley and more time in Warsaw or Istanbul. As Wojciech Sobieraj, CEO and founder of Poland's Alior Bank, put it in an interview with the author of this chapter, when Polish or Turkish bankers speak at an innovation conference, the room is full and everybody wants to listen. When British or other developed countries' bankers begin their presentations, people leave to do their e-mails and phone calls. This is also confirmed by recent reports by Forrester Research (Wannemacher and Walker, 2014) and Ensor et al. (2014). Forrester Research analyzed the functionalities of online and mobile banking and published its 2014 rankings. In global mobile banking ranking, the Turkish Garanti Bank came first with 80 points and Polish mBank third with 72 points. The first three places were taken by European banks. The average in this ranking was 61 points in a 0–100 range. In European online banking the Polish mBank came first, with a score of 82 in the 0–100 range, followed by the Turkish Garanti Bank with 71 points. The average was 62, and the global online giant ING scored just 52. In 2012 Alior Bank from Poland received the BAI–Finacle Global Banking Innovation Award in the disruptive solution category. The award was for developing Alion Sync, a new-generation bank that implemented a series of innovations, including fully virtual branches and fully paperless services. In 2013 Alior Bank received three other global awards[2] and in 2014 it was also selected as the best European bank by *Retail Banker International*.

There is a puzzle. Poland is ranked very low in overall innovativeness, and has been deteriorating over the past decade. It was ranked 72nd in the global World Economic Forum innovation ranking in 2014 (see Table 11.1), is among the laggards of innovation in the European Commission's (2014) ranking, and holds 45th position in the Global Innovation Index

Table 11.1   Poland's position in World Economic Forum innovation ranking, 2006 and 2014

	Grade		Ranking*	
	2006	2014	2006	2014
Quality of scientific research institutions	3.8	**3.9**	58	63
Company spending on R&D	3.8	**2.8**	31	98
University–industry collaboration in R&D	3.6	**3.5**	38	73
Government procurement of advanced tech. products (1 = lowest price, 7 = advanced technology)	3.6	**3.2**	76	89
Availability of scientists and engineers	4.2	**4.2**	75	62
Capacity for innovation (1 = firms buy new technology, 7 = firms create new technology)	4.1	**3.8**	30	67
Innovativeness rank overall	3.5	**3.3**	44	72

*Ranking is based on more than 14,000 surveys sent to C-level managers globally; responses are in the range from 1 (lowest innovativeness) to 7 (highest innovativeness).
Source: World Economic Forum.

ranking. And yet its financial institutions are winning global innovation competitions, justified by the supreme overall quality of services and the user experience. This requires special attention, as such a situation is typical of many emerging markets where innovation ecosystems are young, weak, and inefficient, and there are company-specific or financial services-specific factors that make such companies global leaders of innovation, despite all local obstacles.

This chapter aims to show that emerging markets now play a very important role as innovation leaders in the field of financial services. It will provide hard evidence and inspiring stories, so that in the area of financial services, the traditional way of dividing the world into the developed North, that innovates, and the underdeveloped South, that consumes innovation invented by the North,[3] will be proven to be wrong and outdated, at least in the financial services industry.

However, this chapter finds some support for the hypothesis put forward in the literature (Acemoglu et al., 2006) that when innovation happens in the South outside mobile financial services, it is not utilized in industrialized countries as much as the innovation potential would suggest. But we argue that it is not because these innovations have no user or market value in developed countries, it is caused by the lack of a proper innovation ecosystem that facilitates globalization of innovation, including poor access to venture capital. Also large physical and cultural distance continues to be an important barrier, despite advancing economic and financial globalization processes, only briefly interrupted by the Lehman moment.

This chapter is organized as follows: section 2 shows that in the past decade or so developed countries devoted massive resources to financial innovation, but it was heavily concentrated in an area that allowed top financial institutions to make billions of dollars and euros in profits and hurt Main Street at the same time. At the same time, developing countries avoided this trap and delivered a broad range of financial innovations that benefited their clients in terms of better access, higher quality, and higher satisfaction. It also led to a reduction of the financial exclusion problem. This section presents a review of such innovations. Section 3 presents the key results of five case studies of Polish companies that developed and implemented disruptive innovations in financial services, changing the Polish market and expanding internationally. Section 4 presents these case studies, which are based on interviews by the author of this chapter conducted with founders/ the CEOs of these companies. This is the first time such a selection of case studies has been presented. Section 5 concludes the discussion.

Diverging paths of innovation in developed and in emerging markets

As presented in the previous section, there are many companies in the developed world that work on disruptive innovations in financial services. There are almost 1 trillion reasons for this activity. More precisely, pretax profits of the banking sector in 2013 reached an all-time high of 920 billion dollars despite sluggish economic growth and heavy losses faced by some banks that kept delaying full recognition of loan provisions after the Lehman moment (*The Banker*, 2014). People are looking at these enormous profits and rightly concluding that there is no reason why banks in the twenty-first century should enjoy such high profits as the majority of services they deliver can be provided outside the banking sector.

In the following section we will show that massive innovation in the banking sector in developed countries led to this unprecedented rise in profits. And that it is not necessarily a good thing.

Developed countries and innovative weapons of mass financial destruction

In order to understand how a bank-led innovation frenzy in the past two decades translated into massive profits, one should first understand, where does money come from? In the old textbooks the story went as follows: Jim brings his savings, say 100,000 dollars, and deposits them with the bank, which in turn grants a loan to John using Jim's savings. Not all of it, as the bank has to place a small fraction, from 2 percent in the West to as much as 15–20 percent in some developing countries, at the central bank as a so-called mandatory or fractional reserve. But John has 98,000 dollars and pays for his new SUV, and this amount lands in the bank serving the SUV dealer.

This is a new deposit that can be used by that bank to create the new loan ... and the story goes on. In this old and nowadays completely wrong story banks had to collect lots of deposits to create lots of new loans. But as shown by McLeay et al. (2014) in the *Bank of England Quarterly Bulletin*, the correct mechanism of creating money starts from the loan decision, and then the loan automatically translates into someone's deposit. Therefore banks create money not out of deposits, but "out of thin air," and later this "thin air" becomes new deposits. This is the true process of money creation.

Once banks understood this true mechanism of money creation they proceeded as follows. Traditional ways of making loans were limited, because banks were making loans out of thin air and regulators paid close attention to the ratio of bank capital to the risk-weighted loan portfolio. If this ratio fell below levels determined by regulators, because of a surging denominator, regulators became worried, or even angry. And in the terminal case they would replace the bank board with a new one. So banks invented new ways of making loans without abusing the minimum capital requirement, and they did it outside banks' balance sheets. To facilitate this process they hired the best brains, math, and physics and finance PhDs from leading world universities and invented a range of new, extremely complicated products. These products were sold to clients around the globe and the latter believed that a new wave of financial innovation had come, that you could increase an expected return to a level typical for risky assets and be exposed to a very low risk at the same time. This was the new normal. It was supposed to be the ultimate moment when diversification led to massive risk reduction. The Lehman moment showed that, instead of diversification the world has moved toward a massive concentration of risk in large financial institutions, a risk that was not properly measured, that was not hedged, and that led a wave of bankruptcies. It took unprecedented government and central bank action, including nationalization and recapitalization with taxpayers' money, to stop a giant domino effect. The innovation in financial services in developed markets in the past 15 years produced the great global recession in 2009.

Emerging markets and customer-need-driven innovation

The situation was very different in emerging markets. They also suffered recession or growth slowdown, as the global demand collapse affected their exports and led to massive capital flight. But it was not caused by their own financial innovation; these markets were simply infected by developed countries, as financial markets are connected and problems in one country easily transmit to another. Complex derivatives have never become popular in emerging markets; financial institutions were busy making money using traditional products and domestic regulators were against such practices. So innovation was focused on real client needs.

In financial services globalization of innovation is at the same time both easier and more difficult than in other sectors. It is easier, because globalization processes in finance have gone much further than in any other sector. For example, news affecting one market instantly transmits into other markets, within milliseconds, even if these markets are thousands of miles away.[4] On the other hand, the banking sector is heavily regulated, and offering services across borders is subject to increased scrutiny by many regulators, while establishing a local presence is subject to lengthy and sometimes painful licensing processes. It creates a powerful asymmetry. When innovations are brought to the market by nonbanks, they can be developed more quickly and can spread very fast, as regulator approval is not needed. However, when banks come up with innovative ideas, often regulators say no, because they do not fully understand what these innovations would mean for their traditional risk metrics.

These factors were very important in developed markets where almost all disruptive innovations in financial services were delivered by nonbanks, with the infamous, already mentioned, exception of toxic financial derivatives, which were invented by banks. However, in emerging markets banks were smaller, and corporate structures were flatter and more conducive to fostering real innovation, focused on real client needs. So in the South innovation in financial services came from both sides: regulated banks and nonregulated start-ups. And in the South regulators were also smarter than in the North; they did not allow development of toxic financial derivatives but they did let the real product and services innovation take place.

The South has two more advantages over the North when it comes to innovation in financial services. According to the EIU (2012) report, almost all adults in developed countries have bank accounts: 98 percent in Germany, 97 percent in the United Kingdom, 96 percent in Japan, and 88 percent in the United States. The financial needs of citizens in these countries are met by the old, fat, and inflexible banking sector. In developing countries the share of the population serviced by banks is much smaller: 64 percent in China, 56 percent in Brazil, 48 percent in Russia, and only 35 percent in India. So there is room for nonbank companies to provide services for those who do not have enough income to qualify for a bank account. And indeed we have seen many companies from the telecom and retail sectors as well as small payment services firms that have moved aggressively in these markets and provided innovative solutions to millions of underbanked or unbanked citizens in these countries.

The EIU (2012) provided a review of such initiatives in many emerging markets. For example, microfinance has grown massively in Peru, where volumes trebled between 2007 and 2012. But the creation of a credit information infrastructure in Peru allowed good loan standards to be maintained despite the lending boom, with microlenders accounting for 5 percent of total loans. Interestingly, the main player in this market – Mibanco – started as an NGO and still operates under this status. Banks noticed that this market took off and developed their own activities. Crediscotia Financiera, a unit of Scotia Bank, and Banco de Credito del Perú, partly owned by the World Bank, are among such key players. Another trend has developed in India. The country's largest mobile operator Bharti Airtel launched a nationwide mobile-wallet service, Airtel Money, in February 2012. This service allows users to load cash into their mobile devices and spend it paying their utility bills, recharge their mobile accounts, sign insurance policies, invest in mutual funds, and make purchases at retailers, small grocery stores, and pharmacies. In 2012 Airtel was used in more than 7,000 merchant outlets. Airtel was the first mobile operator worldwide to achieve 200 million subscribers in 2014. This service has provided much-needed payment functionalities to the heavily underbanked Indian society, with only 55 percent of the population having access to a bank account and 10 percent having life insurance policies. The Reserve Bank of India (RBI), after a period of reluctance, has in recent years been an active promoter of such initiatives and had granted 17 mobile wallet licenses by 2012. But the RBI prefers banks to offer such services and by 2014 more than 170 banks had been granted licenses to offer mobile payment services. Consequently these payments skyrocketed. According to RBI data, in October 2014 alone customers made mobile transfers worth more than 47 billion rupees, while in October 2012 it stood at 19 billion rupees. Handset sellers also recognized the enormous potential of the Indian mobile payment services market. Nokia, which accounts for 40 percent of the Indian mobile phone market and has more than 200,000 dealers across the country, in December 2011 launched the Nokia Money nationwide service. It allows dealers to act as Nokia Money agents. Clients can load cash onto their mobile phones at Nokia dealers and pay for various services without a bank account. It is independent of any particular mobile network, and targets semiurban and rural areas and below-average-income urban areas. Its network of more than 200,000 dealers dwarfs the network of bank branches that has only 75,000 outlets. Payment services are also developed in partnerships. For example, Vodafone launched its M-Pesa payment and money deposit service in cooperation with local lender the HDFC Bank.

But mobile financial services are not limited to simple payments. In Afghanistan a partnership was formed between Vodafone and Roshan to offer financial services to a society where 70 percent of people are illiterate, only 9 percent of adults have access to bank accounts, and all transactions are done with cash. In Afghanistan, Kenya, and the Philippines people have already started using more sophisticated financial services available on their mobile phones (see Figure 11.1).

The most successful mobile financial services application in emerging markets is M-Pesa in Kenya, operated by Safaricom. In 2013 *The Economist*[5] reported that Safaricom had 19 million subscribers out of 43 million adults living in Kenya and 15 million of them used mobile payment services. About 60 percent of Kenyans have no bank account while 80 percent have mobile

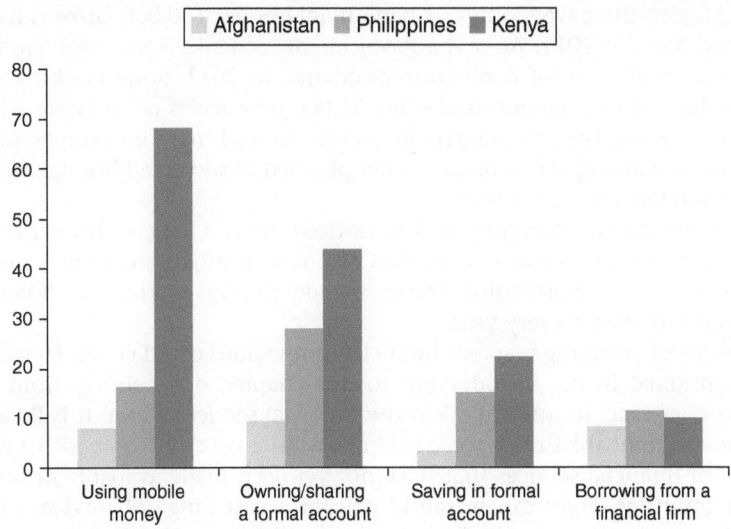

Figure 11.1 Percentage of adults using four types of financial services
Source: Economist Intelligence Unit.

phones. M-Pesa began by enabling remittances from individuals working in cities such as the capital, Nairobi, and Mombasa to family members in rural villages. But it has since morphed into what is nearly a full-service bank branch. In the Philippines more than 8.5 million people use mobile financial services, the second highest number after M-Pesa.

But mobile banking in emerging markets sometimes has a much broader meaning than in the developed world. Banco Bradeso, one of Brazil's top banks, launched its "mobile" service in 2009 by opening bank branches on vessels that deliver goods and passengers along a 1600 km route along the Amazon river. It allowed Banco Brandeso to become the first lender that provides financial services in all 5,564 municipalities in Brazil. These bank boats can open new accounts, accept deposits, pay pensions and benefits, and offer withdrawals.

In emerging markets very unusual businesses have also moved into financial services. For example, in Colombia financial services from various banks are available at more than 3,000 shops owned by Loteria Bartolo, a betting company.

Another big trend in emerging markets was a move by retailers into financial services in Latin America. Chilean retailers began a wide range of banking operations in the late 1990s and since then have moved to Colombia, Peru, and Argentina. The Chilean firm Fagabella operates financial services in more than 150 of its retail outlets under a banking license. The company has more than 2 million cardholders in a country of 17 million citizens.

In Mexico there were only 1.3 bank branches per 100,000 citizens living in rural areas in 2011. After the government recognized this problem and allowed for the use of bank correspondents, in 2011 alone banks signed more than 600 agreements and some 21,000 new access points were added to the network. The scale of expansion was astonishing; for example, BBVA Bancomer had only 2,000 branches but planned to add an additional 20,000 correspondents to its network.

Governments in emerging markets noticed the vast benefits being offered to their citizens by these new models of financial inclusion. So new policy initiatives are mushrooming across the developing world and financial innovations appear every year.

The list of emerging markets' financial innovations could go on. However, as mentioned in the introduction to this chapter, one country from the South does excel in innovation in the financial services sector. It is Poland. Banks and nonbank firms succeeded in providing Polish citizens with a wide range of financial services that have not yet been made available in developed countries. Poles can instantly pay for a vast range of services using mobile phones, instant transfers are easily available, and banks and technology companies are about to provide image recognition capabilities that will transform shopping. And the world's largest Internet bank was created in Poland already in the early 2000s, at a time when Poles had very poor online access. Therefore Poland deserves special attention in our journey of understanding financial innovation in emerging markets. In the next section we will present five financial innovation cases and the results of five interviews with founders and CEOs of these very innovative companies.

How Poland, an overall innovation desert, breeds disruptive innovations in financial services

As shown earlier, Poland is not a very innovative country. And yet it has become a source of many global innovations in financial services, as documented in the minicases below. The author conducted a series of interviews with founders/CEOs of the most innovative financial institutions in order to understand how it was possible: Alior (bank), BZWBK (bank, member of Santander Group), Cinkciarz (Internet foreign exchange and transfers platform), mBank (bank, member of Commerzbank group), and XTB (Internet trading platform).

Several factors were mentioned during these interviews. One factor is common to all emerging markets. Because of a long period of socialism that ended in 1990 Poland had few legacy assets. Because Poland was a latecomer to the Western-style banking universe, few processes were developed and the old-type infrastructure was not established. Therefore bankers in Poland started to invest in modern infrastructure while developed countries had large legacy assets that hindered the innovation process. For example, paper-based banking has never been developed in Poland; checks appeared in Poland only very briefly and were immediately replaced by debit and credit cards. This latecomer effect was used very effectively. The early stage of financial services development was based on credit/debit card technology which stimulated innovation, as these instruments allowed for adding new functionalities and new products and created the possibility to develop a client-centric approach at a time when traditional banking was still bank-centered.

Secondly, Poland had and still has a large pool of highly qualified specialists, especially in the IT area. Polish teams regularly win or take top positions in global IT competitions, including coding, problem solving, and modern marketing.

Thirdly, Polish society traditionally was very entrepreneurial. So there were many young people not satisfied with typical corporate jobs, who wanted to innovate and wanted to start their own companies. It worked poorly in many industries due to large legislative obstacles, but the financial services sector created an innovation-enabling ecosystem, and it worked very well indeed. Interestingly, many innovations were developed in banks that were daughter companies of established foreign banks. It defies a common perception that innovation, research, and development can take place only at corporate headquarters. On the contrary, Polish daughter companies in financial services were much more innovative than their parent institutions.

Corporate structures were much flatter in Poland than in developed countries. Banks introduced various schemes that helped the innovation culture to grow and thrive, including recognition of innovators and various incentive schemes.

Finally, lack of financial resources also contributed to this success. Bank clients in developed countries have large cash reserves, and banks make enough money managing these funds in the traditional way. There was very limited pressure to improve. In Poland, bank clients were relatively poor, and pressure to provide new services, new products, and to cut costs was far greater. Therefore lack of funding, often seen as a major obstacle to innovation, in the Polish case stimulated creativity and innovation. Having too much money is not a good thing; on the contrary, it often makes people lazy and obstructs innovation.

Polish companies that excel in innovation in financial services share certain characteristics that can be grouped into several categories: leadership, vision and strategy, client-centric approach, innovation-friendly culture, flat structure and flexibility, and trigger (a triggering event).

Leadership

All leaders are experts in their fields – they have spent years in the field and understand their business very well. Wojciech Sobieraj of Alior worked for many years at C-level positions in the banking sector. Mateusz Morawiecki also worked for many years in the banking sector before becoming an innovative CEO. Slawomir Lachowski (2012) had a shorter banking history, but he gained experience at the very top – as a member of the Visa International Advisory Board. Marcin Pioro, before founding the Cinkciarz Internet currency exchange and transfers platform, owned a network of real-world currency exchange booths and graduated from the computer science department. XTB's Jakub Zablocki was a salesperson on the trading floor in one of the most sophisticated banks in Poland at that time, BRE Bank. All the leaders are very passionate about their businesses. They work 24/7 and have important goals ahead.

Leadership style differs, but it depends on the stage of the company's development. Cinkciarz is still a family-owned start-up, but with a whopping 3 billion euro-plus of revenues planned in 2014. Its owner and CEO participates personally in all important activities; for example, he invented a new brand that will be used for international expansion, he personally follows the main rivals and scans market trends. He does it often between midnight and 3 a.m. XTB's founder is no longer its CEO, although he remains a key decision-maker. Over time he learned to adjust his management style to the changing stage of his business. In the early stage he was the key source of new ideas, and personally programmed many new functionalities and drafted necessary documentation. Those days made him a true expert in the field. But now he focuses on the creation of a proper, flat structure that allows a culture of innovation and professionalism to remain strong and continue supporting XTB's international growth. He also moved to London where the process of building relationship capital with institutional clients is much more effective than if done out of Warsaw. Wojciech Sobieraj is a natural born leader. Charismatic, he makes great speeches, and inspires and empowers people. And he knows very well that teamwork is very important. Very unusually, he met personally with everyone who was recruited before the formal Alior launch. It may seem impossible to meet several hundred people, not in a big hall, but personally, but it did work. It created a unique atmosphere, and people shared his passion and were highly motivated.

Slawomir Lachowski created a unique team; there was even a "Lachowski team" brand created in Poland. They developed and ran mBank as a single department within the larger BRE bank, where Lachowski served as its CEO. The team has developed a unique set of values that united them, and annual assessments were conducted on the basis of these values, which was an important management innovation. In an interview Lachowski cited four leadership characteristics that make a great leader: passion, openness, courage, and humility.

Mateusz Morawiecki runs a large, well-established organization, BZ WBK Bank, that is part of the international financial group Santander. So his major leadership challenge was to engage people around the mission and vision of BZ WBK. In this endeavor he reached beyond internal stakeholders;

he succeeded in bringing clients and even people unrelated to the bank into the continuous improvement and innovation process.

Vision and strategy

All leaders and organizations had a clear and powerful vision and strategy. Slawomir Lachowski knew that the Internet would revolutionize financial services, and his "Lachowski team" shared this vision. At that time banks were offering Internet access to bank accounts as a premium service, and they charged for it. mBank's idea was to make the Internet a major channel that will allow three basic services: saving, loans, and payments, without needing to visit the bank branch. While today it seems obvious, back in 2000 it was a disruptive and revolutionary vision, especially in Poland which was a country with poor online access for citizens. Nowadays, Mr Lachowski runs another innovative bank, Smart Bank, that offers fully paperless operations and voice recognition.

Mateusz Morawiecki and BZ WBK knew that they could offer a much better and more personalized service to their clients by making better use of the vast amount of data that the bank stored. They had limited capacity to do it internally so they successfully reached out to the academic community. It was a unique achievement in a country that was notorious for very poor cooperation between business and academia. Mr Morawiecki deeply believes that SMEs (small- and medium-sized enterprises) are the backbone of the Polish economy, generating a large part of GDP. In order to help SMEs thrive, the bank succeeded in creating a very efficient ecosystem that channeled funds much more effectively to SMEs than a typical bank, by developing advanced Business Intelligence and Big Data tools.

Wojciech Sobieraj transformed Alior from a bank into a networked technology company that would offer its services globally thorough cooperation with leading companies in various sectors in many countries. They already cooperate with mobile operators and supermarket networks. Cinkciarz's founder plans to become a global foreign exchange player and his goal is to repeat his massive success in Poland in other markets, including the United States. Jakub Zablocki strives to challenge the supremacy of Russia as the provider of an Internet trading platform and now offers his own product. It should also support XTB's transformation from around the 15th largest world broker into one of the world's leaders. All five leaders have long-term goals and all make innovations and client focus the pillars of their long-term strategy.

Client-centric approach

All five companies have a very strong client-centric approach. Alior was basically designed by clients, while mBank and BZ WBK run focus groups to find out what functionalities should be offered. Cinkciarz dwarfed other Internet currency exchange and transfer platforms by opening accounts at all significant banks in Poland, which shortened the time it took to transfer money.

Cinkciarz, XTB, and all three banks focus on user experience; for example, Jakub Zablocki of XTB from time to time tries other trading platforms and makes sure that XTB's user experience is superior to that of his competitors. He said in an interview that XTB is very complex inside, while the user interface is as simple as possible. This client-centric approach translates into very high Net Promoter Scores, in the range of positive 35–50, into low churn, and into large inflows of new clients.

Innovation-friendly culture

All five institutions devote a lot of energy to creating and maintaining an innovation-friendly culture. They constantly watch market developments, consumer preferences, and new frontiers. Alior Bank is part of a network of innovating banks from Korea, Malaysia, Turkey, Brazil, South Africa and, of course, Poland, where members share experience and inspire each other. Alior launches some 10 innovations and 30 improvements every year. It holds regular innovation Mondays where insiders can present their new ideas. Wojciech Sobieraj plans to expand these Monday meetings to the external world, including universities. mBank's culture is strongly supporting of innovation; team members are regularly invited to present their ideas to the board. The BZ WBK board makes sure that the corporate culture is conducive to information flow; they run regular multidepartmental brainstorming sessions and employ quants from academia to stimulate the innovation process. They share ideas within the Santander network of banks, which includes 14 different geographies. BZ WBK also actively supports universities within the Santander Universidades network which leads to many innovations in the CSR space. For example, BZ WBK together with Vistula University developed a new methodology to teach entrepreneurship to high school students, using e-learning, gamification, and real-life projects. The BZ WBK network of branches is involved in the project and more than 20 high schools took part in the pilot launch in 2014.

XTB has had a very innovative culture, but with the number of branches and foreign geographies growing, maintaining this culture remains a challenge that demands managers' attention. It works well as, for example, many innovative ideas are sourced from the Spanish branch to headquarters located in Warsaw. Cinkciarz works on the frontiers of innovation, its platform works on all systems, it offers all types of currency exchange products, and it was even the first currency exchange platform in the world to release the Google glass application. Cinkciarz has a strong team of its own software engineers and all core applications are internally written, so they do not face any limitations and can very quickly add functionalities or modify their systems.

Flat structure and flexibility

Another common characteristic of Polish innovators in financial services is flat structure and flexibility. Cinkciarz is the 30th largest Polish firm by revenue but still remains a family start-up, with a very strong role played by its founder. Decisions are taken quickly and effectively. XTB grew out of the start-up stage a few years ago, and its founder moved to London to expand business there and delegated the CEO role to a Czech citizen, but the XTB structure remains flat and the decision process efficient. Alior Bank cultivates a project culture; many projects are run every year, and they have names, the most successful one being called Kill Bill. Alior Bank's middle-ware system is very flexible: it allows for adding new services from any exter-nal provider, while maintaining strict safety. It works so well that when Alior took over another bank, nobody noticed when they plugged in the acquired bank to their middleware. mBank for the entire period of eight years while Mr Lachowski was in the leading role was simply one department within the larger BRE bank. The structure was as flat as it gets. But it was so power-ful that today the entire bank, known as the BRE bank, has been renamed mBank. At BZ WBK, which is a large bank by Polish standards, there is a very strong and agile project culture and silos are eliminated by frequent interdisciplinary brainstorming sessions.

Trigger

In every analyzed case of major financial innovation there has been a trigger, an event or events that inspired the company founder or CEO. Cinkciarz's founder took out a mortgage denominated in Swiss francs. When he came to a bank branch to pay the first installment he noticed that the bank had applied a very large currency spread, much larger than at the net-work of currency exchange booths owned by Mr Pioro. It inspired him; if someone offered clients with Swiss franc mortgages currency exchange at a much smaller spread it would make a sound business case. It was possible by matching buyers and sellers in a dark pool and by aggregating unmatched orders. XTB's founder heard complaints from his clients at BRE bank that they were not able to trade 24/7, and that spreads were too wide. He knew there was no reason to stop trading at 17:00 Warsaw time, when US markets are still open. So he launched an Internet-based trading platform. Wojciech Sobieraj after having worked at C-level positions in the banking sector for a number of years felt that banks did not offer quality services to clients. He and his two colleagues did a detailed analysis and estimated that with 400 million euro they could launch a bank that would offer premium ser-vices to Main Street. It took 18 months and a lot of persistence but they found a proper investor. Each time Slawomir Lachowski came back from Silicon Valley in his capacity as a VISA International Advisory Board member he knew that the Internet would change the banking landscape completely. And he wanted to be in that game. One day he decided to go for it and 100 days later mBank was launched. Finally, BZ WBK's Mateusz Morawiecki for many years observed that the backbone of the Polish economy – SMEs – faced tough credit constraints. He knew that it was an information asymmetry problem and banks had a lot of stored data that could be used to reduce this problem, but they did not use it. And he launched a project that allowed better credit access for Polish SMEs. It was the first Big Data project of its type, long before it became a buzzword.

The case of BZ WBK

A number of Polish banks have implemented disruptive innovations developed in Poland. These disruptions won a huge market share, changed the market structure and behavior of millions of clients, and are now expanding in many other countries. The most notable examples are Alior, Cinkciarzl.pl, mBank, XTB, and BZ WBK. This last bank is used as an illustration.

BZ WBK, a member of the Santander group

BZ WBK is the third-largest Polish bank by assets (see Figure 11.2) with 6 million clients, including 3.5 million mobile banking clients. It is also a member of the global Santander group. Despite being a large and complex organization in one of the most regulated sectors it is highly innovative, which has been recognized internationally. This section, which is based on an interview with BZ WBK CEO, Mateusz Morawiecki, explains why and how this was achieved. There were two moments in the bank's history that rapidly accelerated the innovation process, which in turn translated into

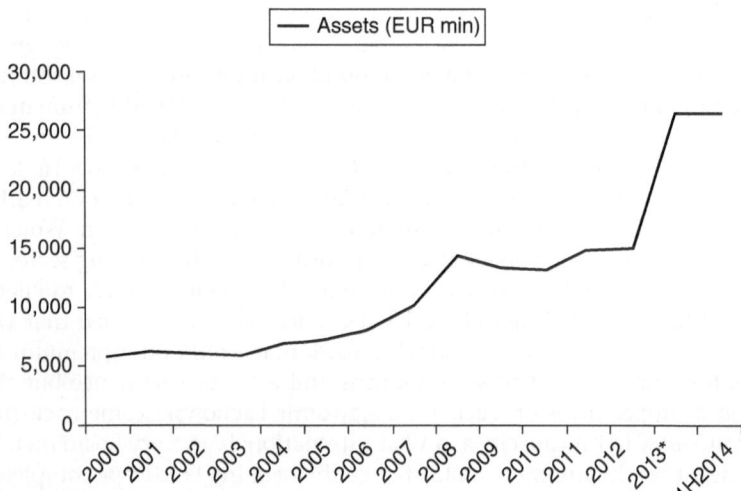

Figure 11.2 Assets of BZ WBK bank
Note: *In 2013 BZ WBK took over another bank in Poland.
Source: BZ WBK bank.

strong business results: starting the Big Data project and putting clients at the center of everything.

After Mr Morawiecki became CEO in 2006, the bank decided that BZ WBK should make much better use of its client and transaction data to do well what banks are expected to do: analyze credit risk. They hired math, physics, econometrics, and statistics PhDs in their bank computing center in Wroclaw and started crunching Big Data sets, after anonymization of the data to make it compliant with data privacy regulations. They started well before the Big Data buzzword hit the headlines. The outcome of this first ever banking Big Data exercise was beyond expectations. The effectiveness of marketing campaigns increased markedly, various conversion and loyalty indicators improved, and data engineers started a very close and fruitful cooperation with business lines. They even started Big Data social network analysis well before Facebook became known in Poland. For example, finding payment regularities between bank clients and their clients allowed the bank to offer target products to firms that were not bank customers, resulting in a credit risk well below the typical level seen in new bank–client relationships. Since then thousands of product campaigns have been carried out; each campaign's effectiveness is measured using Big Data tools and this process is producing superior results.

The second milestone was the exercise run by BZ WBK a few years ago. Bank management decided to put clients at the center of all bank processes, and asked its 12,000 staff to reach out to clients and ask them what functionalities, what products, they would welcome. There were two types of common answers: clients wanted mobile services and integration of banking with other activities, especially e-commerce. And the bank decided to act. Within a few years it launched a very advanced mobile application that runs on all platforms and offers an easy way to pay for a wide range of services: from parking, shopping, public transport (agreements were signed with more than 130 cities in Poland), and buying cinema tickets to flower service. The Santander group runs regular benchmarking exercises for all 14 geographies where they are present, and BZ WBK consistently comes first in terms of reach and quality of mobile banking. Figure 11.3 illustrates the evolution over time. Polish solutions are now implemented in other geographies where the Santander group has a presence.

BZ WBK is also aware that there are more innovators outside the bank than inside it. So they launched a crowd-sourcing website called the "Bank of ideas" where people can submit their ideas. Tens of thousands of such ideas were submitted and a dedicated team analyzes all of them. More than 600 were implemented, leading to improved service and lower costs. The most promising innovators were invited to join the bank, adding to the bank's innovation culture.

These powerful innovation processes were facilitated by reducing silo culture, which is so strong in a typical bank. IT staff work closely with business

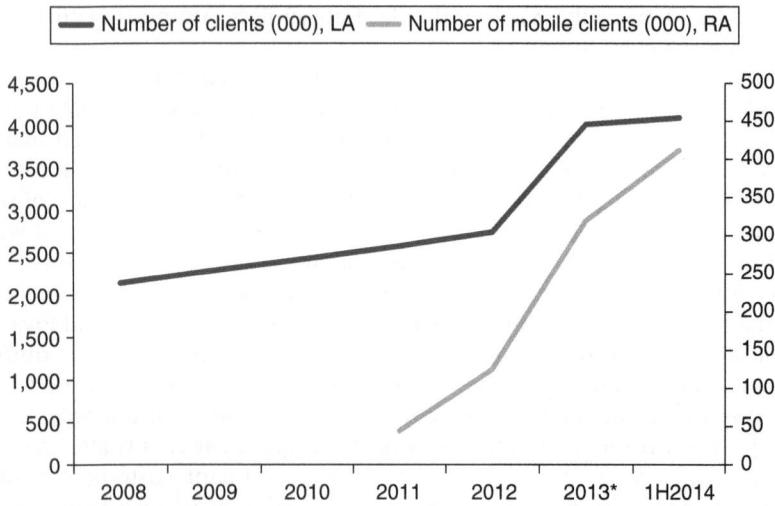

Figure 11.3 Number of BZ WBK clients and number of mobile clients
Note: *In 2013 BZ WBK took over another bank in Poland.
Source: BZ WBK.

lines, interdisciplinary brainstorming sessions are held regularly, and new ideas are submitted to the board without the delays that are typical of other organizations that have more hierarchical structures.

Finally, BZ WBK works closely with Polish tech companies. For example, mobile banking was developed by 100 BZ WBK IT engineers together with the Polish technology firm eLeader that added another 100-plus engineers to this innovation process. The bank also cooperates with more than ten technology start-ups, many of them Polish university spin-offs. Again this is an excellent example of strong cooperation between business and research institutes in a country that faces large problems in making this interface work and deliver results. It is a lesson for other emerging markets that is worth studying.

Conclusions

In this chapter we have shown that many disruptive innovations in financial services are created and implemented in emerging markets. While the developed North put enormous resources into the production of innovative products called toxic derivatives, the developing South put client needs at the center of the innovation process and delivered better service and better access, reducing the financial exclusion problem and contributing to economic and social development. Many innovations in financial services that originated in the South were later adopted by the West, although this rate of adoption was not very high due to the large cultural and perceptual distance and amid the still poor quality of innovation ecosystems in emerging markets that makes the international scaling of business models difficult.

We showed that Poland is a very interesting case on the financial services innovation landscape. It is a country that scores very poorly in the overall innovation rankings and yet it is home to many very innovative financial institutions, and these disruptive innovations not only massively changed the Polish market, but also expanded to others. The Polish case should be carefully analyzed by other countries which also have weak innovation ecosystems and exhibit poor cooperation between business and universities. Even with such barriers in place it is possible to breed local innovative companies that deliver disruptive innovation on a global or regional scale. Poland has proved it.

Notes

1. Miller (2014) listed in her *Fast Company* article the following companies: Nice Systems (Israel), Square (USA), Bitcoin (unknown), GiveDirectly (USA), Dwolla (USA), TransferWise (USA, Estonia), OneID (USA), MasterCard (USA), Estimize (USA), and eToro (USA, and Cyprus for tax reasons). Wile (2014) has the following list of top ten in his *Business Insider* article: Paga (Nigeria), M-Pesa (Kenya), InVeture (USA), Venmo (USA), Splitwise (USA), Cover (USA), Simple (USA), Ripple (USA), Stripe (USA), and Balanced (USA).
2. These were: Retail Banker International Award in the category Best Online Banking Strategy, in May 2013, BAI–Finacle Global Banking Innovation Award in the category Credit Online Process, and Best Use of IT in Retail Banking in the Banking Technology Awards.
3. See Krugman (1979), Coe and Helpman (1995), Flam and Helpman (1987), and Grossman and Helpman (1991).
4. A flash-crash on the NASDAQ led to an index drop by almost 10 percent within less than an hour. The Tokyo Stock Exchange was closed but futures contracts on this index instantly fell by more than 10 percent.
5. See http://www.economist.com/news/finance-and-economics/21574520-safaricom-widens-its-banking-services-payments-savings-and-loans-it

References

Acemoglu, D., Aghion, P., and Zilibotti, F. (2006). "Distance to Frontier, Selection, and Economic Growth." *Journal of the European Economic Association*. Retrieved from http://economics.mit.edu/files/4472, 4(1), p. 37–74.

The Banker. (2014). "The Top 1000 World Banks." *The Banker*. Retrieved from http://www.thebanker.com/Top-1000-World-Banks (accessed October 12, 2014).

Boor, P., Oliveira, P., and Veloso, F. (2014). "Users as Innovators in Developing Countries: The Global Sources of Innovation and Diffusion in Mobile Banking Services." *Research Policy*, 43 (9): 1594–607.

Chipchase, J. (2009). "Mobile Phone Practices and the Design of Mobile Money Services for Emerging Markets." Retrieved from http://www.ict.kth.se/courses/

ID2216/wiki/lib/exe/fetch.php?media=janchipchasedesigningmobilemoneyservice svfinal-091108204720-phpapp01.pdf (accessed October 12, 2014).

Coe, D. T. and Helpman, E. (1995). "International R&D Spillovers." *European Economic Review*, 39: 859–87.

The *Economist*. (2013). "Is It a Phone, Is It a Bank?" *The Economist*. Retrieved from http://goo.gl/tMtEK

EIU. (2012). "Beyond Branches. Innovations in Emerging Market Banking." Economist Intelligence Unit. Retrieved from http://goo.gl/R96DDp

Ensor, B., L'Hostis, A. with Berdak, O., Walker, S. and Causey, A. (2014). "European Mobile Banking Functionality Benchmark." Forrester Research. Retrieved from http://goo.gl/CIc6U3 (accessed October 28, 2014).

European Commission. (2014). "Innovation Union Scoreboard 2014." European Commission. Retrieved from http://ec.europa.eu/enterprise/policies/innovation/policy/innovation-scoreboard/index_en.htm

Flam, H. and Helpman, E. (1987). "Vertical Product Differentiation and North–South Trade." *The American Economic Review*, 77: 810–22.

Grossman, G. and Helpman, E. (1991). "Quality Ladders in the Theory of Growth." *The Review of Economic Studies*, 58: 43–61.

Krugman, P. (1979). "A Model of Innovation, Technology Transfer, and the World Distribution of Income." *The Journal of Political Economy*, 87 (2): 253–66.

Lachowski, S. (2012). *Disruptive Innovation in Banking. A Business-Case in Low-Cost Finance*. Warsaw: Studio Emka.

McLeay, M., Radia, A., and Thomas, R. (2014). "Money Creation in the Modern Economy." Bank of England. Retrieved from http://www.bankofengland.co.uk/publications/Pages/quarterlybulletin/2014/qb14q1.aspx

Miller, N. (2014). "The World Top 10 Most Innovative Companies in Finance." *Fast Company*. Retrieved from http://www.fastcompany.com/most-innovative-companies/2014/industry/finance

Petalcorin, A. (2011). Personal Communication.

Wannemacher, P. and Walker, S. (2014). "2014 Global Mobile Banking Functionality Benchmark." Forrester Research. Retrieved from http://goo.gl/EH6zws

Wile, R. (2014). "10 Financial Innovations that Are Changing the World more than Bitcoin." *Business Insider*. Retrieved from http://www.businessinsider.com/10-financial-innovations-more-exciting-than-bitcoin-2014-1

12
Education 3.0: Facing the Challenge of Human Capital Building in Emerging Economies

Gabriel Sanchez Zinny

The face of the educational challenge

Educational systems in a large number of countries are facing a critical state of affairs in many aspects. Poor-quality schools, persistent social inequality, and local administrations that struggle to stretch their funds in the most efficient way possible mean that large chunks of our student-age population receive insufficient education – if not drop out.

In the face of these issues, for a few years now, a new generation of reformers has realized that without serious changes, students will be tragically unprepared for the demands of the global economy in the twenty-first century. These changes and challenges need to harness the entrepreneurial energy of the nongovernmental sector together with the promise of new technologies, to deliver high-quality education to traditionally underserved populations.

The fact that students are failing to gain the skills that lead to success in the highly competitive, dynamic global market is a problem for the individuals but it also has long-term consequ ences for national growth and competitiveness. Education is, more than ever, central to national prosperity, impacting productivity, wages, and GDP growth.

The OECD, which includes the world's most advanced economies, concluded in its 2012 evaluation of education in the world's most advanced economies that more than half of the GDP growth of its member countries could be traced to the income growth of those with higher education.[1]

Moreover, across the OECD, those with more education are not only earning increased wages but are more consistently employed. From 2008 to 2011, those with a postsecondary degree saw their unemployment number rise[2] by 1.5 percent, while that number increased by 3.8 percent for those with a lower degree or no degree. The latter faced unemployment[3] of 16.2 percent in 2011, almost tripling that of workers with college degrees or higher (4.9 percent). According to expert Rebecca Strauss at the Council on Foreign Relations (CFR):

> Human capital is perhaps the single most important long-term driver of an economy. (…) It is an economist's rule that an increase of one year in a country's average schooling level corresponds to an increase of 3 to 4 percent in long-term economic growth. Most of the value added in the modern global economy is now knowledge based.[4]

In many societies, especially in emerging markets, this dynamic may reproduce sharp inequalities of outcomes between those who are able to access quality education, and those who are not. Reasons are complex and interconnected – including poverty, family dysfunction, or failing neighborhood schools. Ethnic and class effects on educational attainment also add to this equation, while low enrollment in early childhood programs and kindergarten is determinant for the trajectory at later educational levels.

Emerging economies have had one of their best decades in memory due to persistently high commodity prices. By 2010, the Latin American growth was booming at an average of 5.9 percent and it cooled only slightly to 4.3 percent in 2011.[5] But as the current boom fades, these countries are realizing that the higher value added, diversified, innovation and technology-oriented economies that create sustained growth, rising wages, and increased productivity will require a revolution in human capital.

For the past several decades, policy makers in these regions have focused on expanding access to education and obtained some good results. According to the *Millennium Development Goals Report* of 2012,[6] "many more of the world's children are enrolled in school at the primary level, especially since 2000. Girls have benefited the most. [In addition,] Many countries facing the greatest challenges have made significant progress towards universal primary education." In the developing regions the net enrollment rate for children of primary-school age rose from 82 to 90 percent between 1999 and 2010.

In Latin America, for example, according to the Inter-American Development Bank (IDB) school access for four to five-year-olds increased from 36 percent in 1990 to more than 60 percent at the present day – including girls, making the region one of the most gender-equal ones of the world in terms of early access to school.[7] And over the past two decades, secondary-school completion rates have climbed from 30 percent to nearly 50 percent.

But these achievements have not changed the fundamental underperformance of education systems, highlighting the need to focus on quality. In the last PISA report, more than half of participant countries fell below the OECD average rating. Indonesia, Turkey, and Hungary[8] ranked far below it, the first with a more than 115-points breach in math, the second some 50 points below, and the last with a 15 points breach. Poland came close but managed to outpace the average by some 24 points, placing itself among the most developed economies. None of the Latin American countries reached the OECD average. Chile, the highest performer, barely scored 450 – 44 points below the average. Uruguay and Mexico averaged 67 and 68 points below, while Brazil's results were down 81 points. The rest of Latin America all score below 400.

A key challenge is that even where access has expanded, dropout rates remain frustratingly high. Nearly 40 percent of students in Latin America drop out, similar to China, where only up to 65 percent of high school students graduate.[9] The numbers are worse in countries like Turkey, where high school graduation rates reach only 45 percent. Others are dealing with significantly low enrollment rates. This is the case in India or Indonesia where according to World Bank data barely 30 percent of their population attend high school.

In part, this phenomenon can be traced to poverty,[10] which forces many youth to leave school and begin working in order to support themselves and their families. According to the Latin American Educational Tendencies Information System (SITEAL in Spanish), an organization affiliated with UNESCO and the Organization of Ibero-American States (OEI), such economic considerations explain 23 percent of student dropouts.[11]

Inequality also strikes hard. UNESCO found, for example, that in Brazil, Mexico, and Uruguay, the children of parents in the highest income quartile achieved results between 25 and 30 percent higher than children growing up in the lowest quartile.[12] And according to the IDB, the difference between the high school graduation rates of the poorest and the richest is often nearly 50 percent.[13]

But an even bigger reason ties directly into the poor quality of the schools. Surprisingly, the biggest factor is not economic at all: SITEAL found that some 40 percent of students drop out because they do not feel that what they are studying is useful or applicable to their future careers.

The last decade of growth resulted in the creation of a middle class which has started to demand better education in some of these emerging countries. Even in Africa, where the middle class's expansion has been less robust, it has nonetheless been noticeable and contributed to increased domestic consumption.[14]

According to an EY report,[15] over the next two decades the middle class is expected to expand by another 3 billion, coming almost exclusively from the emerging world. China's and India's contributions will be substantial. In Asia alone, 525 million people can already count themselves middle class – more than the European Union's total population.

The World Bank,[16] in time, found that the middle classes grew by a full 50 percent over the past decade in Latin America – led by Brazil, Colombia, Chile, Peru, and Costa Rica; 152 million people are now middle class, 30 percent of the region's total population.

This expansion will be difficult to sustain while sharp differences in quality between schools persist. Emerging markets must find a way to leverage the political and economic aspirations of its new middle classes to push policy makers to deliver real progress in improving education for all.

Harnessing the school-to-work transition

The low quality of education and the inequality of access are linked to a third phenomenon: a major disconnect between what is taught in classrooms and the skills demanded in the labor force.

Global competition has increased dramatically. It is no longer enough that only a small portion of the population is able to innovate and create new goods and services. It is vital to scale that capacity to society as a whole.

The OECD's *Skills Outlook*[17] report provides evidence that, over the past four decades, the decline in employment in the manufacturing sector (except for the high-technology industry, which continued to grow) has been offset by growth in the service sector, especially those requiring the highest levels of skills. These services are based on the analysis and transformation of information and, as such, are highly dependent on computers and ICTs. According to the study, in over half of all OECD countries, at least one-third of economic activity is concentrated in high-tech manufacturing, communications, finance, real estate, and insurance.[18]

One of the drivers of this change is linked to the fact that the growing middle classes are bringing in new buying habits and expanding nontraditional markets, making innovation central to creating new products and services and, by doing so, new jobs. The global consultancy PWC pointed out back in 2012 that improving the effectiveness of innovation in a disruptive market was a major strategic priority for organizations, with three out of four CEOs planning to upgrade their R&D capacities.

In addition, Deloitte found in its *Human Capital Trends 2012* report that two of the top three things investors are looking for are dynamic emerging economies and an educated workforce that can sustain high-value growth.

However, nearly two-thirds (63 percent) of the world's CEOs[19] were revealed in a survey conducted by PWC to be concerned that they will not be able to find people with the skills to fill the required new positions.

These challenges are particularly acute for emerging markets' multinational companies – like the so-called "multilatinas,"[20] based in Latin America. Multilatinas are now responsible for 2.1 million employees in the region and approximately $780 billion in annual revenue.[21] They, however, opt to invest most of their money elsewhere in the world, increasing foreign direct investment outflows[22] mostly because of the difficulty in finding the requisite human capital to hire.

The IDB revealed that the quality of the most important skills required for the modern workforce has been steadily decreasing.[23] While employers value what are known as "socioemotional skills," related to personality traits like punctuality, politeness, work ethic, responsibility, empathy, and adaptability, a full 80 percent cited these as the "most difficult to find" among candidates.

McKinsey reported that in the world "half of youth are not sure that their postsecondary education has improved their chances of finding a job, and almost 40 percent of employers say a lack of skills is the main reason for entry level vacancies."[24]

The common denominator in nearly all cases is an asymmetry in information between the education system and the labor market. At least 46 percent of students who enter college do not finish within six years.[25] Among those who do, college does not ensure the same quality jobs as it did before, or the solid middle-class income that it once did. McKinsey reports that "nearly half of college grads say they are in a job that doesn't require a four year degree, and four in ten graduates of the nation's top 100 colleges are unable to get jobs in their chosen field."[26]

Many emerging markets are still largely unprepared to compete in this context. In the 2013 *Doing Business* report,[27] among Asian economies, China was one of the best ranked with Taiwan filling second place, but other subregional areas come in only 90th. Indonesia, in time, ranked 120th and India, 134th. Moreover, on the other hand – Chile, Peru, Colombia, and Mexico – only Chile scored in the top 40 worldwide, coming in at 37. Venezuela brings up the rear at 180th, Argentina is 124th, Brazil comes in at 130th, and Mexico is 48th. Europe's Hungary and Poland rank 54th and 55th, while Turkey comes in at 71st.

But the most dramatic sign of how unprepared many of these economies are to compete with the rest of the world – and perhaps the most tragic manifestation of the education system's failings – is the rising share of youth who fall into the "NEET" category, short for youth "not in employment, education, or training."

The OECD estimates that over 16 percent of its member countries' population aged 15–24 fall into the NEET category or, in Spanish, "ni-nis" (those that *ni estudia ni trabaja*). The problem became global after the last economic crisis, with the highest rates reached in 2008, nearing 41 percent.[28]

The situation is worse, on average, in countries with low education levels and among immigrants, compared with the well-educated. One example is Mexico, with the proportion reaching 22 percent of 15–24-year-olds. In Brazil, that number is 20 percent, and reaches 27 percent for women. Turkey,[29] in turn, has the highest proportion of NEETs in the OECD with 35 percent, with the ratio being twice as high among women as in men. By 2011, 16 percent of tertiary-educated young Hungarians[30] were NEETs.

Along with NEETs, an investigation led by the pan-American magazine *Americas Quarterly*[31] found that some 60 percent of young workers are stuck in informal jobs with low productivity, low income, and precarious working conditions, a phenomenon that also threatens social mobility and economic advancement, dragging down the entire economy's trajectory with a downward spiral effect.

A sign that employers recognize the crisis of human capital – and are starting to take proactive steps to address it – is the increase in philanthropic private spending on educational programs. In a Brookings Institution[32] study coauthored with education expert Justin W. van Fleet, we found that firms are indeed taking matters into their own hands in the absence of robust government reform efforts – but there are still serious holes in their approach.

The largest firms are making investments of up to between \$224 and 569 million annually; 46 percent of those investments exceeded three years in duration. But much of this money is spent with little coordination or collaboration with government organizations: a full one-half of corporations reported not coordinating contributions with anyone outside of the company.

With many factors at work, addressing the skills gap will require a multifaceted approach. The war for talent is only going to heat up, and it is time for our business and policy leaders to catch up.

Introducing "Education 3.0"

As we have seen, the current world economy is defined by a paradigmatic shift from growth based on mass production to growth through innovation fueled by knowledge creation. This shift has had profound effects for education systems that find themselves unprepared to produce graduates with the necessary technological skills and the problem-solving, creative mentalities that the new economy demands. It is in this context that the presence of ICT in classrooms and curriculums offers such promise.

New players and models have suggested several paths forward in the struggle to reconnect school and work. The most promising examples all include expanding options through the involvement of the private sector, both nonprofit and for-profit, and leveraging the potential of technology to revolutionize the way in which students learn.

This new era in education has been referred to as "Education 3.0" because of the way that it has seen the mobilization of numerous stakeholders – from parents to business to advocacy groups and not from any government policy – to work towards better education.

The concept itself was first introduced by the researchers at the Innovation Unit of the Global Education Leaders' Program (GELP).[33] At heart, the message is simple: leaders, institutions, and citizens need to acknowledge that the learning arrangements that have produced results in the past no longer work for the vast majority of young people, and it is time for change.

Just like the original Industrial Revolution required a total overhaul of education systems in order to train a new generation of workers in unprecedented industries, the technological revolution requires preparing our children for a global workforce based on new types of skills which are critical to being successful in the coming decades.

Today, as, according to *The Economist*, "firms are constantly experimenting with new technologies and production processes,"[34] jobs requiring high-tech skills – and thus, a good education – provide increasingly high returns, while wages for low-skill jobs are stagnating or falling.[35]

Bringing ICT into the education system can mean many different things: expanding teacher access to students in remote areas via video links, training teachers to use online resources to upgrade their lesson plans, or providing students with their own low-cost computers.

Private and for-profit groups have led the way by taking advantage of the Internet's ability to reach millions of students at low cost, as well as its potential to expand the interactive nature of the modern classroom. They can serve as complements to existing brick-and-mortar schools, especially in low-budget or otherwise underresourced areas. Online software can also be a force multiplier for overburdened teachers, allowing them to streamline their curriculums and their evaluation mechanisms. It also provides relatively user-friendly means of tailoring coursework to individual students, allowing them to learn at their own pace and according to their own aptitude, which has been shown to strongly improve educational outcomes.

These advances can represent a substantial improvement in educational options, especially for disadvantaged groups who generally have access only to poorer quality public schools.

This way, a critical factor for middle-income countries is that by connecting their students to the broader world with these technologies, the positive spillover effects reach beyond the educational system and go to the heart of social inequality and exclusion. This is because connecting students to online technologies often means connecting families and entire communities that had previously been isolated and marginalized.

Mobilizing the private sector

Changes imposed by the new digital age are not just happening in the so-called developed world. "Today, people with connected smartphones or tablets anywhere in the world have access to many (if not most) of the same communication resources and information that we do while sitting in our offices at MIT ... In short, they can be full contributors to the world of innovation and knowledge creation," argue MIT scholars Mark Brynjolfsson and Andres McAffee in their recent book *The Second Machine Age*.[36]

Moreover, changes to the broader system are coming from below, rather than being imposed in a top-down process, as students, families, communities, and entrepreneurs begin to create their own learning models. They are "voting with their feet" and choosing alternatives that guarantee educational quality – from homeschooling, and blended learning academies, to other private or independent options.

Technology and blended learning tools are driving a process of customization or tailoring of education for the needs of individual students. Simultaneously, it is creating more data-driven analytics, which allow for real-time calibration of methods based on a large-scale statistical understanding of what is working and what is not.

Altogether, these changes are causing an unprecedented and generalized disruption of the education status quo. Not only is it empowering parents, students, and teachers to take more control over outcomes, it is introducing new players into the system to become education providers. And these disruptions are taking place on every level, from pre-K to university to the remedial school sector.

The success of private education models holds particular promise as new education technologies have the potential to dramatically alter children's educational experiences while at the same time lowering costs – a combination that is a boon for underfunded, understaffed, or otherwise underperforming schools and districts. The private sector is often the best positioned to innovate, apply, and scale up these technological solutions. In Latin America, while there is strong variation between and within countries, an average of 19 percent of students attend private schools, according to SITEAL, rising to a full 26 percent in the case of high school.[37]

The involvement of the for-profit sector is also a paradigm shift in education. Indeed, venture capital is migrating to start-ups and entrepreneurs in the for-profit education sector. This capital is largely allocated by seed incubators, which channel investor funding into new companies through a competitive application process. Start-ups chosen through this process are given seed money as well as mentorship and networking support, in exchange for a share of equity.

The number of such start-up accelerators has grown exponentially over the past half-decade. While in the US in 2007 there were just four, according to technology news leader *Techcrunch*, there are now over 100 throughout the country.[38]

In addition, according to *Inc. Magazine*, the share of venture capital invested in education companies was only 1 percent as recently as 2011 – compared with 38 percent in technology companies and 19 percent in health-care firms.[39] But this is changing rapidly: US private investment in ed-tech companies reached $1.1 billion in 2012, compared with just $52 million in 2005.[40] Leaders in this burgeoning space include Imagine K12, an ed-tech start-up incubator led by cofounder and partner Tim Brady, and the New Schools Venture Fund (NSVF), led by CEO Ted Mitchell and which has been raising venture funds to invest in education entrepreneurs for 15 years.

Philanthropy is another way that private actors are channeling their energy and wealth into seeking immediate improvement in education. Fortune 500 companies have contributed nearly $500 million to education projects in developing countries.[41] Individual leaders are also putting their personal wealth on the line for this goal. *Forbes* reminds us that America's richest men and women are giving more to education-related causes than to any other issue.[42]

The work of social entrepreneurs who are out "in the field" experimenting, implementing, and always seeking to upgrade their products or systems to deliver better results, is another path to innovation. Social entrepreneurs combine the trial-and-error attitude of the business start-up world with the social development goals of the nonprofit and NGO sector.

Brazil is serving as an example for education tech entrepreneurs.[43] Perhaps the most compelling example is Descomplica, an online platform that serves as a "full service online classroom," with a wide range of study guides and instructional videos. But there are other examples such as Duolingo,[44] a language teaching app founded by the Guatemalan Luis von Ahn; or Open English, founded by the Venezuelan Andres Moreno.[45]

In the higher education field, apart from the major groups like DeVry, Pearson, or Laureate,[46] a new wave of entrepreneurs is also getting involved at the ground floor of education. David Stofenmacher founded UTEL, an online university in Mexico with more than 6,000 students;[47] and Dionisio Garza promoted the expansion of Regiomontana University. Julio Noriega, of Peru, is leading investors into the technical and vocational sector with Vigenta Educacion. And in Brazil, Carlos Souza has replicated Veduca's Massive Open Online Course (MOOC) online course model, a platform with 3 million visits and $1.3 million in outside investment.[48]

As a result, in the latest survey of the Latin American Venture Capital Association (LAVCA), which profiles 105 of them from around the world, education appears as the third most attractive sector, after only consumer/retail and financial services.[49]

Several trends have contributed to making the education market attractive for investors. First, the increasing demand for access and improved education quality. Second, the growth of the middle classes with the necessary resources to consume higher-quality education in the struggle to move up the economic ladder. And finally, the education markets realizing the need for better and more efficient management.

These realities are combining with deep changes in school systems in developing areas: in particular the explosive growth of classroom technology and the drive towards personalized and blended learning alternatives.

As demand for quality education to compete in a globalized economy will continue to rise throughout the world, tighter public budget constraints and frustration over poor services, especially for low-income communities, will push for increased public–private dialogue and partnership. These are especially important given the fundamental role governments play in helping to scale experimental investment models to reach as many people as possible. Governments in general do not have a comparative advantage in innovation, but they can help expand good ideas, while the private sector social investors can help them fill their capacity gap.

Mexico with the INADEM institute, Chile with CORFO, or Colombia with its InnPulsa program, are only a few examples of governments that have already jumped into efforts to incentivize social entrepreneurship.

Multilateral organizations are another major actor, as they can cross borders and marshal both funding and expertise in the service of social goals. The IDB Multilateral Investment Fund (MIF) is a perfect example, being the largest provider of technical assistance to the private sector in Latin America. Since 1993, the MIF has launched over 1,700 projects with funds totaling nearly $2 billion, and its institutional support for social entrepreneurship has included the creation of LAVCA, Brazil's governmental venture capital promotion agency INOVAR, and the promotion of global industry best practices and proinvestment regulatory reform throughout the region.[50]

Countries are also pouring resources through development aid. The 2012 Index of Global Philanthropy showed the US at the front with an Official Development Assistance (ODA) budget of over $30 billion yearly,[51] nearly tripling that of the UK, Germany, France, and Japan, the next largest donors.

In addition, in 2006 foreign assistance from emerging donors reached nearly 10 percent of total global aid flows, but by 2011 it rose as high as 30 percent.[52] Brazil and India's foreign aid spending grew by more than 20 percent between 2005 and 2010, while China and South Africa's increased by 10 percent.[53]

The implications of new technologies for education quality

One key aspect of technology's role in Education 3.0 is the potential it has for the personalization of education. Digital tools that can be integrated with traditional classroom structures – known as blended learning – bring the computing power and immediate responsiveness necessary to tailor lessons and evaluations to each student's specific needs, goals, and style of learning.

Blended learning technologies can help educators better focus their efforts and enhance their capabilities. With options such as interactive learning software, or programs that provide real-time tracking and feedback on individual student performance, the school's access to quality teaching can shift dramatically in a positive direction.

At the same time, the responsive and interactive nature of these tools brings its own benefits. The personally tailored nature of content drives new enthusiasm for learning, and the capacity of education technologies to provide students with immediate feedback targets an individual's strengths and weaknesses. This is because online exercises and testing allow students to receive real-time insight into how well they did and where they need to focus their energies. We may be, in other words, seeing the beginning of the end of "one-size-fits-all education."[54]

Data storage and analysis of capabilities made possible by digital technologies build on and expand the benefits of personalized learning. They allow not only faster and more comprehensive evaluation of student and teacher performance, but blended learning and related programs also enable real-time adjustments based on the instant feedback provided by big data. At the same time, cloud-based computing dramatically increases the power and reach of learning technologies.

This, of course, does not mean that all school interaction needs to be web-based. In fact, face-to-face instruction is equally, if not more, important when introducing students to new forms of learning. Student progress can be monitored on an ongoing basis by the teacher, who oversees their behavior and performance, serving as an expert facilitator of the web-based educational content.

Under a successful blended learning model, the two modes – digital and traditional – reinforce each other and make up for each other's weaknesses. The digital aspect provides data on learning and the incorporation of critical concepts or abilities, while the in-person component allows the teacher to observe the data along with the behavior and personal evolution of students, serving as a powerful complement to the digital learning. The role of teachers and tutors remains critical, while technology enhances the learning experience.

Rather than a substitution of teachers, then, online education is emerging as an opportunity for teachers to focus on those parts and processes of education which can only be taught by a human being or which are significantly improved by in-person guidance. Technology is also giving them a powerful set of new tools that allow them to play an even more involved role in guiding their students. New technologies streamline and make intuitive the ways that teachers give students both positive and negative feedback. One of the prime examples of this is ClassDojo, a cloud-based software platform that both tracks student behavior, as well as allowing teachers to take a proactive role in improving that behavior.

The fact that parents can easily access the data generated on their kids' behavior is a crucial bonus. Parents can now have a first-hand view into what is going on with their children inside the classroom. This enables them to get involved at a much deeper level, and work with teachers to identify and address potential behavioral problems before they get routinized.

In other ways as well, start-ups are creating new venues and forums for teachers to communicate and learn from each other. And others are using "gamification" to improve student performance and enhance the reach of teachers.

In sum, blended learning enables students to get the best of both worlds: the expanded access, quality, and personalization of online content with the feedback, community, and human touch of traditional schooling.

There are platforms that eschew the physical classroom experience completely and give students from all over the world a personalized educational path, like the case of the massively popular Khan Academy.[55] At the same time, a number of online programs and applications are allowing students to "leapfrog" their failing local schools and receive free or low-cost personalized education with nothing but an Internet connection. The previously mentioned Descomplica, Duolingo, Julio Profe, or Open English are only a few examples.

Higher education is providing some good examples. In 2003, more than four out of five US colleges offered at least one online class. Between 2003 and 2008, the number of students taking online classes almost tripled, and by 2010 nearly a third of college students were participating in at least one online course.[56] By 2013, *New York Times* columnist Thomas Friedman was arguing that "nothing has more potential to enable us to reimagine higher education than the massive open online course, or MOOC, platforms that are being developed by the likes of Stanford and the MIT and companies like Coursera and Udacity."[57] Together, Udacity, edX, and Coursera – another similar platform – together reach over 2 million students in almost 200 countries.[58]

If we look outside of the classroom, we can see the impact that the shift to education technology is having in businesses. Major data analytics platforms are increasingly being deployed in the service of technical training, workforce development, and lifelong learning goals at major multinational corporations, including Microsoft, Oracle, and IBM.

A final key plank of digital education's impact is its ability to provide options specifically for lower-income or socially marginalized students, who are often shut out from higher levels of education due to cost or the need to work to support their families.

Education as a philanthropic, anti-poverty measure is nothing new to the developing world. But there is still a deep need for affordable options for acquiring what are known as "middle skills" – in other words, for students to complete the equivalent of high school, or vocational and language training programs – that improve their job prospects.

This "school-to-work" segment is woefully underserved by several education systems in emerging economies. Online programs and Associates degrees in everything from accounting and business administration to nursing and paralegal training can dramatically improve the job prospects of those low-income students who have been habitually underserved by their education systems.

Lessons for the future of education technology

Technology, by itself, is no silver bullet. If innovative platforms are not supported by and integrated with an overarching strategy – which differs on a case-by-case basis – then technology alone will do little to reverse the decline in education quality.

Indeed, to enable their students to take full advantage of the ed-tech revolution, schools need not only connectivity and infrastructure, but also the ability to implement a coherent plan for merging digital technologies with the learning environment, as well as measuring its impact and making course corrections where necessary. The effective use of digital technologies in the classroom depends on *how* they are used – which role they play in the so-called "interactive triangle" formed by the content and the functions assigned to both students and teachers.

While this represents a significant challenge, the fact that students are now accustomed to the use of technology in all aspects of their lives – and expect the same at school – is becoming a driving force of change to be reckoned with. Teachers are also growing more comfortable with technology as schools are beginning to equip them with the necessary hardware.

One of the clearest outcomes of the introduction of ICT is that schools are better able to prepare their communities to face the challenges of the twenty-first century. By encouraging "soft skills" such as project-based learning, problem-solving logic, and teamwork, among others, digital education has become crucial for helping students prepare for the real world in a way that was not possible even two decades ago.

It is quite evident that innovative approaches to education are contingent upon the expansion of connectivity and computer use.[59] According to the 2013 report from the Americas Society/Council of the Americas (ASCOA) on Internet use, Latin America is seeing some progress in this area: Internet use has reached 43 percent region-wide – although high-speed broadband access remains considerably lower. Chile, unsurprisingly, leads the region in broadband service, with 10 percent of its population covered. Still, Latin America's Internet connectivity lags behind many parts of the globe, which places a ceiling on its ability to adopt new education technologies.[60]

And as long as the digital divide is not reduced in schools – as long as technology is not effectively used in the classrooms – it will be hard for schools to catch up to their modern counterparts.

Effective implementation of a blended learning school model also demands that teachers acquire new skills, not only to use the technological platforms but also to succeed at their new role as facilitators of knowledge and skills. As with any other human resources development process, this is both time- and resource-consuming.

Clearly, while we argue for the inevitability of a new wave of outside participation in our moribund education systems this must happen in addition to – rather than in opposition to – robust government involvement. The more our leaders understand the true potential of new technologies in education, the faster the democratization – through universal access and personalization – will occur.

Conclusion

Global economies have been changing and are increasingly dependent on technology and high-skill knowledge workers that are able to push for innovation as a source of added value. In this context human capital and education become vital in order to keep those economies competitive and to continue to drive global development.

This cannot be done in isolation. It is not the problem of governments and educational bureaucrats, or corporations, or individuals, on their own. Rather, this challenge demands a coherent, combined, cross-border effort.

The stagnating educational performance is already leading to serious gaps in many emerging markets' development that could undermine their future potential. And while many have made remarkable progress, especially in terms of expanding access to primary school, many are also still far behind regarding the quality of education and the critical transition from college to work. Despite increased government budgets[61] for education in the last decade, emerging countries are still sitting in the bottom third of global education scores.

Given this reality, maintaining the status quo will not cut it. Instead, education desperately needs innovation, and every important stakeholder (governments, teachers, but also parents, private sector entrepreneurs, or other actors in the NGO or nonprofit worlds) will need to participate in the reform.

Committing to building multilateral alliances to promote world-class human capital also becomes crucial. Mexico and China are starting to walk this road with the US with their 100,000 Strong or FOBESII programs. Universities are also increasingly promoting student exchange, corporations are becoming engaged in building diverse workforces that leverage the skills they need to remain competitive, and education is growing as a market for successful and sustainable investment by entrepreneurs.

Improving the quality of education worldwide, especially in developing nations, requires the participation of the private sector, which has always been out-front in developing new processes, tools, and systems. These are necessary to work on the most pressing problems of the future, such as developing new ways to teach next-generation skills, improving and implementing technological platforms that bring new teaching tools to the classroom, fostering innovation, and scaling up successful local models.

The proliferation of the so-called twenty-first-century skills that combine technological prowess with creativity and leadership capabilities also requires a new look at the recruitment and training of teachers capable of teaching such skills, or who possess them themselves. Online learning might be able to produce excellent content, streamline teaching processes, and reduce costs overall, but by itself it will not be enough to teach "soft" skills like teamwork, curiosity, persistence, and tolerance of other cultures.

In the words of the Brookings Institution: "The future prosperity of our global economy depends on our ability to recognize our shared responsibility in providing quality education and act with new energy to invest in its provision in emerging market economies and the developing world."[62]

Acknowledgement

Coauthored by Cristina Autorino. Based on Sanchez Zinny, Gabriel (2014): *Educación 3.0. The Struggle for Talent in Latin America*, Washington, DC: Books & Book Press.

Notes

1. OECD, *Education at a Glance 2012*, p. 38. http://www.oecd.org/edu/highlights.pdf
2. Elizabeth Redden, "'Amplifying' Education's Value," *Inside Higher Ed*. June 26, 2013. https://www.insidehighered.com/news/2013/06/26/oecd-education-glance-report-considers-relationship-between-recession-education-and
3. Redden, "'Amplifying' Education's Value."
4. Rebecca Strauss, "US Education Slipping in Ranks Worldwide, Earns Poor Grades on CFR Scorecard," Council on Foreign Relations. June 17, 2013. http://www.cfr.org/education/us-education-slipping-ranks-worldwide-earns-poor-grades-cfr-scorecard/p30939
5. UN News Center, "Latin American Growth Predicted to Slow Down due to Economic Crisis – UN Report," The United Nations. December 21, 2011. http://www.un.org/apps/news/story.asp?NewsID=40817&Cr=latin+america&Cr1=#.VAdTMWPYEXJ
6. *The Millennium Development Goals Report 2012*. United Nations. http://www.un.org/millenniumgoals/pdf/MDG%20Report%202012.pdf
7. Marcelo Cabrol and Miguel Székely (eds), *Educacion Para La Transformacion*, The Inter-American Development Bank (IDB). October 2012. http://publications.iadb.org/handle/11319/392?locale-attribute=en
8. OECD, PISA 2012 Database. http://www.oecd.org/pisa/keyfindings/PISA-2012-results-snapshot-Volume-I-ENG.pdf
9. http://www.aneki.com/oecd_countries_high_school_graduation_rates.html?number=all
10. "IDB Launches GRADUATE XXI, an Initiative to Prevent High-School Dropout in Latin America," Inter-American Development Bank (IDB) News Release, Dec 10, 2012.
11. Néstor López; Vanesa D'Alessandre and Silvina Corbetta, "Informe Sobre Tendencias Sociales y Educativas en América Latina 2011," May 12, 2011. http://www.siteal.iipe-oei.org/informe_2011
12. Irina Bokova et al., "Llegar a los Marginados," UNESCO, 2010, p. 11. http://unesdoc.unesco.org/images/0018/001878/187865S.pdf
13. "High School Graduation Rates In Latin America," Graduate XXI, Inter-American Development Bank. http://www.iadb.org/en/topics/education/infographic-high-school-graduation-rates-in-latin-america,7110.html
14. Mario Pezzini, "An Emerging Middle Class," 2012, *OECD Observer*. http://www.oecdobserver.org/news/fullstory.php/aid/3681/An_emerging_middle_class.html#sthash.666uqMqS.dpuf
15. EY, "Hitting the Sweet Spot. Middle Class Growth in Emerging Markets." http://www.ey.com/GL/en/Issues/Driving-growth/Middle-class-growth-in-emerging-ma.rkets%20
16. *World Bank News*, "Latin America: Middle Class Hits Historic High," World Bank Group. November 13, 2012. http://www.worldbank.org/en/news/feature/2012/11/13/crecimiento-clase-media-america-latina
17. This report presents the initial results of the OECD Survey of Adult Skills (PIAAC), which evaluates the skills of adults in 24 countries. It provides insights into the availability of some of the key skills and how they are used at work and at home. A major component is the direct assessment of key information-processing skills: literacy, numeracy, and problem solving, in the context of technology-rich environments. See: http://skills.oecd.org/skillsoutlook.html
18. This rate, authors anticipate, is likely to underestimate the impact of new technologies on traditionally low-skilled sectors, such as primary production and extractive industries, which are also incorporating advanced technologies into their operations. Agriculture, for example, is being transformed by biotechnology

and computerization (e.g. GPS technology; the use of IT to manage sales and monitor markets).

19. Over 1300 CEOs in 68 countries were surveyed for PWC's report. "The Talent Challenge," May 20, 2014. http://press.pwc.com/global/skills-gap-is-hampering-businesses-recruitment-efforts/s/6d07c69e-c1a2-4ba0-b13f-bbc9c2d6bbe4

20. Javier Santiso, "Here Come the Multilatinas," Beyondbrics, *The Financial Times*. August 17, 2011. http://blogs.ft.com/beyond-brics/2011/08/17/guest-post-the-decade-of-the-multilatinas/?

21. "Rankings Multilatinas: Las Empresas Mas Globales de America Latina," *America Economia*. 2010. http://rankings.americaeconomia.com/2010/multilatinas/ranking_multilatinas.php

22. Alicia Barcena, Álvaro Calderón, Mario Castillo, René A. Hernández, Jorge Mario Martínez Piva, Wilson Peres, Miguel Pérez Ludeña and Sebastián Vergara, "Foreign Direct Investment in Latin America and the Caribbean, 2010," The Economic Commission for Latin America and the Caribbean (ECLAC). 2010. http://www.cepal.org/publicaciones/xml/0/43290/2011-138-LIEI_2010-WEB_INGLES.pdf

23. Busso, Matías, Bassi, Marina, Urzúa, Sergio and Vargas, Jaime, "Disconnected: Skills, Education, and Employment in Latin America," The Inter-American Development Bank (IDB). March 6, 2011. http://www.iadb.org/en/topics/education/disconnected-home,5928.html

24. Dominic Barton, Diana Farrell and Mona Mourshed, "Education to Employment: Designing a System that Works," McKinsey Center for Government. http://mckinseyonsociety.com/downloads/reports/Education/Education-to-Employment_FINAL.pdf

25. Joy Resmovits, "College Dropout Crisis Revealed in 'American Dream 2.0' Report," *Huffington Post – College*. February 1, 2013. http://www.huffingtonpost.com/2013/01/24/college-dropout-crisis-american-dream-20_n_2538311.html

26. Susan Adams (2013): "Half Of College Grads Are Working Jobs That Don't Require A Degree" http://www.forbes.com/sites/susanadams/2013/05/28/half-of-college-grads-are-working-jobs-that-dont-require-a-degree/

27. World Bank, "Doing Business 2013: Smarter Regulations for Small and Medium-Size Enterprises," World Bank Group. 2013, p. 3. http://www.doingbusiness.org/~/media/GIAWB/Doing%20Business/Documents/Annual-Reports/English/DB13-full-report.pdf

28. OECD, "Society at a Glance: Youth neither in Employment, Education, nor Training." Accessed September 5, 2014. http://www.oecd-ilibrary.org/sites/soc_glance-2014-en/04/03/index.html;jsessionid=24guhqmhpfem.x-oecd-live-01?contentType=&itemId=%2Fcontent%2Fchapter%2Fsoc_glance-2014-14-en&mimeType=text%2Fhtml&containerItemId=%2Fcontent%2Fserial%2F19991290&accessItemIds=%2Fcontent%2Fbook%2Fsoc_glance-2014-en

29. Andreas Schleicher, "OECD Country Note Turkey: Education at a Glance," Organization for Economic Cooperation and Development. 2013. http://www.oecd.org/edu/Turkey_EAG2013%20Country%20Note.pdf

30. Andreas Schleicher, "OECD Country Note Hungary: Education at a Glance," Organization for Economic Cooperation and Development. 2013. http://www.oecd.org/edu/Hungary_EAG2013%20Country%20Note.pdf

31. José Manuel Salazar-Xirinachs, "Generation Ni/Ni: Latin America's Lost Youth," *Americas Quarterly*. Spring 2012. http://www.americasquarterly.org/salazar

32. Justin van Fleet and Gabriel Zinny, "Corporate Social Investments in Education in Latin America and the Caribbean: Mapping the Magnitude of Multilatinas' Private

Dollars for Public Good," The Brookings Institution. August 13, 2012. http://www.brookings.edu/research/papers/2012/08/investments-latin-america-van-fleet

33. Innovation Unit for GELP, "Redesigning Education: Shaping Learning Systems around the Globe," Booktrope Editions. April 30, 2013. http://www.amazon.com/Redesigning-Education-Shaping-Learning-Systems/dp/1620151456/ref=tmm_hrd_title_0

34. *The Economist*, "The Future of Jobs: The Onrushing Wave." January 16, 2014. http://www.economist.com/news/briefing/21594264-previous-technological-innovation-has-always-delivered-more-long-run-employment-not-less

35. S. Craig Watkins, "Rethinking the 'Race between Education and Technology' Thesis," Connected Learning Research Network. December 4, 2013. http://clrn.dmlhub.net/content/rethinking-the-race-between-education-and-technology-thesis

36. Erik Brynjolfsson and Andres McAfee, "The Second Machine Age: Work, Progress, and Prosperity in a Time of Brilliant Technologies," W. W. Norton & Company. January 20, 2014. http://www.amazon.com/Second-Machine-Age-Prosperity-Technologies/dp/0393239357/ref=tmm_hrd_swatch_0?_encoding=UTF8&sr=&qid=

37. Ana Pereyra, "Boletin No. 8: La fragmentacion de la oferta educative: la educación publica vs la educación privada," Sistema de Informacion de Tendencias Educativas en America Latina (SITEAL). 1999. http://www.siteal.iipe-oei.org/sites/default/files/educacion_publica_vs_educacion_privada.pdf

38. Rip Empson, "Economic Impact of Startup Accelerators: $1.6B+ Raised, 4,800+ Jobs Created, 2,000 Startups Funded," *Techcrunch*. November 27, 2012. http://techcrunch.com/2012/11/27/economic-impact-of-startup-accelerators-1-6b-raised-4800-jobs-created-2000-startups-funded/

39. Eric Markowitz, "Bill Gates: Education System Needs More Entrepreneurs," *Inc. com*. March 7, 2013. http://www.inc.com/eric-markowitz/bill-gates-education-system-needs-more-entrepreneurs.html

40. Ellis Booker, "Education Tech Investments Surpassed $1 Billion in 2012," *Information Week*. January 25, 2013. http://www.informationweek.com/software/education-tech-investments-surpassed-$1-billion-in-2012/d/d-id/1108366?

41. Justin W. van Fleet, "Increasing the Impact of Corporate Engagement in Education: Landscape and Challenges," Brookings Institution. July 12, 2011. http://www.brookings.edu/research/speeches/2011/07/12-corporate-engagement-vanfleet

42. Nicole Perlroth, "NewSchools CEO Ted Mitchell: My Best Idea for K-12 Education," *Forbes*. September 19, 2011. http://www.forbes.com/sites/nicoleperlroth/2011/09/19/newschools-ceo-ted-mitchell-my-best-idea-for-k-12-education/

43. Rip Empson, "Descomplica Lands $5M from Social+Capital, AngelList's First International Syndicate to Become Brazil's Go-To Online Classroom," *TechCrunch*. February 6, 2014. http://techcrunch.com/2014/02/06/descomplica-lands-5m-from-socialcapital-angellists-first-international-syndicate-to-become-brazils-go-to-online-classroom/

44. Duolingo website, accessed September 5, 2014. https://www.duolingo.com/

45. Camila Souza, "More Money, More Problems: Open English CEO Andres Moreno Talks Funding," Tech Cocktail Miami. May 5, 2014. http://tech.co/andres-moreno-funding-2014-05

46. Laureate University, "Our Network," accessed September 5, 2014. http://www.laureate.net/OurNetwork

47. Sanchez Zinny, Gabriel (2014), "Education innovation: Lessons from Latin America" Available at http://edexcellence.net/articles/education-innovation-lessons-from-latin-america

48. Holly Else, "Brazil's Home Grown MOOC, Veduca, Has High Hopes," *Times Higher Education*. January 16, 2014. http://www.timeshighereducation.co.uk/news/brazils-home-grown-mooc-veduca-has-high-hopes/2010440.article

49. http://lavca.org/wp-content/uploads/2015/09/Coller-Capital-LAVCA-Latin-America-Survey-2015-English.pdf

50. Multilateral Investment Fund (MIF), "The MIF and Early Stage Financing," The Inter-American Development Bank. February 2013. http://idbdocs.iadb.org/wsdocs/getdocument.aspx?docnum=37479609

51. Hudson Institute, "Strong Gains in Private Giving to Developing World on Heels of Global Recession, Reports 2012 Index of Global Philanthropy and Remittances," April 2, 2012. http://www.hudson.org/research/8851-strong-gains-in-private-giving-to-developing-world-on-heels-of-global-recession-reports-2012-index-of-global-philanthropy-and-remittances-

52. "New World: Emerging Donors and the Changing Nature of Foreign Assistance," The Center for Global Development. November 21, 2011. http://www.cgdev.org/blog/brave-new-world-emerging-donors-and-changing-nature-foreign-assistance

53. Elizabeth Dickinson, "Look Who's Saving the World: BRICS Pump up Foreign Aid," *The Christian Science Monitor*. March 26, 2012. http://www.csmonitor.com/World/Global-Issues/2012/0326/Look-who-s-saving-the-world-BRICS-pump-up-foreign-aid

54. Cathy Davidson, "Going From One-Size-Fits-All Education to One-Size-Fits-One," *Fast Company*. May 31, 2012. http://www.fastcoexist.com/1679921/going-from-one-size-fits-all-education-to-one-size-fits-one

55. Salman Khan, "Let's Use Video to Reinvent Education," Technology, Education, and Design (TED). March 2011. https://www.ted.com/talks/salman_khan_let_s_use_video_to_reinvent_education/transcript

56. Ryan Lytle, "Study: Online Education Continues Growth," US News and World Report Education. November 11, 2011. http://www.usnews.com/education/online-education/articles/2011/11/11/study-online-education-continues-growth

57. Thomas Friedman, "Revolution Hits the Universities," *The New York Times*. January 26, 2013. http://www.nytimes.com/2013/01/27/opinion/sunday/friedman-revolution-hits-the-universities.html?_r=0

58. Tarun Mitra, "The Big Three: Udacity, Coursera, and edX," LearnQ Alpha. May 2012. http://lurnq.com/lesson/EdTech-Startups-to-Watch/section/The-Big-3-Udacity-Coursera-edX/

59. Pew Foundation, "Mobile Technology Fact Sheet," PewResearch Internet Project. Accessed September 5, 2014. http://www.pewinternet.org/fact-sheets/mobile-technology-fact-sheet/

60. Rachel Glickhouse, "Explainer: Broadband Internet Access in Latin America," Americas Society/Council of the Americas. February 13, 2013. http://www.as-coa.org/articles/explainer-broadband-internet-access-latin-america

61. To mention a few examples, countries like Argentina have been investing around 6 percent of GDP in education. Poland and Hungary's investment as a proportion of their GDP has neared 5 percent, while India and Indonesia have averaged some 3.5 percent of their GDP. See: http://data.worldbank.org/indicator/SE.XPD.TOTL.GD.ZS/countries

62. http://www.brookings.edu/experts/winthropr \t "_blank"Rebecca Winthrop, Gib Bulloch, Pooja Bhatt and Arthur Wood (2013), "Investment in Global Education: A Strategic Imperative For Business". Available at http://www.brookings.edu/research/reports/2013/09/investment-in-global-education

13
Health-Care Innovation in Emerging Markets

Françoise Simon

Major emerging markets such as India, China, and Brazil, in addition to presenting rising middle-class populations, are now playing a part in driving the transformation of the health-care sector, from R&D and clinical trials to local manufacturing.

However, significant challenges remain, from intellectual property (IP) issues to manufacturing quality and drug pricing and reimbursement. The health-care sector follows the general structure of three clusters of emerging markets:

- The BRICMT group, comprising Brazil, Russia, India, China, Mexico, and Turkey, have shown strong growth, and some are reaching market sizes comparable to those of mature Western countries.
- The "second-tier" markets include a group of Eastern Europe economies, as well as some countries in Southeast Asia and Latin America.
- The third group includes African markets with high populations but smaller market sizes. Although the region has high potential, only a few of its countries, such as South Africa, Egypt, Algeria, and Nigeria, have a pharmaceutical market volume exceeding $1 billion (Booz & Company, 2011).

As a consequence of economic growth and shifting demographics, epidemiology patterns in the higher-income emerging markets are evolving from communicable to chronic illnesses such as diabetes, and cardiovascular and oncological diseases.

However, economic growth is not, by itself, a predictor of innovation success or of the market potential for high-cost innovation products.

Innovation depends on a complex set of factors including government financing for health-care services, the rise of private insurance to supplement often inadequate public funds, patient access to services, and good regulatory systems for manufacturing. In addition, the positive role of the

233

private sector, and especially of foreign investment and technology transfer, is directly linked to IP protection (see Figure 13.1).

Since 2005, when all WHO (World Health Organization) member states committed to universal health coverage, progress has been made towards the United Nations Millennium Development Goals (MDGs), but the gap remains large between these and market realities. For instance, nearly half of all HIV-infected patients requiring antiretroviral therapy still were not receiving it by 2011, and an estimated 150 million people suffer great financial harm because they must pay out-of-pocket for health services (WHO, 2013).

In India, with a population of over 1.2 billion, the addressable segment for an oncology therapy with a $10,000 yearly cost would only be 23 million people. By contrast, for an anemia treatment cost of $270 a year, as many as 290 million people could have access and affordability (Booz & Company, 2011).

In response, some markets are moving towards value-driven drug evaluation and pricing. Brazil has created CONITEC (National Commission for Incorporation of Technologies in the Unified Healthcare System),

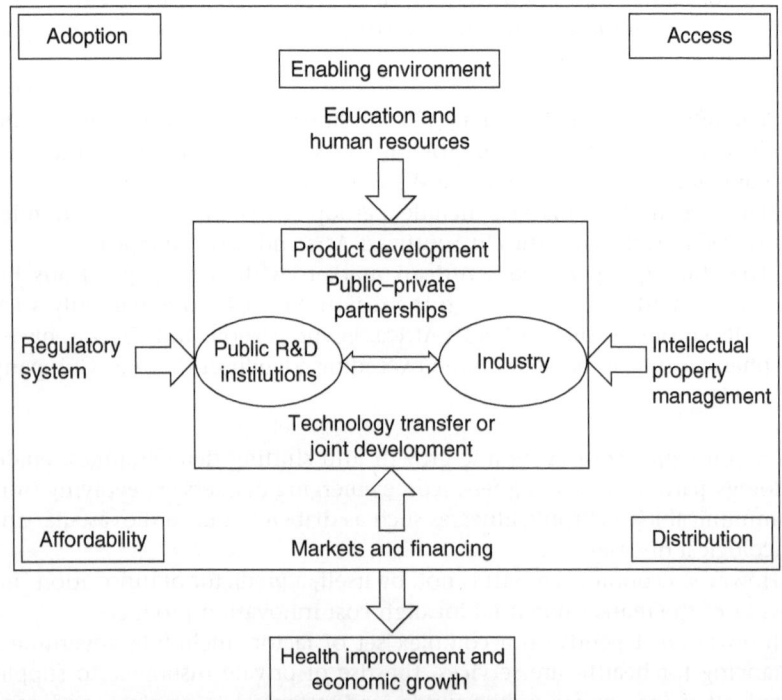

Figure 13.1 Health delivery systems and services
Source: Adapted from Morel et al. (2005).

an economic evaluation agency, and China is collaborating with the British National Institute of Health and Clinical Excellence (NICE) (Booz & Company, 2011). These policies, as well as inadequate IP protection, may have a negative impact on innovation. Since 2009, China has had a National Drug Reimbursement List (CNDRL) and also capped retail prices for its Essential Drug List. This had two unintended consequences: drug shortages, as manufacturers stopped production of some low-profit products, and quality issues in a context of drastic cost reduction. In response, China announced in April 2014 the easing of price caps on 520 low-cost essential drugs (IMS Institute for Healthcare Informatics, 2014).

A key barrier to private-sector-driven innovation remains the inadequacy of IP protection. After the Patent Act of 1970, IP applied only to process manufacturing, not products, which helped make India the world leader in the manufacturing of generic drugs, but blocked market entry for foreign firms. After India joined the World Trade Organization (WTO) in 2005, IP was introduced for pharmaceutical products, but this has been offset in part by some cases of compulsory licensing such as the granting to local firm Natco the rights to manufacture and market Bayer's oncology drug Nexavar (sorafenib) in 2012 (Booz & Company, 2011).

Despite these challenges, emerging markets are expected to represent a growing share of medicines, including biologic products, produced by Western as well as local firms such as Biocon in India. Global spending in medicines is projected to reach over $1 trillion by 2017, and biologic agents will continue to outpace overall growth and reach up to 20 percent of total market value by 2017.

Although China's growth was revised down, following a decrease in GDP prospects, its pharmaceutical market is expected to grow between 14 and 17 percent in the 2013–18 period. Volume-based growth will be driven by government efforts to expand the Essential Drug List, as well as to improve access to services, insurance coverage, and use of private hospitals (IMS Institute for Healthcare Informatics, 2013).

In Brazil, the government has supported health research with its Biotechnology Development Policy and a ten-year, $4 billion development program. An upper-middle-income country, Brazil has a population of about 200 million people, most of whom live in poverty despite a GDP per capita that reached over $11,000 at purchasing power parity (PPP) by 2011. Over 70 percent of the disease burden for the poor is contributed by chronic noncommunicable illnesses such as diabetes, cancer, respiratory and cardiovascular conditions, thereby showing similarity with the disease profile of mature markets. Brazil has a strong R&D capacity, but the private sector contributes only about 20 percent of the total investment. A notable strength is its well-developed vaccine industry, with one of the best national immunization programs among developing countries. The country has over 180 biotech firms, and it received a $2.5 million grant in 2007 to set up manufacturing capacity for the influenza vaccine (Abuduxike and Aljunid, 2012). Given the diversity of research and market conditions across emerging markets, this chapter will cover in more depth India, China, and Brazil, to show their different approaches to health-care innovation.

India

India now accounts for over 17 percent of the world's population, and is projected to become the most populous country by 2050, with 1.6 billion people. It shows rapid urbanization, with over 30 percent of the population in massive metropolises of more than 10 million such as Delhi, Mumbai, and Kolkata. It is very diverse, with at least six major religions, several officially recognized languages, and 28 states with large differences in income. Despite GDP growth averaging 6.6 percent in 1990–2010, the country has not increased public spending on health care correspondingly, and the literacy rate is only 74 percent.

In addition to chronic diseases such as diabetes and cardiovascular illnesses, India has seen the growth of communicable diseases such as HIV/AIDS. Tuberculosis in India is the number one cause of death, with a rate double that of China, accounting for over one-quarter of all cases worldwide (with a total of 8.8 million). Despite rising per capita income, reaching nearly $3,700 by 2011 at PPP, India is divided between a large share of its population (69 percent) in rural areas with inadequate access to health services, and a growing middle class of about 250 million that can afford Western allopathic medicine (Burns, 2014).

When it comes to balancing the outcomes of patient access, high quality care, and cost efficiency, India therefore faces challenges on all fronts. It rates low on indicators such as infant mortality or life expectancy at birth, it has inadequate regulation of providers and medical product quality, and nearly 70 percent of all health-care costs are borne out of pocket by the population (see Figure 13.2).

National policies

Since 2000, India has increased the government's role in funding health care, engaged in initiatives to develop the biotechnology industry, and supported the rise of private-sector health insurance. Biotechnology was defined by the 1992 Convention on Biological Diversity as any technological application that uses biological systems, living organisms, or derivatives thereof, to make or modify products or processes for specific use (MOITI, 2008).

Following the creation in 1986 of the Department of Biotechnology (DBT), India approved in 2007 an extensive National Biotechnology Development Strategy (NBDS), with several recommendations, including a Small Business Innovation Research Initiative to fund early-stage research, public/private partnership support, a national task force to set up model academic

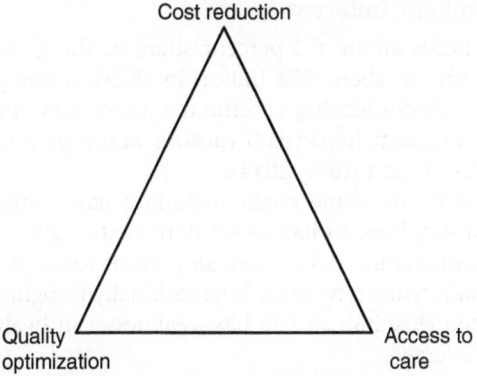

Figure 13.2 The triple aim of health care
Source: Adapted from Lawton (2014).

curricula, the reengineering of university departments to promote interdisciplinary research, and the creation of biotechnology parks. In addition to the well-established technology clusters in Bangalore and Hyderabad, the DBT has set up other clusters near Delhi and in Punjab. It has encouraged international collaborations such as the Stanford–India Biodesign Program, and a substantial allocation from the UK's Wellcome Trust to fund biomedical research over five years (Frew, 2014).

The Indian government is also seeking to extend insurance coverage. Under private and public insurance, only about 25 percent of the population is covered. The newer central and state government schemes aim to increase protection of the poor against catastrophic health events, focusing on surgical procedures and secondary care. The government aims to cover 300 million people in the near future. Further recommendations include the introduction of quality-based purchasing, better utilization management, and expanded autonomy for public hospitals (La Forgia and Nagpal, 2012).

A key step toward private-sector innovation was the entry in 2005 of India into the WTO, whereby it had to comply with the agreement of Trade-Related Aspects of Intellectual Property Rights (TRIPS). Their main elements were:

- Enforcement of product patent protection in all branches of technology, including drugs.
- 20 years of protection, instead of 14 or 7 in the case of the Indian Patent Act.
- No discrimination between imported and domestic products.

However, the promotion of private-sector innovation remains challenging, since India has imposed compulsory licensing in recent cases.

Indian biotechnology industry

Indian biotech holds about a 2 percent share of the global industry, and is expected to grow to about $74 billion by 2020. It comprises about 400 companies and includes leading vaccine manufacturers. India is the largest producer of recombinant hepatitis B vaccine, and is gaining importance as a clinical trial destination (IBEF, 2014).

Along the health-care value chain including payers/insurers, providers, distributors, and suppliers, India has substantial strengths in product development, low manufacturing costs (50–55 percent lower than in the West), and provider quality, shown by some internationally recognized organizations such as the Apollo Hospitals. It still has weaknesses in health infrastructure and insurance coverage.

Diagnostics is a competitive subsegment with double-digit growth, reaching nearly $500 million in revenue by 2010, split between multinationals such as Roche, Abbott, and Siemens, and domestic companies.

The therapeutics sector is led by Biocon, followed by the Serum Institute, Panacea, and Shanta Biotech. The Serum Institute was the first Indian company to be approached by the WHO in 2008 to develop and manufacture the H1N1 flu vaccine (Frew, 2014).

Although firms such as Biocon have a strong portfolio of biologics, R&D as a whole remains limited in India, and product development is dominated by generics. The top five Indian companies still account for only 1 percent of R&D spend by the top five global multinationals, and over 75 percent of innovations carried out in India are incremental (Tyagi et al., 2014).

Indian company strategies

Leading Indian domestic firms are practicing networked innovation, with a combination of organic growth, geographic expansion, and partnerships with multinationals.

Biocon's strategy

Aspiring to US$1 billion in revenues by 2018, and reaching nearly US$500 million in 2014, Biocon has a broad portfolio including small molecules, biosimilars, and contract research services. Founded in 1978 by Kiran Mazumdar-Shaw, it divested in 2007 its historic enzymes business to Novozymes and began to develop lovastatin, a cholesterol-lowering drug whose patent expired in 2001. Following its launch of recombinant insulin Insugen, it moved to biologics in 2006, with the first humanized monoclonal antibody for head-and-neck cancer developed in India.

Between 2005 and 2010, Biocon entered into numerous R&D licenses and other partnerships. In 2014, it launched CANMAB for a type of HER2-positive breast cancer, positioned as "the world's most affordable breast cancer drug."

Biocon's collaboration with Mylan, started in 2009, includes generic insulin analogs and biosimilar monoclonal antibodies. Both companies share development and capital costs, and have a profit-sharing arrangement in regions where Mylan has exclusive commercialization rights (the US, Canada, Europe, Australia, and New Zealand for insulin, targeting a global market size of $16 billion, and yielding a $20 million upfront fee for Biocon). Other partnerships have included a 2004 alliance with a US antibody technology partner, Vaccinex, and a 2007 agreement with Abraxis Bioscience to outlicense a biosimilar GCSF (granulocyte colony stimulating factor) to North American and European markets.

Biocon's R&D approach focuses on the entire development pathway, from process development to clinical research. Its patent portfolio totals over 900 applications worldwide with over 180 granted patents, covering fermentation, protein purification, drug delivery systems, and biologics. Importantly in the Indian context, the company's manufacturing meets cGMP standards (good manufacturing practices), and Biocon was the first Indian biotech to receive ISO 9001 certification. Beyond its Bangalore site, Biocon has planned a $200 million investment in a Malaysia site. Biocon also leverages its manufacturing with research services; it is India's largest CRO (contract research organization) with over 2,000 scientists, a capital investment of about $130 million, and agreements with 16 of the top 20 pharmaceutical companies worldwide. Its key customers include:

- Bristol-Myers Squibb – largest R&D Big Pharma center in Asia, started in 2009, extended to 2020, for novel molecule research.
- Baxter partnership in 2014 with Biocon subsidiary, Syngene, centered in part on preclinical evaluation in parenteral nutrition and renal therapy.
- Abbott Nutrition's R&D center in India allied with Syngene in 2012 to develop a nutrition products line with emerging market needs.

Biocon's stepwise strategy, moving from small molecule generics to biologics through organic growth, manufacturing optimization, and R&D partnerships, has been rewarded by the marketplace: on the first day of listing for its 2004 IPO (initial public offering), Biocon was only the second Indian company to cross the $1 billion mark, and it continues to lead the Indian industry (Biocon, 2015).

Other Indian biotechs

Besides Biocon, other domestic biotechs have had varying strategies, most often with a focus on vaccines. The Serum Institute was the first Indian firm to be approached by the WHO to manufacture the H1N1 flu vaccine.

Shanta Biotechnics was founded in 1993, and also focused on vaccine manufacturing. It developed international partnerships including, in South Korea, the International Vaccine Institute, and managed an exit strategy, with its 2009 purchase by Sanofi Pasteur at an estimated valuation of €550 million (Frew, 2014).

Panacea Biotec was set up in 1984 and developed collaborations including one with the Netherlands Vaccine Institute. It had an IPO in 1995 and gained its first product patent two years later. However, it encountered quality issues in its manufacturing in 2011, but these were later resolved.

Besides leaders like Biocon, many Big Pharma firms have engaged in partnerships with Indian companies, but these have met with varying success, at least in part because of IP challenges. Representative cases include Gilead's collaboration with Indian generic firms in HIV/AIDS, and Bayer's contrasting experience in oncology.

Gilead partnerships in India

Founded in 1987, Gilead had become the global market leader in HIV treatment, with its first antiretroviral product Viread (tenofovir) or TDF in 2001, followed by Emtriva (emtricitabine) and combination therapies Truvada, Atripla, and Complera by 2011.

Global estimates of HIV patients were more than 34 million that year. By 2000–1, two Indian generic firms launched antiretrovirals (ARVs) at respective prices of $800 and $295 per patient per year, vs an average US cost of $10,000–$15,000, which limited its access to only 2 percent of patients in developing countries. Gilead filed the Viread patent application in India in 1997, but by 2006, while it was still under review by the Indian Patent Office, it faced pre-grant oppositions from generic manufacturers including Cipla and some NGOs.

To increase access to its ARVs, Gilead adopted a new strategy in 2006, by then extending voluntary, nonexclusive licenses to Indian companies to manufacture generic TDF-based ARVs for developing countries. It subsequently partnered with 14 Indian firms, including Mylan, Hetero, Ranbaxy, and Strides Arcolab. The deals allowed them to manufacture ARVs in India and sell them there and to 94 other countries, as well as to codevelop drug combinations. By 2011, 1.8 million patients were receiving Gilead HIV drugs in the developing world, and the price for generic Viread had fallen to 19 cents per day. Given this success, Gilead expanded its agreements with its four partners, and extended them to the Medicines Patent Pool (MPP), and Natco that could itself sublicense to any qualifying Indian firm. Royalties to Gilead were reduced from 5 to 3 percent and generic TDF was also allowed to be sold for chronic hepatitis B. Most importantly, the partners were granted nonexclusive rights to three of Gilead's pipeline HIV medicines.

By contrast, the official policy was negative. In 2009, the Indian Patent Office rejected TDF patents on the grounds that they lacked an inventive step. Gilead then appealed the decision. In the US, the NGO PubPat filed a request for a reexamination of the Viread patents, claiming that the TDF compound was "obvious." Following its review, the US Patent and Trademark Office (USPTO) concluded that the patents were not obvious and reaffirmed their validity (Sachan et al., 2013).

By leveraging innovation in advanced manufacturing by Indian firms, Gilead had created long-term partnerships that balanced incentives for local innovations and global access at low cost.

Bayer and the Nexavar experience in India

Founded in 1863, Bayer has had a long history in India, where it was first set up in 1896. Bayer India was targeting $1.3 million in sales by 2015, from medical care, animal health, and pharmaceuticals, comprising women's health and cardiovascular drugs, as well as specialty medicines.

The USPTO granted a first patent for kidney cancer to Bayer's Nexavar (sorafenib) in 2007, and it was a blockbuster by 2012, with $1.04 billion in global sales. Competitors included Pfizer's Sutent, Novartis's Afinitor, and GSK's Votrient. Although not a cure, Nexavar could prolong a kidney patient's life by four to five years, and could be taken orally. Bayer applied for an Indian patent in 2001, but Nexavar was sold at $5,500 for a month's supply, at a price equivalent to that in the US.

The Indian generic firm Natco had filed for a voluntary license for Nexavar, but was turned down by Bayer. In 2011, Natco then applied for a compulsory license (CL), proposing to sell generic sorafenib at a discount of over 97 percent ($160 per month).

In March 2012, the Indian Controller General granted the first ever CL in India to Natco, committing it to a 7 percent royalty to Bayer. The ruling was based on low availability (mostly in top metropolitan hospital pharmacies), high price, and a lack of "working" of the patent in India. Bayer then filed a petition to block Natco's generic; it had also launched an access program in cooperation with Indian providers, selling Nexavar at 10 percent of the market price. By 2012, Bayer estimated that 73 percent of eligible patients were covered by this program. Nevertheless, the WHO had reacted positively to the ruling, and its potential expansion to other Western drugs in that category remained a threat for foreign manufacturers.

These contrasting strategies by Gilead and Nexavar illustrate the remaining barrier to innovation and technology transfer posed by compulsory licensing, despite the entry of India into the WTO in 2005. Some of these opportunities and challenges also apply to Brazil, within a very different market situation.

Brazil

Brazil's economy is the second largest in the Western Hemisphere and the eighth largest in the world, with a largely urbanized population of about 200 million, a 2013 GDP of $2.3 trillion at PPP, and a growth rate of over 5 percent in the 2000–12 period. This growth, however, slowed in 2013 and was expected to drop further by 2015 (Lima, 2015).

While there have been improvements in living standards, life expectancy at birth was only 73.5 years in 2010, and the under-five child mortality rate was still 16 deaths per 1,000 live births (World Bank, 2013). A major problem remains the disparity in income and access to health services between southern states and the poorer northeast region.

Health system evolution

A major policy event was the creation of the Sistema Unico de Saude (SUS) or Unified Health System in 1996, which sought to embody the principle of a universal right to health established in the Federal Constitution of 1988. The SUS seeks to shift responsibility for its administration to municipal entities, with technical and financial cooperation from the federal government and the states.

The SUS unified disparate subsystems (Social Security, the Ministry of Health, states, and municipalities) and invested heavily in primary care with its flagship program, the Family Health Strategy (FHS). The current system has a decentralized structure, with a tripartite administration by the federal, state, and municipal Ministry of Health secretariats, and with financing at all three levels (see Figure 13.3).

While the Agencia Nacional de Saude Suplementar (ANS) supervises health insurance plans, health providers and pharmaceutical products are the responsibility of the Agencia Nacional de Vigilencia Sanitaria

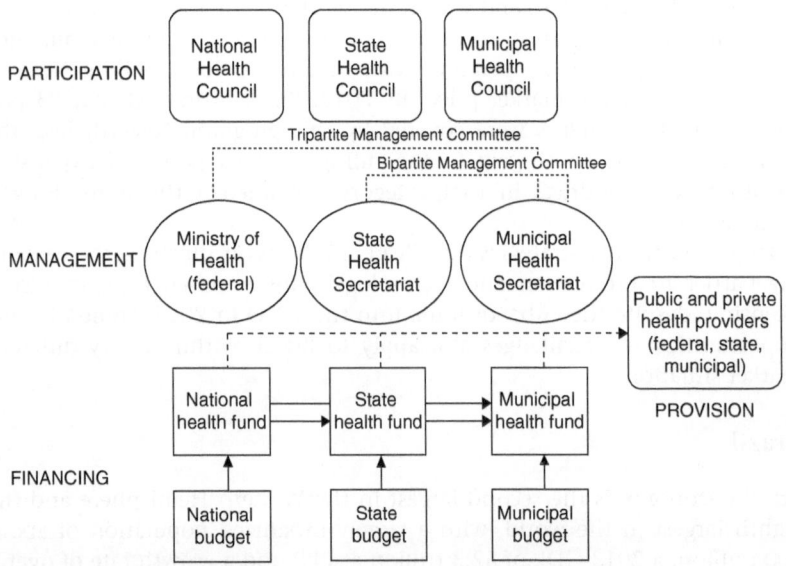

Figure 13.3 The decentralized national health system in Brazil
Source: Adapted from Mori Sarti et al. (2012).

(ANVISA), functioning as the equivalent of the US' FDA (Food and Drug Administration).

Under the SUS, many health indicators such as the under-five child mortality rate improved (from 58 to 16 deaths per 1,000 live births, from 1990 to 2009), but inefficiencies remain in the hospital and insurance sectors. Nearly 70 percent of the population uses the SUS, but private insurance grew by almost 50 percent in 2000–10, to reach nearly 46 million. Of the 9 percent of GDP spent on health in 2011, 47 percent came from the government, but 53 percent was privately funded. The current health system is mixed:

- Publicly funded and provided care (65 percent of medical consultations)
- Publicly funded but privately provided care (10 percent of consultations)
- Privately financed and funded care (25 percent of consultations)

General hospital care is largely supplied by the private sector, under contract with the SUS. Specialist care is concentrated in larger urban centers, and tertiary care is provided mainly by the SUS. The ministries of Health and Education retain about 100 large referral and teaching hospitals (World Bank, 2013).

There are wide differences in the outcomes of this system across therapeutic areas. While the SUS-administered HIV/AIDS program has been very successful, specialist care shows access problems.

A Federal Audit Tribunal found that, due to low access to specialist care and diagnostics, 60 percent of cancer patients were diagnosed at a late stage (3 or 4). There are general delays in accessing treatment, with median wait times in 2010 from 76 to 113 days, depending on the type of treatment. Paradoxically, Brazil purchases advanced medical equipment, but a substantial proportion is allocated to municipalities that do not have the size to utilize it appropriately. The country has over 6,700 hospitals, but over 65 percent have fewer than 50 beds and operate at a very low level of efficiency (45 percent mean bed occupancy rate) (Gragnolati et al., 2012).

This is being alleviated by a consolidation trend: in 2009, Brazilian insurer Amil acquired Medical Health, with a combined coverage of over 6 million members, and it also started to buy hospitals, with a network of 40 throughout Brazil. In 2012, United Health moved to buy Amil for about $4 billion, with further foreign investment in hospitals pending regulatory approval.

National policies – intellectual property

The National Industrial Property Institute (INPI) regulates patent applications. Although Brazil signed the WTO/TRIPS agreement, major obstacles remain:

- Long approval period (backlog of about nine years by the Biotechnology Patent Division)

- Restrictive patentability criteria, with a long list of nonpatentable items including nucleotide and peptide sequences derived from living organisms
- Restricted access to biodiversity, requiring approval to access the genetic heritage (Resende, 2012)

A notable case of Brazil's policy is its contentious compulsory licensing in 2007 of the Merck AIDS drug, efavirenz, to secure supply at a lower cost. The 1997 Brazilian Patent Law requires that foreign products be made in Brazil within three years of receiving a patent. If the foreign company does not comply, Brazil can issue a compulsory license to a local firm. This was done as early as 2001, when the Ministry of Health authorized FarManguinhos to produce nelfinavir, a Pfizer drug licensed to Roche in Brazil. Negotiations ensued, resulting in a 40–65 percent reduction by Roche and Merck for the price of five drugs.

This supported a very effective AIDS program, but alienated foreign firms, several of which threatened to withdraw from Brazil. Several program milestones have occurred from 1986 to the present time:

- 1986: National Program on HIV/AIDS is established (1,537 cases)
- 1991: Ministry of Health (MOH) starts distributing free antiretroviral drugs (18,487 cases)
- MOH launches a large education campaign and begins to reimburse treatment under the SUS (25,186 cases)
- 1996–98: Free drug distribution is established by law in 1996, and mandated for private insurers by 1998 (91,916 cases)
- 2001: Brazil threatens to break patents and negotiates price reductions (139,473 cases)
- 2007–8: Survival rates improve significantly (474,273 total cases since 1980). Brazil invests $10 million in a Mozambique factory for antiretroviral drugs.

Thanks to behavior changes resulting from information campaigns, as well as accessible treatment, mortality peaked in 1994 (12.2 deaths per 100,000 population), then dropped by half by 1998 (Gragnolati et al., 2012).

Brazil's biopharmaceutical industry

Although the Brazilian market is significant, with approximately $40 billion in sales, and a growth of 10.5 percent annually since the mid-2000s, its innovation output is small (only 0.45 percent of the international database of biotech patents). Moreover, only 7 percent of these originate from Brazilian companies, whereas 73 percent come from universities and research centers, 13 percent from foreign firms, and 7 percent from foreign universities.

In addition to patent threats for foreign firms, this may be related to the historical dominance in Brazil of large generic firms such as Ache, Hypermarcas, Eurofarma, or EMS. The biotech sector is more recent, with initial R&D in biofuel, agribusiness, and environmental domains as well as health care. Out of 143 biotechs, 33 percent focus on human health. Most are highly dependent on foreign imports, especially reagents and laboratory equipment; 75 percent are clustered in São Paulo, Minas Gerais, and Rio de Janeiro, with others in the southern states of Rio Grande do Sul and Parana.

Business models are varied, with 66 percent relying on sales of internally developed products, and others adding services, sales of third-party products, and outlicensing of technologies (Resende, 2012). The market is dominated by foreign firms and large local generics companies.

Biotech financing

Capital in Brazil is generated by corporate partnerships, venture capital, and private equity, as well as public funds. Public funding includes CRIATEC, a seed capital fund, the BNDES (National Social and Economic Development Bank), and FINEP, an agency linked to the Ministry of Science, Technology, and Innovation. There has also been significant funding from private nonprofits such as the Gates Foundation (Resende, 2012).

Company strategies: alliances and acquisitions

Acquisitions by foreign firms have been increasing. Sanofi was one of the pioneers in Brazil, first establishing a presence in 1955. Sanofi maintains four manufacturing sites, with a staff of over 5,000. In 2009, it bought Medley, Brazil's third-ranked firm, raising its sales to $1.8 billion. Other acquisitions include Pfizer's purchase that same year of a substantial stake in a generic firm, Laboratorio Teuto Brasileiro. Public–private partnerships are also increasing, and a notable case is the GSK/Fiocruz alliance.

GlaxoSmithKline/Fiocruz partnership

GSK announced in 2009 a major collaboration with Fiocruz, leveraging Brazil's know-how in vaccine manufacturing, and the rising importance of vaccines in the GSK portfolio, accounting for 15 percent of sales. Vaccines complemented therapeutics well, allowing the company to demonstrate value to payers through local manufacturing. In Brazil, SUS purchased 90 percent of vaccines, 50 percent of medical equipment, and 25 percent of pharmaceuticals. GSK's competitors included Novartis and Johnson & Johnson, who also had expansion objectives in emerging markets.

The Fundaçao Oswaldo Cruz (Fiocruz) was set up in 1900 with a mandate to improve public health, and currently operated a large site in Rio de Janeiro and four other scientific hubs. It employed 4,500 people, with an annual budget of $800 million.

Initially supplying active ingredients to Fiocruz's biologics division, Bio-Manguinhos, GSK started in 1985 to share production technology for a polio vaccine. In 1998, a five-year deal allowed for GSK to transfer Hib (haemophilus influenza type B) vaccine technology. Brazil's government purchased the vaccine from GSK at a negotiated price during that time, and by 2009, the country reached comprehensive immunization.

In 2009, GSK and Fiocruz announced a major new public–private partnership for a vaccine against pneumococcal infections, and the joint development of a dengue fever vaccine. GSK agreed to transfer technology for its Synflorix vaccine, including a complex manufacturing sequence of eight processes and over 200 quality control tests. GSK was to sell the vaccine to the Health Ministry over a ten-year period, at an initial price of €11.50 per dose, expected to drop to €5 in later years (vs a European price of €35–40 per dose). It was expected that GSK could sell over $1.5 billion of Synflorix vaccines, as Brazil aimed to vaccinate all newborns.

A second component of the GSK/Fiocruz deal was joint R&D for a vaccine for dengue fever, which infected between 50 and 100 million annually worldwide. GSK and the Brazilian government each pledged $51 million to research and test the vaccine (Daemmrich, 2012).

Domestic Brazilian firms

The industry is polarized between large, diversified generic firms and small biotech innovators with links to universities. Generic manufacturers include EMS, founded in the mid-1950s and reaching revenues of $2.8 billion by 2012, with five divisions: similar, generics, branded drugs, hospital drugs, and over-the-counter (OTC) products.

EMS's competitors include Hypermarcas (consumer goods conglomerate founded in 2001, with 2013 revenues of $1.8 billion, and OTC products including antiseptics like chlorhexidine). Another generic firm with nearly 50 years of history in Brazil, Ache has licensing agreements in 11 countries for its portfolio in areas including hypertension, atherosclerosis, depression, and inflammation. A partnership with Hypermarcas, EMS, and Uniao Quimica aimed to create a biotech joint venture, Bionovis.

In biotechnology itself, many companies have state and university links. Instituto Butantan is affiliated with the São Paulo State Secretary of Health. It was founded in 1901 according to the Pasteur Institute model, focusing research on venomous animals to develop vaccines such as rabies, hepatitis, diphtheria, and tetanus. It has since advanced to molecular biology and immunology, produces monoclonal antibodies, and operates the Hospital Vital Brazil, specializing in the treatment of poisonous animal stings.

International academic links are shown by other biotechs such as FK Biotecnologia, founded in 1999 and focusing on oncology with links with Dana Farber, and the Axis Biotec Group, set up in Rio de Janeiro's Biotechnology Development Park. Through its companies Cellpraxis and

Pharmapraxis, the group focuses on regenerative medicine and cell therapy, and has links with the Federal University of São Paulo (UNIFESP) and the University of South Florida. By contrast with Brazil, a very different innovation model is developing in China.

China

Although technology was a key part of the "four modernizations" program launched in 1978, the real impetus came in China after 2000. The 2006 National Medium and Long-Term Plan for Science and Technology Development aimed by 2010 to derive 60 percent or more of economic growth from technological progress. Between 2000 and 2010, China's R&D expenditure doubled as a share of GDP, to 1.75 percent, and China's world share of researchers was equal to that of the US (20 percent, with 1.4 million scientists).

Following China's entry into the WTO, its applications for international patents more than tripled in 2006–11, representing 9 percent of the world total. Of these, 21 percent were for chemicals and biopharmaceuticals. The 2006 National Plan aimed to increase R&D intensity to 2.5 percent of GDP by 2020, but also to promote "indigenous innovation," decreasing reliance on foreign technology by 30 percent or below existing levels. In the 12th Five Year Plan (2011–15), the emphasis was put on the life sciences, especially drug discovery and infectious diseases.

A key element of China's innovation system is the rising role of enterprises. Government research institutes account for less than 20 percent of R&D spend. Given that producing in China is a key way for foreign firms to access the market, multinationals have set up numerous R&D and production centers (over 350 R&D centers in Shanghai alone by 2010) (Fabre, 2014).

Health system in China

To provide health services to its population of 1.3 billion, China has over 900,000 medical facilities, most of them state-run. Although much progress has been made with basic coverage for 95 percent of the population, problems remain, such as a fragmented infrastructure, underfunding of preventive services, and wide regional variations in access to health care.

Public hospitals have a three-tier structure: tier 1 is made up of small community centers, tier 2 covers regional facilities, and tier 3 includes large provincial hospitals with 500 or more beds. Given the low incidence of private health insurance, part of medical expenses is paid out-of-pocket; this amounted to 36 percent of expenditure in 2010 (Herzlinger and Kindred, 2014). Due to the low reimbursement rate for hospitals, these expect a share of income from drug price mark-ups, since hospital pharmacies dispense a large proportion of drugs. Medicines from foreign and domestic suppliers are distributed by a fragmented network of wholesalers, and sold mostly by hospitals, with a smaller share sold in retail pharmacies (Yu et al., 2010).

National policy – intellectual property

The governance of the biotech industry includes the National Development and Reform Commission (NDRC), the State Food and Drug Administration, and the ministries of Health and Commerce. NDRC is in charge of strategic planning and regulates drug prices, the Ministry of Commerce regulates the export and import of medical devices and equipment, and the Ministry of Health (MOH) guides reform and monitors clinical trials (Wang et al., 2009).

The Patent Office of the State Intellectual Property Office (SIPO) is responsible for the processing of patent applications. In 2000, China's Patent Law was amended in order to comply with the TRIPS agreement of the WTO. However, there were significant enforcement issues, illustrated by the notable case of Pfizer's Viagra patent (Li et al., 2010).

After Pfizer's launch of Viagra (sildenafil) for erectile dysfunction (ED), it had a rapid global uptake, reaching over $1 billion in sales by 1999. In China, about 74 percent of men aged 60–69 suffered from ED, and by 2004, the market was estimated at $7–12 billion. In July 2000, Pfizer was approved to sell Viagra in China, and it was granted a patent by SIPO in 2001. Before approval, illegally imported Viagra pills and local counterfeits flooded the market.

As early as 1998, Guangzhou Viamen, a local firm, had registered its own drug under a trademark ("Weige") similar to the Viagra name, and it started selling it in 2003. In addition, within a month after Pfizer's patent approval, 12 local firms challenged its validity, arguing that it failed to fulfill a "novelty" requirement of China's Patent Law. Pfizer then filed a court case against SIPO. In 2006, SIPO's ruling was overturned and the patent upheld. However, despite a final appeal regarding the trademark, Pfizer lost that litigation in 2009. The company had invested $500 million in China for four production sites and an R&D center. Independently of the patent issue, due to the dominance of counterfeits and local generics, China only accounted for less than 0.5 percent of Viagra's global sales of $1.3 billion in 2000. The launch of branded competitors in 2003 (Bayer's Levitra and Lilly's Cialis) further eroded Viagra's share, before its China patent was due to expire in 2014.

This episode left a lasting impression on foreign companies, and the Office of the US Trade Representative released in 2005 a Special 301 report, including China on the Priority Watch List due to "serious concerns about [its] compliance with its WTO TRIPS obligations" (Li et al., 2010).

China's biopharmaceutical industry

China includes over 5,000 manufacturers, virtually all of which produce generics. Despite the government's efforts to consolidate it, partly by requiring compliance with GMP standards, the industry remains fragmented, with the top ten companies holding less than 11 percent of the market.

About 70 percent of production is Western-type drugs, with the rest comprising traditional Chinese medicines (TCMs). Most leading players, such as Sinopharm or Shanghai Pharmaceuticals, are state-owned enterprises (SOEs) that are often vertically integrated, owning pharmacy chains. Distribution is fragmented, with over 13,000 distributors, who reap most of the drug price margins (Tao, 2014).

Price controls and reductions have long been practiced in China. In 1996–2007 alone, the government reduced the price of 1,500 drugs 19 times. It also set up an Essential Drug List of over 300 medicines, for which hospital pharmacies could not raise prices. The National Reimbursement Drug List (NDRL) split over 2,000 products into two groups: one of about 500 low-cost drugs, and the rest comprising higher-priced, patented Western drugs (Herzlinger and Kindred, 2014). Within the industry, biotech had started to develop as early as 1984, and counted over 900 firms by the mid-2000s. The main coordinating body is the China Center for Biotechnology Development (NCBD), set up in 1983 under the Ministry of Science and Technology.

While the SOE status of many biotechs has advantages including government relationships, they have grown by increasing volume and prices, and have been largely limited to incremental innovations. However, some remarkable leaders have recently emerged, such as BGI (Beijing Genomics), founded in 1999 and now counting divisions in the Americas and Europe.

Foreign company strategies

In order to gain market access, and despite patent issues, most multinationals have been long established in China. Pfizer has several GMP manufacturing facilities, and set up in 2012 a joint venture with Hisun to market branded generics. Germany's Merck announced in 2011 a $1.5 billion investment for its Asian R&D center in Beijing. Similarly, Novo Nordisk invested an additional $100 million in a China R&D complex that opened in 2012.

Western companies also continue to be very active with Chinese acquisitions. In 2010 alone, Sanofi purchased an OTC distributor for $521 million, Nycomed acquired for $210 million a majority share of a local manufacturer, and Sumitomo paid $96 million for a minority share in a domestic distributor. GSK entered China in 1984 with a joint venture in Tianjin, and has since invested about $500 million in an R&D center and six manufacturing sites, with a research staff of 7,000. Its Chinese portfolio includes hepatitis and respiratory drugs, antibiotics, dermatology, depression, and oncology products (Tao, 2014).

Bayer operates several companies in Greater China (including Hong Kong and Taiwan) and is engaged in several collaborations with the Chinese Academy of Science. It also supports research programs at Tsinghua University and the China European International Business School (CEIBS) in Shanghai.

Domestic company strategies

The industry is dominated by large SOEs with greatly diversified portfolios. Some optimize their innovation through technology transfer from joint ventures. Sinopharm, jointly owned by the China National Pharmaceutical Group and Fosun International, reached revenues in 2013 of $33 billion and has a broad portfolio, ranging from TCMs to devices, reagents, and pharmaceuticals. Its joint ventures include deals with Otsuka, J&J, and BMS. The Harbin group products include antibiotics and OTC brands, with 73 percent of its portfolio devoted to Western medicines, and the rest to TCMs. It is expanding its R&D through investment from banks such as Citic Capital in Hong Kong.

A notable example of an evolving SOE is Shanghai Pharmaceuticals.

Shanghai Pharmaceuticals

Shanghai Pharmaceuticals (SPH) is a vertically integrated conglomerate with a wide range of products, from active ingredients to small molecules and biologics as well as TCMs. It was formed by the merger of several SOEs and had an IPO in Hong Kong in 2011. By that time, its revenue was nearly $9 billion. Major initial investors were Temasek Holdings (Singapore), the Hong Leong Group (Malaysia), the Bank of China, and Pfizer.

Unusually for an SOE, SPH has spent about 4.5 percent of revenues on R&D. It focuses on five therapeutic areas: cardiovascular, metabolism, central nervous system, anti-infectives, and immunology. Its strategy is to prioritize products with high margins, market shares, and entry barriers. SPH also owns China's second-largest distribution network, engages in direct sales with hospitals, and has a network of about 1,700 retail pharmacies. All of the company's manufacturing sites meet or exceed GMP standards.

Before its IPO, SPH set up an agreement with Pfizer for distribution and cooperation on drug approval and commercialization, which could be followed by further R&D collaboration (Herzlinger and Kindred, 2014).

At the other end of the industry spectrum, cutting-edge biotechs have emerged, and one of the most dynamic and internationally recognized is BGI.

BGI's strategy

Founded in 1999, the group has now gained a global footprint, with divisions in the Americas and Europe. It was created as a nongovernmental independent research institute in order to represent China in the Human Genome Project. By 2002, BGI had sequenced the rice genome. The following year, it decoded the SARS virus genome and developed a diagnostic kit. It also set up a genomics collaboration with Zhejiang University. By 2008, it had published the first human genome of an Asian individual.

In addition, BGI is certified as meeting ISO 9001 standards for high-throughput sequencing services. It set up in 2010 BGI Americas in

Cambridge, MA, and BGI Europe in Copenhagen. It has alliances with most of the top global pharmaceutical firms, and has made acquisitions such as California-based Complete Genomics, a supplier of DNA sequencing technology, for $118 million (Specter, 2014).

Conclusion

In conclusion, the analysis of biotechnology innovation in India, Brazil, and China shows a very diverse set of health systems, with different proportions of government regulation and financing, but with general progress toward a goal of basic universal health-care coverage. The industry situations are also varied, with a dominance of SOEs in China, vs long-established generic manufacturers in Brazil and India.

Across these countries and in emerging markets in general, local talent and university resources, linked to foreign investment, are fueling the rapid development of remarkable biotechnology companies such as Biocon and BGI. To reach the optimal potential of their technology, governments in emerging markets will need to address the remaining challenges, from inequities in access to health care, to IP and manufacturing quality issues.

References

Abuduxike, A. and Aljunid, S. (2012). "Development of Health Biotechnology in Developing Countries: Can Private-Sector Players Be the Prime Movers?" *Biotechnology Advances*, 30: 1589–601.

Biocon (2015). Biocon company data, Company History. Retrieved from http://www.biocon.com/biocon_invrelation_com_history.asp.

Booz & Company. (2011). "Pharma Emerging Markets 2.0: How Emerging Markets Are Driving the Transformation of the Pharmaceutical Industry." Booz & Company. Retrieved from http://www.strategyand.pwc.com/global/home/what-we-think/reports-white-papers/article-display/pharma-emerging-markets

Burns, L. (ed.). (2014). *India's Healthcare Industry*. Cambridge: Cambridge University Press.

Daemmrich, A. and McKown Cornell, I. (2012). "GlaxoSmithKline in Brazil: Public/Private Vaccine Partnership," *Harvard Business School*, Case 9-712-049, June.

Fabre, G. (2014). "The Real Leap Forward: China's R&D and Innovation Strategy." In Taylor, R. (ed.) *The Globalization of Chinese Business. Implications for Multinational Investors* (pp. 3–25). Oxford, UK: Elsevier/Chandos Publishing.

Frew, S. (2014). "India's Biotechnology Sector." In Burns, L. (ed.) *India's Healthcare Industry* (pp. 477–99). Cambridge: Cambridge University Press.

Gragnolati, M., Lindelow, M., and Couttolenc, B. (2012). *Twenty Years of Health System Reform in Brazil – An Assessment of the Sistema Unico de Saude*. Washington, DC: World Bank.

Herzlinger, R. and Kindred, N. (2014). "Shanghai Pharmaceuticals." Harvard Business School, Case 9-313-016, 2014.

IBEF (India Brand Equity Foundation). (2014). "Biotechnology Industry in India." Retrieved from http://www.ibef.org/

IMS Institute for Healthcare Informatics. (2013). "The Global Use of Medicines: Outlook through 2017." Retrieved from http://www.imshealth.com/portal/site/ imshealth/menuitem.762a961826aad98f53c753c71ad8c22a/?vgnextoid=9f819e46 4e832410VgnVCM10000076192ca2RCRD&vgnextchannel=a64de5fda6370410Vgn VCM10000076192ca2RCRD

IMS Institute for Healthcare Informatics. (2014). "Harbingers of Change in Healthcare." Retrieved from http://www.imshealth.com/portal/site/imshealth/menuitem.762a96 1826aad98f53c753c71ad8c22a/?vgnextoid=af47009812568410VgnVCM10000076 192ca2RCRD&vgnextchannel=a64de5fda6370410VgnVCM10000076192ca2RCRD

La Forgia, G. and Nagpal, S. (2012). *Government-Sponsored Health Insurance in India – Are You Covered?* Washington, DC: World Bank.

Lawton R. B. (ed.) (2014), *India's Healthcare Industry.* Cambridge: Cambridge University Press.

Li, Y., Lu, J., Tao, Z., and Wei, S-J. (2010). *Viagra in China: A Prolonged Battle over Intellectual Property Rights.* Asia Case Research Centre, University of Hong Kong, Case HKU902.

Lima, M. (2015). "Brazil Analysts Raise Inflation, Cut GDP Forecast for Third Week." *Bloomberg News.* Retrieved from http://www.bloomberg.com/news/2015-01-19/ brazil-analysts-raise-inflation-cut-gdp-forecast-for-third-week.html

MOITI (Massachusetts Office of International Trade and Investment). (2008). "Brazil Biotechnology Industry." Retrieved from http://kooperation-international.de/file-admin/public/cluster/Belo_Horizonte/BrazilBiotechnologyIndustry_Mass.pdf

Morel, C. M., Tara, A., Denis, B., et al. (2005). "Health Innovation Networks to Help Developing Countries Address Neglected Diseases." *Science,* 309 (5733): 401–4.

Mori Sarti, F., Ivanauskas, T., Montoya Diaz, M., and Coelho Campino, A. C. (2012). *Towards Universal Health Coverage in Latin America and the Caribbean: Case Study Measuring Inequalities in Health in Brazil.* Washington, DC: World Bank.

Resende, V. (2012). *The Biotechnology Market in Brazil.* Washington, DC: US Department of Commerce.

Sachan, N., Tathambothia, A., Nehru, R., and Dharanaj, C. (2013). *Collaborative Commercialization at Gilead Sciences: Resolving the Innovation vs. Access Tradeoff.* Indian School of Business, Case ISB 025, 2013.

Specter, M. (2014). "The Gene Factory." *The New Yorker.* Retrieved from http://www. newyorker.com/magazine/2014/01/06/the-gene-factory

Tao, Z. (2014). *The GSK Scandal: When Questionable Global Practices Met Imperfect Institutions in Emerging Markets.* Asia Case Research Centre, University of Hong Kong, Case HK 1049.

Tyagi, S., Mahajan, V., and Nauriyal, D. K. (2014). "Innovations in the Indian Drug and Pharmaceutical Industry: Have They Impacted Exports?" *Journal of Intellectual Property Rights,* 19: 243–52. Retrieved from http://nopr.niscair.res.in/ handle/123456789/29286

Wang, K., Hong, J., Marinova, D., and Zhu, L. (2009). "Evolution and Governance of the Biotechnology and Pharmaceutical Industry in China." *Mathematics and Computers in Simulation,* 79 (9): 2947–56.

World Bank. (2013). *Brazil Report, 2013.* Washington, DC: World Bank.

World Health Organization. (2013). *World Health Report 2013.* Geneva: Research for Universal Health Coverage.

Yu, X., Li, C., Shi, Y., and Yu, M. (2010). "Pharmaceutical Supply Chain in China: Current Issues and Implications for Health System Reform." *Health Policy,* 97 (1): 8–15.

14
Catch-Up Innovation and Shared Prosperity

Mark A. Dutz

This chapter provides an economic development perspective on the role of catch-up innovation for shared prosperity. Innovation happens when entrepreneurs commercialize new ideas through markets to improve business productivity and raise profits. In developing countries, most businesses operate quite far from the global "best practice" technological frontier. Innovation in these contexts should be defined to include not only frontier new-to-the-world innovation but also catch-up innovation, namely the adoption and commercialization of existing products and technologies that are new to the firm. These "imitation" catch-up activities are legitimately considered as innovations since their adoption involves adaptations to the local context. They typically involve lower risk to the firm than frontier innovations (Dutz, 2015). And catch-up possibilities are the main reason emerging economies grow faster on average than high-income ones. The chapter focuses on how catch-up innovations can affect the income and consumption opportunities of poor people, mainly through the linkages between innovation and jobs. The chapter also examines selected policy actions for boosting shared prosperity – promoting the income growth of the bottom 40 percent of people in each country – by helping lower-income people contribute to and benefit from catch-up innovation.

Broadly defined, innovation is synonymous with business productivity upgrading: it is the growth in output not from greater use of existing types of physical capital and labor but from better uses of them.[1] Importantly, innovations require investments in intangible knowledge capital, including learning about and adopting "soft" technologies such as better branding, managerial practices, and business models. Aggregated across all enterprises, knowledge capital investments are an important addition to traditionally measured GDP, accounting for over 10 percent of expanded GDP in the US and over 7 percent in China. These expenditures drive an important type of structural change (SC), namely within-firm SC. And recent studies suggest that the within-firm component of SC is very important. In India, productivity increases of the largest firms drove its manufacturing miracle, accounting for almost four-fifths of the rise in productivity of manufacturing plants between 1993 and 2007 (Bollard et al., 2013). And differences in business productivity dynamics between the US, Mexico, and India, driven by higher investments in knowledge capital assets over time in the US, could plausibly account for most of the gap in total factor productivity (TFP) (Hsieh and Klenow, 2014).

Innovation matters for both growth and shared prosperity. We know that long-run growth relies on innovation (Aghion and Howitt, 2009). Changes in the patterns of technology diffusion linked to catch-up innovation reportedly account for 80 percent of the increase in cross-country differences in per capita income over time (Comin and Ferrer, 2013). Adoption lags have been converging over time across countries. However, penetration rates of new technologies within countries have been diverging – with large gaps within countries between the best and average technological practices within each industry. The relation between innovation and shared prosperity, on the other hand, is more nuanced. Three main channels from investments in innovation capabilities to shared prosperity are: (1) directly through more and better jobs for lower-income entrepreneurs, managers, and workers; (2) directly through higher consumption by lower-income people; and (3) indirectly through technological and other positive spillovers within and across industries. Innovation, however, can also stifle shared prosperity: if the benefits of productivity gains are appropriated by owners of physical capital and senior managers of businesses without commensurate income increases by lower-income workers; if innovation is primarily directed to products only affordable by higher-income consumers; and if innovation occurs mainly in export-oriented enclaves where the positive spillovers largely flow to firms and consumers abroad rather than to the rest of the local economy.

The role of government policy actions to boost shared prosperity through catch-up innovation is critical, but not well enough understood yet. Policy actions may need to be "tilted" or oriented to improve the income and consumption opportunities of poor people. This policy tilting can occur at three levels: (1) at the level of individuals through skills upgrading and labor market intermediation support; (2) at the level of firms, mainly through technology adoption support; and (3) at the level of the broadly defined business environment through social mobility support.

The chapter is organized as follows. The first section presents evidence across countries on the role and importance of innovation as a driver of overall SC. The second section presents the main channels through which innovation can result in greater shared prosperity. The third section discusses some policy actions that support investments in innovation capabilities that are likely to boost shared prosperity. A final section concludes. The limited experience of countries that have been experimenting with "tilting" policy actions to favor the productive employment of lower-income people suggests that more policy experimentation and scaling up successful approaches could yield dividends if shared prosperity is a valued policy goal.

Innovation and business productivity

Innovation is the most important driver of sustainable growth, but is not as well understood as it should be. This section has two main messages. First, business productivity upgrading is an important type of SC that is constantly transforming the economy. Second, more detailed measurement of firm expenditures on the different types of capabilities required for business productivity upgrading is a prerequisite to better understand how catch-up innovation occurs, and the appropriate role of government in best promoting it.

The importance of business productivity upgrading for aggregate productivity growth

Business investments in knowledge capital contribute to an important type of SC, namely within-firm SC, which in turn contributes to aggregate productivity dynamics. A three-way typology of the process of industrial development helps clarify the role of investments in knowledge capital and within-firm SC within aggregate economy-wide productivity dynamics:[2]

1. *Between-industry SC,* as resources are reallocated across industries from some activities (e.g. lower-productivity agriculture) to others (e.g. higher-productivity, typically more physical and knowledge capital-intensive agriculture, manufacturing, and services).
2. *Within-industry SC,* as resources are reallocated across enterprises within industries, both on the extensive margin (Schumpeterian "creative destruction" selection effects as less productive firms exit and more productive firms enter) and on the intensive margin (as less efficient firms contract and more efficient ones expand), altering each industry's structure.
3. *Within-firm SC,* as resource reallocation, new investment, and innovation within firms alter each firm's structure and productivity, linked to the adoption and use of new-to-the-firm technologies and to shifts to higher-productivity product lines.

The knowledge capital approach opens up the black box of conventional TFP by explicitly measuring the knowledge elements of the business component of overall TFP dynamics. It allows their impact and spillovers to be measured across businesses and industries.

One of the most direct empirical explorations of the relative importance of business productivity upgrading for overall productivity growth is a recent analysis of India's manufacturing productivity performance. Average plant growth rates increased in value added from 4.2 percent per year (1980–92) to 9.6 percent per year (1993–2007). Bollard et al. (2013) decompose value-added growth into input growth (accounted for jointly by labor and physical capital) and productivity growth, and then decompose the latter into the three-way SC typology. They trace the largest productivity growth to within-plant TFP as opposed to cross-industry or within-industry reallocations: within-plant TFP growth increases from 2.4 to 7.1 percent across the two periods, with by far the largest part of that increase accounted for by large plants of 200 workers or more. Harrison et al. (2013) report similar findings: most of the productivity improvements in Indian manufacturing between 1985 and 2004 occurred through "learning" (improvements in average firm productivity or within-firm SC) rather than through "stealing" (reallocating market shares towards more efficient firms).

Another recent exploration of the importance of within-firm SC is a comparison of employment growth and TFP over the plant life cycle by Hsieh and Klenow (2014). They examine the importance of plant-specific, intangible knowledge capital accumulation over time for understanding differences in aggregate manufacturing TFP between the US, India, and Mexico. In a prior empirical study, Hsieh and Klenow (2009) had shown evidence that resource misallocation between plants accounts for about one-third of the gaps in aggregate manufacturing TFP between the US and China and India. One way to interpret their more recent finding, as they emphasize, is that although the gains from SC across plants are important, the differences in within-plant productivity account for most (two-thirds) of the remaining gap in aggregate TFP between poor and rich countries.

The importance of investments in knowledge capital for innovation

Innovation typically requires investments by businesses in different types of "soft," intangible assets. Historically, expenditures on knowledge capital were treated as current intermediate expenditures – thereby assuming that all their benefits are reflected in the current year's output of goods and services. In terms of company valuation, this was done at least in part because external market transactions were not available for accountants to validate these flows of value created within the company. Capitalizing these expenditures based on imputations inferred from the cost of capital makes corporate profits look larger, as businesses would no longer count, for instance, net R&D after depreciation as a cost. When knowledge capital is added to the balance sheets of a sample of R&D-oriented US companies in 2006, it becomes the single most important source of shareholder value: while conventional equity accounts for 42 percent of the market capitalization of the firms, adding knowledge capital more than doubles this to 86 percent of market capitalization. So including knowledge capital assets in financial statements largely removes the puzzle that the market typically puts a value on shareholder equity of more than twice the reported book value of a company (Hulten and Hao, 2008).

Seminal work applying direct expenditure methods to measure investments in knowledge capital was done by Corrado et al. (2009; see also OECD, 2013). This method correctly capitalizes those outlays that contribute to production and value beyond the taxable year and treats them as longer-lived knowledge investments – using the same cost-based accounting that is used for physical capital. The Corrado et al. (2009) classification divides investments in knowledge capital into three types:

1. *Digital capital* (or "computerized information"): including software and databases.
2. *Intellectual capital* (or "innovative property"): including R&D, creative assets, architectural, engineering, and other designs, and mineral exploration and evaluation.
3. *Human–organizational capital* (or "business competencies"): including *managerial capital* (individual skills upgrading through worker and management training); *marketing capital* (market research, branding, and advertising); *organization capital* (organizational improvements and new business models); and *collaboration capital* (an additional subcategory suggested by Dutz et al. (2014) that is likely most important for businesses more distant from the global technological frontier, namely spending on networking and on peer-to-peer learning from local clusters, foreign buyers and sellers, consultants, "technology capture" study tours, and other mechanisms for knowledge transfer).

Table 14.1 presents available data on aggregate knowledge capital accumulation in Brazil, China, and India relative to the US. By treating capitalized investments in knowledge as separate measured inputs to GDP growth, the economy's gross investment rate and GDP are increased: the inclusion of investments of the main types of knowledge capital expands measured GDP in 2006 relative to its conventional counterpart by 10.4 percent for the US, 7.1 percent for China, 4.3 percent for Brazil, and 2.7 percent for India.

Measurement of knowledge allows the investment efforts across businesses and countries in building catch-up innovation capabilities to be compared. As highlighted in Table 14.1, spending on business competencies is the most important category in the US, but less important in emerging economies. Reported spending on training is particularly low, especially in India. This may be linked to underreporting what is actually spent, but also to insufficient incentives for enterprise-based training linked among others to informational barriers and insufficient market competition. Bloom et al. (2014) suggest that investment in managerial capital as a form of catch-up innovation is significantly lower in many developing countries. Caliendo et al. (2015) provide evidence that spending on organizational

Table 14.1 Knowledge capital investments as percentage of expanded GDP, 2006

	US		China		Brazil		India	
1. *Computerized information*	*1.24*	*12%*	*1.88*	*27%*	*1.10*	*26%*	*0.22*	*8%*
2. *Innovative property*	*4.07*	*39%*	*3.40*	*48%*	*2.01*	*47%*	*1.48*	*56%*
Research and Development (R&D)	1.69	16%	1.02	14%	0.56	13%	0.80	30%
Mineral exploration and evaluation	0.78	8%	0.21	3%	0.03	1%	0.00	0%
Copyright and license costs	0.55	5%	0.08	1%	0.11	3%	0.05	2%
Development costs in financial ind.	0.55	5%	0.47	7%	1.10	26%	0.38	14%
Designs, incl. architectural and engineering	0.50	5%	1.62	23%	0.21	5%	0.25	9%
3. *Business competencies*	*5.04*	*49%*	*1.80*	*25%*	*1.17*	*27%*	*0.95*	*36%*
Reputation and Branding	1.35	13%	0.38	5%	0.56	13%	0.11	4%
Advertising expenditure	1.24	12%	0.38	5%	0.51	12%	-	-
Market research/Branding	0.11	1%	-	-	0.05	1%	-	-
Training and Development	1.05	10%	0.29	4%	0.34	8%	0.01	0%
Continuing vocational training	-	-	0.29	4%	-	-	-	-
Apprentice training	-	-	-	-	-	-	-	-
Business process improvements	2.64	26%	1.13	16%	0.27	6%	0.83	31%
Purchased	-	-	-	-	0.03	1%	-	-
Own-account	-	-	1.13	16%	0.24	6%	-	-
Total knowledge capital investment	*10.35*	*100%*	*7.08*	*100%*	*4.28*	*100%*	*2.65*	*100%*

Source: Dutz et al. (2012b), Hulten and Hao (2012), and Hulten et al. (2012).

capital through changes in the internal organization of firms is critical to understanding how firms expand and contract. They find that firms that add organizational layers expand output much more. Although such firms pay lower average wages, in part by hiring less experienced workers in all preexisting layers, this expansion could provide a stepping stone for lower-income workers to be hired for the first time and then progress to better-paying jobs. There is also suggestive evidence that spending on collaboration capital may be important for the diffusion and adaptation of existing technologies to local contexts. Dutz et al. (2014) present such evidence for the Chilean wine industry: regression results show that spending on foreign consultant services is statistically significantly correlated with export sales. They also report that few firms actually spend on R&D: training, software, and reputation and branding were the top three asset categories across both Chile and the UK (based on a similar survey applied across manufacturing industries). Importantly, the average life lengths across all asset types in Chile and in the UK are greater than one year, supporting the case for capitalizing knowledge assets.

Investment growth in physical capital shows no statistically clear relation with TFP growth. By contrast, there is a strong positive correlation between TFP and knowledge capital growth, consistent with spillover effects – when one firm invests in software or adopts a new business process, not only does that firm become more productive but other firms also benefit over time (Corrado et al., 2013). This is good for economy-wide productivity and provides a rationale for policy intervention.

Pathways from innovation to shared prosperity

This section discusses the importance of three main channels from investments in innovation capabilities to shared prosperity. First, and most importantly, innovation has the potential to improve the productivity and remuneration of jobs held by poor people. Second, innovation can directly improve the consumption opportunities of lower-income households. Third, innovation can indirectly stimulate shared prosperity to the extent that there are technological and other positive spillovers that improve the income and consumption opportunities of the poor.

Innovation impacts through more and better jobs

Innovation impacts employment through multiple channels of varying timescales and complexity. This subsection has three main messages. First, product innovations often generate more jobs than process innovations, but process innovations are also important if cost reductions expand demand sufficiently to require expansions in jobs. Second, innovating firms often employ more low-skilled and female workers than noninnovating firms. Third, as innovation increases productivity, it creates winners and losers, typically increasing the real income of owners of physical and knowledge capital and the earnings of higher-skilled workers over others – and absent government intervention, the types of technologies that could spur income growth for poor people may not be adopted.

The overall effect of innovation on jobs is sensitive to the character of the innovation and its setting. Process innovation can lead to productivity gains which enable firms to produce the same level of output with fewer inputs, including direct labor-saving substitution impacts or "displacement" effects. However, these direct negative effects can be counterbalanced by indirect output expansion or "compensation" effects when the cost reductions from the innovation spur price reductions that in turn stimulate higher demand and greater output. Product innovation, on the other hand, generally leads to "market-expansion" output effects when it stimulates domestic and foreign demand for the firm's outputs, thereby enhancing labor demand for the innovating firm. However, innovation typically causes demand diversion from substitute products of other firms in the same or a similar industry or "business-stealing" effects and thereby has an uncertain impact on aggregate employment. The balance of these countervailing impacts on jobs is an empirical question, depending among others on the technology employed, the substitutability of input factors, the own- and cross-price elasticities of demand, and the nature of the business environment.[3]

Dutz et al. (2012a) debunk a conventional view that the benefits of innovation flow only to skilled workers and shareholders of technically sophisticated companies. They provide support for a contrasting view based on the broader "new-to-the-firm" interpretation of catch-up innovation. Analyzing firm-level data from over 26,000 manufacturing enterprises across 71 countries (ranging in per capita income from Germany to Niger), they show that businesses that innovate in this sense employ a higher share of unskilled workers, a higher share of female workers, and attain higher TFP and more rapid employment growth than firms that do not innovate. Importantly, they find no evidence of corresponding offsetting declines in the output and employment of domestic competitive rivals. However, unskilled workers may still be receiving significantly lower income growth; on the other hand, they may constitute new hires that eventually transition to skilled workers through on-the-job learning. These are questions that their data unfortunately did not allow them to address. Importantly though, their findings, coupled with the increasing empirical support in the literature for the view that low-wage jobs are a stepping stone for the integration of the jobless into employment and better-paid work in the future, do provide a key underpinning to innovation-driven inclusive growth.

The effect of innovation on jobs importantly depends on the types of technologies that are adopted. The introduction of new high-skill-biased and labor-saving technologies can lead to widening employment and earnings differentials between higher- and lower-skilled workers, and to technological (structural) unemployment – creating three overlapping sets of losers (Brynjolfsson and McAfee, 2011, 2014): (1) low-skilled workers (technologies often displace routinized tasks and increase the value of more abstract tasks); (2) "non-superstar" workers (in some winner-take-all industries, a few highly talented people get the lion's share of the rewards); and (3) owners of own-labor only (to the extent that technology reduces the relative payments to labor, owners of capital will capture a bigger share of income from production). Brynjolfsson and McAfee argue that mismatches between accelerating technologies and stagnant skills and organizations can be addressed by fostering two types of knowledge capital: skills upgrading and organizational capital. Entrepreneurs have the lead role in inventing new business models that leverage evolving technologies and make productive use of available pools of labor.[4]

Innovation impacts through improved consumption opportunities

Innovation can directly improve the consumption opportunities of lower-income households to the extent that the resulting products, quality mix, and service delivery mechanisms either are more affordable or better meet their needs. However, absent appropriate policy actions, businesses may direct their innovation investments primarily towards higher profit-margin products only affordable by higher-income consumers – as there may be insufficient market incentives for investments in the production and delivery of products such as vaccines for the neglected diseases of the poor and seeds for arid soils where only poor people live. It is noteworthy that recent initiatives by government, business, foundations, and other entities seeking to have a positive impact on the welfare of lower-income households through support of "inclusive" or "base-of-pyramid" innovation have focused largely or even solely on consumption. This focus is important; among others, better health and education services targeted to the poor are critical in increasing their on-the-job productivity. However, government support of inclusive innovation can have its largest impact through the jobs that the adoption, adaptation, and use of innovations can generate – which in turn enable poor people both to improve their own productivity (including through the learning networks accessed as part of employment) and to purchase needed products through their progressively higher incomes.

Innovation impacts through spillovers

In addition to its direct impact on jobs and consumption, innovation also affects shared prosperity through its indirect impact on learning and follow-on innovations by individuals outside the firm and by other firms, both through markets and through spillover effects not reflected in market prices. There are two main messages. First, there are many spillovers that, through their impact on jobs and consumption opportunities, can help promote the well-being of lower-income people. Trajtenberg (2009) broadens the notion of spillovers beyond traditional technological ones by emphasizing a range of innovation-related positive externalities that are particularly relevant for locations where markets are less dense and less well connected. These include: (1) "post-innovation competition effects within markets" spillovers, namely the benefits not captured through markets spurred by an entrepreneur who breaks the mold of a static market by introducing an innovation, thereby forcing rivals to respond and triggering further innovations; (2) "demonstration effects in the diffusion of innovation" spillovers, namely the noncaptured benefit of early adopters that influence the decisions of later adopters, which may take the form of either network externalities, informational effects, or providing a new role model for entrepreneurial individuals to emulate; and (3) other forms of learning spillovers, including employees who share learning when they move to other enterprises.[5]

Second, such technology spillovers do not always result in significant gains to lower-income people. Lach et al. (2008) and Trajtenberg (2009) highlight how policies supporting the development of the Israeli software industry failed to promote increased software usage by other local industries: while ICT industries grew at 16 percent per year during the 1990s, TFP actually declined on an annual basis between 1996 and 2004 in other industries such as transport, construction, retailing, and other business services.

Public policy actions: supporting learning from experimentation

This section examines three types of policy interventions for supporting shared prosperity outcomes from catch-up innovations: (1) at the level of individuals through skills and labor market intermediation support; (2) at the level of firms through technology adoption support; and (3) at the level of the business environment through "creative transformation" and social mobility support.

Supporting innovation at the individual level

The most direct channel to boost shared prosperity is through the productivity upgrading of poor individuals. This should be complemented with better-functioning labor market intermediation support policies to enable them to convert their capabilities into income via the best-matched jobs for them, and move if needed to where these jobs are located. As Piketty (2014) recognizes,

> over a long period of time, the main force in favor of greater equality has been the diffusion of knowledge and skills ... [but] the principal force for convergence [of wealth] – the diffusion of knowledge – is only partly natural and spontaneous. It also depends in large part on educational [and TVET, on-the-job skills upgrading, and labor market intermediation support] policies.

Shared prosperity may benefit from a "tilting" towards lower-income people of four human capital support services: (1) the general education of individuals (from early childhood development to quality secondary and tertiary education), with a particular focus on strengthening each country's engineering capacity; (2) TVET (technical and vocational education and training) services including prejob technical education and apprenticeships, professional training for specific job profiles, and technical skills upgrading; (3) on-the-job experience and within-firm training; and (4) labor market intermediation services that inform and align general education and TVET with the current and anticipated future needs of businesses and raise the likelihood of good employer–employee matches.

Supporting innovation at the firm level

There are a number of ways that policies supporting investments in innovation capabilities at the level of the firm can support shared prosperity. First, supporting innovation at the firm level should be more about catch-up innovation and the diffusion, adoption, and adaptation to local context of already existing technologies than about frontier innovation.[6] For all countries, the cost of not adopting better technologies can be high in terms of forgone growth. Two key sets of policies are facilitating access to, and stimulating adoption of, better technologies: the former through open trade, FDI (foreign direct investment), licensing, technology extension services, and matching grant support;[7] and the latter through standards, regulations, government procurement, and benchmarking programs.

Second, catch-up innovation driven by technology diffusion is often surprisingly slow and can benefit from policies to stimulate uptake by addressing informational problems. Even in the US, it took more than 10 years for half of the major iron and steel firms to adopt by-product coke ovens or continuous annealing lines (Mansfield, 1961). Bloom et al. (2013) found that Indian textile firms were not using cost-effective management practices that have been adopted widely elsewhere. After examining possible reasons, they suggest that informational barriers were the primary factor, in addition to insufficient pressure from market competition. For instance, owner-managers frequently argued that their own product quality was so good they did not need to record defects. However, this turned out to be a mistaken view given that quality was poor by international standards. It suggests that owner-managers had incorrect information on benefit–cost calculations within their own firms. A key question to guide policy action is to understand what works to incentivize firms to adopt better practices in a cost-effective manner that can be scaled across many firms – not only through regulatory reforms promoting greater product market competition but through policies that help reduce information frictions. Relevant issues include how best to reduce barriers to a well-functioning consulting market for advice, how to create more effective benchmarks so that firms understand how far behind they are and therefore are interested in tapping the advice market, and whether it is cost-effective to support the setting up of advice shops and management demonstration projects (see Bloom et al., 2014).

Third, an important area of firm-level pro-poor innovation policy involves support to global collaboration and translational research – to translate existing basic science into commercial solutions adapted and verified to meet local needs of poor people. Investments in joint learning through collaboration, including outlays on peer-to-peer networking with other enterprises and on participation in global value chains to capture existing but new-to-the-firm knowledge, can improve productivity and earnings in low-productivity activities. For example, interviews of leading businesses in

Ethiopia, Ghana, Tanzania, and Zambia highlight that a key determinant of business productivity upgrading was prior investment by these companies' leading entrepreneurs to learn what to produce, where to source inputs, and where to sell outputs through participation in import–export trading networks (Sutton, 2010, 2012, 2013). Dutz and Vijayaraghavan (2014) highlight how global collaboration between Indian researchers and firms and foreign closer-to-the-frontier universities and firms led local biotech firms to adopt structured research protocols as a critical process technology. A notable outcome of the collaboration in translational research was the development of an oral vaccine against rotavirus-caused diarrhea that is expected to be sold to poor people at $1 a dose. And Correa and Schmidt (2014) describe how global collaboration between Brazil's Embrapa and the US Agricultural Research Service led to the development of two-crop-per-year soybeans, more than doubling land productivity.

Supporting innovation at the business environment level

There are two main messages to create an enabling business environment that spurs productivity upgrading in ways that support more inclusive growth. First, the traditional policy emphasis to create a business environment that stimulates "creative transformation" remains critical so that firms are pressured to upgrade existing technologies.[8] Two complementary types of policy actions are required: regulatory simplification and regulatory quality upgrading. Regulatory simplification actions should focus on removing distortionary rules that prevent successful firms from growing and that prevent experimentation and learning. And additional public support actions to enhance regulatory quality are required to complement markets where externalities warrant public interventions. These include the promotion of competition law enforcement and competition advocacy, such as ensuring level-playing-field access to finance and other essential business services,[9] and the public–private provision of common logistics infrastructure and other business services subject to coordination failures.

Distortionary regulations that punish growth by imposing inordinate "costs of success" inhibit productivity upgrading. Based on evidence from India and Mexico, Hsieh and Klenow (2012) suggest that the return on business investments in knowledge capital may be lower in Mexico and India than in the US due to a range of distortions punishing enterprise expansion – such as higher tax enforcement and corruption, difficulties in obtaining skilled managers, bigger contractual frictions in hiring nonfamily labor, difficulties in buying land to expand, higher costs of shipping to distant markets, finance frictions, and other inappropriate size-based thresholds that disincentivize young firms from growing. According to Garicano et al. (2013), the welfare losses of regulations that "tax" firm growth, such as a 50-employee regulatory threshold for labor laws, can be substantial – up to 5 percent of GDP. Such size-based policies as well as policies that "tax" formality significantly more than informality encourage firms to stay informal and small and not invest sufficiently in knowledge capital. They should be avoided.

Distortionary rules that impose inordinate "costs of failure," by inhibiting risky experimentation, also stifle productivity upgrading. Distortionary costs are imposed, among others, by excessively strict bankruptcy regulations and the inability to borrow again in formal financial markets after bankruptcy, high labor hiring and firing costs, the inability to share high-fixed-cost facilities, the absence of sufficiently deep lease, rental, or resale markets, and the negative social stigma associated with failure (not viewing failure as a learning mechanism and as a by-product of high-risk activities). Governments should address these barriers, as well as support information dissemination such as the documentation and widespread diffusion of role models of successful high-risk-taking local entrepreneurs.

The impact of these regulatory reforms may not, absent additional policy actions, boost shared prosperity. Domestic and international product market liberalization and competition, without additional facilitating policies, may result in increased income inequality.[10] Recent empirical work on the unbalanced effect of trade liberalization in Brazil emphasizes the high and long-lasting mobility costs for some workers who lose from productivity-related adjustments (Dix-Carneiro, 2014; Dix-Carneiro and Kovak, 2014). Workers in hi-tech manufacturing prior to the trade liberalization, especially those older, higher-skilled workers (with higher initial wages), faced the largest losses. And workers in regions more negatively affected face continuously deteriorating labor market outcomes even 15 years after the end of the trade liberalization, with significant worker job transitions from tradable to lower-paid nontradable and informal jobs. More broadly, supporting the dynamic adjustment of lower-income workers to permanent technology-related shocks is essential as workers rather than firms bear the brunt of adjustment costs: while a firm's adjustment costs to trade and technology shocks are 0.05 times the wage bill, the average worker's mobility costs are 13 times the annual average wage (Hollweg et al., 2014).

The second message is that additional policies are essential to make "creative transformation" a faster and lower-cost transition for poor people. In addition to policies helping make less well-off individuals more productive and employable, a mix of policy actions are required to ease poor people's mobility, and increase the likelihood that regulatory reform and additional public support actions promoting productivity also boost shared prosperity. Based on the recent work by Dix-Carneiro (2014) and Dix-Carneiro and Kovak (2014) on the unbalanced effect of trade liberalization, there is a rationale for policies to help negatively affected workers adjust more quickly to productivity-related reforms, with moving subsidies performing better in compensating the losers relative to retraining policies. Finally, older workers face substantially higher costs of mobility, so more attention should be given to them following trade reform or other technology-related shocks.

Conclusions

This chapter provides a framework and some illustrative examples on catch-up innovation, and the possible role of government policy actions to boost shared prosperity. The framework and examples highlight how little is known about the linkages between innovation and shared prosperity, and about the design of policy actions that improve outcomes. A number of the issues at the interface between innovation and shared prosperity are quite complex and require a more detailed general framework to be developed as part of a larger work program of analytical, empirical, and policy research. The chapter is intended as a starting point to stimulate further work in this area.

The framework is in three parts. First, to understand how innovation interacts with shared prosperity, we need to better measure investments by enterprises in different types of knowledge capital. There is a wide-open research agenda here to collect and analyze data to better understand which types of knowledge capital are more important for catch-up innovation, how investments in different types of knowledge capital interact with each other and with other physical and human capital assets, and how this varies depending on the stages in an enterprise's life cycle. And we need to better understand how important catch-up innovation is relative to gains from between- and within-industry SC at different stages of economic development.

The second part of the framework seeks to better understand the pathways through which investments in catch-up innovation lead to shared prosperity as opposed to greater income inequality. The chapter puts forward three main channels: (1) directly through more and better jobs for lower-income entrepreneurs, managers, and workers; (2) directly through higher consumption by lower-income people of goods and services meeting their needs; and (3) indirectly through technological and other spillovers. Here too there is an important outstanding work agenda. It appears that recent government and business initiatives seeking to have a positive impact on the welfare of lower-income households through support of base-of-pyramid innovation have focused largely or even solely on consumption rather than the more important jobs channel. How important is each of these channels in practice, and how do they vary across industries? What are the conditions when innovation is more likely to boost shared prosperity rather than exacerbate income inequality? And what are the key interactions between these and other channels linking innovation and shared prosperity?

Finally, the third part of the framework seeks to better understand the role of government policy actions to boost shared prosperity through catch-up innovation. The chapter argues that policy actions may benefit from a tilting at three levels: (1) at the level of the individual; (2) at the level of firms; and (3) at the level of the business environment. The first channel, directly targeting skills upgrading and labor market intermediation support to lower-income individuals and to lower-income regions of each country, seems the most straightforward in practice. Regarding the second channel of firm-level and industry-specific technology adoption support, recent studies show that technology extension support initiatives to upgrade managerial capital have boosted productivity and worker earnings. However, we do not know whether wage inequality among employees has decreased or increased, and whether and how to tilt policy actions to help create more and better jobs for lower-income workers over time. And regarding the third channel, there are many unanswered questions on the effects of different business environments on shared prosperity, including if, and if so how, to tilt policies to help lower-income individuals – as illustrated among others by the insufficiently understood impact of product market liberalization on labor income inequality. Importantly, what we do know from this framework is that there are likely to be considerable gains from spending more resources on learning about the effects of different policy actions across individual, firm, and business environment levels.

Notes

1. Haskel (2011) makes a useful distinction between growth from duplication (investment in more of the same types of inputs and outputs) versus from innovation (investment in new technologies and outputs).
2. The three-way typology of SC presented here broadens the use of the term. McMillan and Rodrik (2011), for example, decompose economy-wide labor productivity changes into the contribution of "SC," defined there as the productivity effect of labor reallocation across sectors (shifts in labor from all agriculture industries to all manufacturing and service industries), and a "within" sector productivity component, without further disaggregation.
3. Based on French, German, Spanish, and UK manufacturing and service firms for 1998–2000, Harrison et al. (2014) find that process innovation has significant displacement effects that are counteracted by compensation effects with no net reduction in jobs, while product innovation is associated with employment growth. Mairesse and Wu (2014) find that the compensation effects of product innovation in China more than counterbalance the displacement effects, with product innovations and exports making a positive net contribution to employment growth.
4. Brynjolfsson and McAfee (2014) provide examples of technologies that are deliberately labor intensive, including online, distributed, problem-solving business models that allow people to offer their labor to the crowd – that bring not only efficiency gains and price declines but also jobs, increasing the value of labor rather than decreasing it. They argue that policy makers should encourage such entrepreneurial creations.
5. Du et al. (2012) show that FDI firms from OECD economies generate positive vertical, interindustry productivity spillovers to domestic firms in China over 1998–2007 both via backward linkages (technological learning by local suppliers from foreign buyers) and forward linkages (learning by local buyers from foreign suppliers).

6. Of course, frontier innovations can also help boost shared prosperity. Brynjolfsson and McAfee (2014) advocate using taxes, regulations, contests, grand challenges, and other incentives to direct technical change toward machines that augment human ability rather than substitute for it.
7. Based on firm-level data spanning 1998–2007, Du et al. (2014) show that tariff reductions associated with China's WTO (World Trade Organization) accession increased the productivity impacts of FDI's backward spillovers, possibly because the additional competition forced domestic suppliers to improve their efficiency and quality. Tax holidays (both corporate income and VAT subsidies) drew FDI into strategic industries that spawned significant vertical spillovers.
8. The term "creative transformation" captures the same concept as the more commonly used "creative destruction," but in a more palatable form from a political-economy perspective – reflecting both Schumpeterian creative destruction (the exit of less productive firms and the entry of more productive ones) and resource reallocation between firms on the intensive margin (the shifting of market shares from less to more efficient incumbent firms, and the associated shifts in workers and other factors of production).
9. According to Aghion et al. (2015), this competition message applies to industrial policy as well: do it in ways that increase competition. Based on a comprehensive data set of all medium and large enterprises in China from 1998 to 2007, they show that investment-specific subsidies and tax holidays can lead to higher productivity growth if they encourage competition – if they are not targeted at a few favored existing champions but spread more broadly, and especially if they are dispersed across a broad range of younger and more productive firms.
10. Grossman et al. (2014) distinguish three mechanisms that govern the effects of trade on income distribution – trade: (i) increases demand for all types of the factor used intensively in exports; (ii) benefits those types of a factor that have a comparative advantage in exports; and (iii) induces a rematching of workers and managers within industries, which benefits the more able types of the factor that achieves improved matches.

References

Aghion, P., Cai, J., Dewatripont, M., Du, L., Harrison, A., and Legros, P. (2015). "Industrial Policy and Competition." *American Economic Journal: Macroeconomics*, 7 (4): 1–32.

Aghion, P. and Howitt, P. (2009). *The Economics of Growth*. Cambridge, MA: MIT Press.

Bloom, N., Eifert, B., Mahajan, A., McKenzie, D., and Roberts, J. (2013). "Does Management Matter? Evidence from India." *Quarterly Journal of Economics*, 128 (1): 1–51.

Bloom, N., Lemos, R., Sadun, R., Scur, D., and Van Reenen, J. (2014). "The New Empirical Economics of Management." NBER Working Paper 20102, Cambridge, MA.

Bollard, A., Klenow, P., and Sharma, G. (2013). "India's Mysterious Manufacturing Miracle." *Review of Economic Dynamics*, 16: 59–85.

Brynjolfsson, E. and McAfee, A. (2011). *Race against the Machine: How the Digital Revolution Is Accelerating Innovation, Driving Productivity, and Irreversibly Transforming Employment and the Economy*. Lexington, MA: Digital Frontier Press.

Brynjolfsson, E. and McAfee, A. (2014). *The Second Machine Age: Work, Progress, and Prosperity in a Time of Brilliant Technologies.* New York: W.W. Norton & Company.

Caliendo, L., Monte, F., and Rossi-Hansberg, E. (2015). "The Anatomy of French Production Hierarchies." *Journal of Political Economy*, 123 (4): 809–52.

Comin, D. and Mestieri Ferrer, M. (2013). "If Technology Has Arrived Everywhere, Why Has Income Diverged?" NBER Working Paper 19010, Cambridge, MA.

Corrado, C., Haskel, J., Jona-Lasinio, C., and Iommi, M. (2013). "Innovation and Intangible Investment in Europe, Japan and the United States." *Oxford Review of Economic Policy*, 29 (2): 261–86.

Corrado, C., Hulten, C., and Siche, D. (2009). "Intangible Capital and US Economic Growth." *Review of Income and Wealth*, 55 (3): 661–85.

Correa, P. and Schmidt, C. (2014). "Public Research Organizations and Agricultural Development in Brazil: How did Embrapa Get it Right?" Economic Premise 145, Washington, DC: World Bank.

Dix-Carneiro, R. (2014). "Trade Liberalization and Labor Market Dynamics." *Econometrica*, 82 (3): 825–85.

Dix-Carneiro, R. and Kovak, B. (2014). "Trade Reform and Regional Dynamics: Evidence from 25 Years of Brazilian Matched Employer–Employee Data." Mimeo, Duke University.

Du, L., Harrison, A., and Jefferson, G. H. (2012). "Testing for Horizontal and Vertical Foreign Investment Spillovers in China, 1998–2007." *Journal of Asian Economics*, 23 (3): 234–43.

Du, L., Harrison, A., and Jefferson, G. H. (2014). "FDI Spillovers and Industrial Policy: The Role of Tariffs and Tax Holidays." *World Development*, 64 (C): 366–83.

Dutz, M. A. (2015). "Resource Reallocation and Innovation: Converting Enterprise Risks into Opportunities." Chapter 10 in Bounfour, A. and Miyagawa, T. (eds) *Intangibles, Market Failure and Innovation Performance.* Cham, Switzerland: Springer Publishing, 241–90.

Dutz, M. A., Kessides, I., O'Connell, S. D., and Willig, R. D. (2012a). "Competition and Innovation-Driven Inclusive Growth." Chapter 7 in de Mello, L. and Dutz, M. A. (eds) *Promoting Inclusive Growth: Challenges and Policies.* Paris: OECD and the World Bank, 221–77.

Dutz, M. A., Kannebley Jr., S., Scarpelli, M., and Sharma, S. (2012b). "Measuring Intangible Capital in an Emerging Market Economy: An Application to Brazil." Policy Research Working Paper 6142. Washington, DC: World Bank.

Dutz, M. A., O'Connell, S. D., and Troncoso, J. (2014). "Public and Private Investments in Innovation Capabilities: Structural Transformation within the Chilean Wine Industry." Policy Research Working Paper 6983. Washington, DC: World Bank.

Dutz, M. A. and Vijayaraghavan, K. (2014). "Supporting Affordable Biotechnology Innovations: Learning from Global Collaboration and Local Experience." Chapter 6 in Dutz, M. A., Kuznetsov, Y., Lasagabaster, E., and Pilat, D. (eds) *Making Innovation Policy Work: Learning from Experimentation.* Paris: OECD and the World Bank, 155–92.

Garicano, L., Lelarge, C., and Van Reenen, J. (2013). "Firm Size Distortions and the Productivity Distribution: Evidence from France." NBER Working Paper 18841, Cambridge, MA.

Grossman, G. M., Helpman, E., and Kircher, P. (2014). "Matching, Sorting, and the Distributional Effects of International Trade." Mimeo, Harvard University.

Harrison, A. E., Martin, L. A., and Nataraj, S. (2013). "Learning versus Stealing: How Important Are Market-Share Reallocations to India's Productivity Growth?" *World Bank Economic Review*, 27 (2): 202–29.

Harrison, R., Jaumandreu, J., Mairesse, J., and Peters, B. (2014). "Does Innovation Stimulate Employment? A Firm-Level Analysis Using Comparable Micro-Data from Four European Countries." *International Journal of Industrial Organization*, 35: 29–43.

Haskel, J. (2011). "Innovation: A Guide for the Perplexed." Mimeo, Imperial College.

Hollweg, C. H., Lederman, D., Rojas, D., and Ruppert Bulmer, E. (2014). *Sticky Feet: How Labor Market Frictions Shape the Impact of International Trade on Jobs and Wages.* Washington, DC: The World Bank.

Hsieh, C. T. and Klenow, P. J. (2009). "Misallocation and Manufacturing TFP in China and India." *Quarterly Journal of Economics*, 124 (4): 1403–48.

Hsieh, C. T. and Klenow, P. J. (2012). "The Life Cycle of Plants in India and Mexico." NBER Working Paper 18133, Cambridge, MA.

Hsieh, C. T. and Klenow, P. J. (2014). "The Life Cycle of Plants in India and Mexico." *Quarterly Journal of Economics*, 129 (3): 1035–84.

Hulten, C. and Hao, J. X. (2008). "What Is a Company Really Worth? Intangible Capital and the 'Market to Book Value' Puzzle." NBER Working Paper 14548, Cambridge, MA.

Hulten, C. and Hao, J. X. (2012). "The Role of Intangible Capital in the Transformation and Growth of the Chinese Economy." NBER Working Paper 18405, Cambridge, MA.

Hulten, C., Hao, J. X., and Jaeger, K. (2012). "The Measurement of India's Intangible Capital." Mimeo.

Lach, S., Shiff, G., and Trajtenberg, M. (2008). "Together but Apart: ICT and Productivity Growth in Israel." The Foerder Institute for Economic Research Working Paper 3-08.

Mairesse, J. and Wu, Y. (2014). "Employment Growth, Export, Product Innovation and Distance to the Productivity Frontier in China: A Firm-Level Comparison across Regions, Industries, Ownership Types and Size Classes." Mimeo, UNU-MERIT.

Mansfield, E. (1961). "Technical Change and the Rate of Imitation." *Econometrica*, 29 (4): 741–66.

McMillan, M. S. and Rodrik, D. (2011). "Globalization, Structural Change and Productivity Growth." Paper 2 of *Making Globalization Socially Sustainable.* Geneva, Switzerland: ILO-WTO.

OECD. (2013). *Supporting Investment in Knowledge Capital, Growth and Innovation.* October, Paris: OECD Publishing.

Piketty, T. (2014). *Capital in the Twenty-First Century.* New York: Belknap Press.

Sutton, J. et al. *An Enterprise Map of Ethiopia* (2010, with N. Kellow), *An Enterprise Map of Tanzania* (2012, with D. Olomi), *An Enterprise Map of Ghana* (2012, with B. Kpentey), *An Enterprise Map of Zambia* (2013, with G. Langmead), London: International Growth Center, LSE.

Trajtenberg, M. (2009). "Innovation Policy for Development." In Foray, D. (ed.), *The New Economics of Technology Policy*, pp 367–96, Edward Elgar Publishing, Cheltenham, UK.

15

Inclusive Innovation: Harnessing Creativity to Enhance the Economic Opportunities and Welfare of the Poor

Carl Dahlman, Esperanza Lasagabaster, and Kurt Larsen

Introduction

In recent decades, technological innovation has led to increased productivity, higher economic growth, and vastly improved living conditions around the globe, lifting millions out of poverty. Yet, access to the fruits of this rapid development has been uneven, with a large share of the world's population not benefiting from these advances. Nearly 2.5 billion people live on less than US$2 a day (see Table 15.1). More than 35 percent of the population does not have access to basic sanitation facilities and 22 percent does not have access to electricity (World Bank, 2014).

Clearly, innovation per se is not sufficient to improve living standards: it must be inclusive, meaning that it must be accessible and affordable to those at the BoP (base of the pyramid) or help create better economic opportunities for them. *Inclusive innovation* refers to the "knowledge creation, acquisition, absorption and distribution efforts targeted directly at meeting the needs of the low-income or the BoP population"[1] in order to improve their welfare and access to better economic opportunities.[2,3] Frequently, inclusive innovation involves broad partnerships of people and organizations from the public and private sectors, research centers, academia, foundations, nongovernmental organizations, and bilateral and multilateral donors, each with its own comparative advantages and tools, to leverage different aspects of the innovative process and bring innovations to those who need them most.

The rationale for focusing on inclusive innovation is becoming ever more compelling. First, despite impressive technological advances and economic growth, more than 40 percent of the world's population – 2.5 billion people – still falls below the international poverty line (Table 15.1) and lack access to basic goods and services. Second, inequality has been increasing within many countries, becoming a source of instability. The fact that the effects of climate change disproportionately affect the most vulnerable populations could further exacerbate inequality.

Table 15.1 Number of poor and poverty rates at the international poverty lines of US$1.25 and 2.00 per day, 1981–2010

	1981	1993	1999	2005	2008	2010
Below US$1.25/day (purchasing power parity)						
Persons in world millions	1,938	1,910	1,743	1,389	1,289	1,215
As share of world population %	51.2	41	34.1	25.1	22.4	20.6
Below US$2.00/day (purchasing power parity)						
Persons in world millions	2,585	2,941	2,937	2,596	2,471	2,396
As share of world population %	69.6	63.1	57.4	46.9	43	40.7

Source: World Bank (2014).

Inclusive innovation is gaining momentum in some emerging markets, such as India and China, whose governments have made it a policy priority. Declaring 2010–20 the "Decade of Innovation," the government of India created the National Innovation Council in 2011 to develop a framework for inclusive innovation and promote pertinent programs at the federal and state levels. China is also formulating a holistic strategy for promoting inclusive innovation.

Some examples of inclusive innovation have been extensively documented. The Green Revolution helped dramatically increase the productivity of small farmers in developing countries. More recently and on a smaller scale, Tsingua Solar has leveraged advanced research and development (R&D) to bring affordable solar energy solutions to the poor in China and a few other countries. Inclusive innovation does not always require new R&D; existing ideas and technologies can be recombined in a novel way to make the innovations affordable, accessible, and applicable.

For example, M-Pesa in Kenya has leveraged information and communications technology (ICT) to deliver financial services to the poor at affordable rates. Today, it has more than 15 million accounts, and variations of this model are in place in other developing countries. The Narayana Hrudayalaya hospital in India leverages efficient operations and economies of scale to provide world-quality cardiac surgery services at one-tenth of the price in the United States. These examples, however, are still far too few. Moreover, most inclusive innovations have not reached a large enough scale to make a significant impact.

Some governments in emerging markets have begun to implement policies designed to stimulate innovation that addresses the needs of the poor and excluded. Analyzing and sharing these experiences are important so that inclusive innovation can be promoted more efficiently and successful programs can be adopted by other countries. Inclusive innovation is also of interest to developed countries, where rising inequality and fiscal constraints generate pressure to deliver more with less. Local companies, especially in large emerging markets, are starting to develop goods and services for low-income groups – a new phenomenon driven by the growing innovation capacity of those firms. At the same time, multinational companies (MNCs) are realizing that if they do not develop affordable products for these markets, they will be preempted by domestic firms, which will also begin to sell some of these products in their own home markets (Prahalad, 2005; Immelt et al., 2009).

This chapter provides a conceptual framework on inclusive innovation, with the aim of stimulating a discussion of how it can be promoted more broadly.

Market and government failures

The literature on innovation points to numerous market failures that might lead to underperformance of innovation outcomes from a social welfare standpoint.

First, the benefits of innovation are not fully appropriated by the innovator due to knowledge spillovers. Second, the problem of information asymmetry in market transactions and other unique features of innovation projects can constrain the financing of innovation. Third, innovation depends on complementarity of assets such as access to human capital and specialized skills, technological infrastructure, and other forms of knowledge. Shortcomings in these markets could hinder innovation.

In the case of inclusive innovation, market failures are exacerbated by the poverty, low levels of schooling, and isolation of the target population. Although the marginal social utility of the innovative effort allocated to the needs of the poor is likely to be higher than that of many conventional innovations, innovators do not often explicitly focus on creating these goods and services because a sufficiently attractive market is not perceived to exist.

Emerging trends in inclusive innovation

A rich diversity of experiences of pro-poor innovations is emerging. These experiences share many common elements with other types of innovations but also present unique features, which are summarized below.

Nature of pro-poor innovation

Inclusive innovation uses both high and low technologies. Moreover, the core of the innovation often lies in the business, organizational, or delivery model. While in some cases the technology is new, in most cases the innovation involves a recombination of existing technologies. Some examples are Drishtee, a distribution system for rural village stores in India, and the Narayana Hrudayalaya Hospital in India (see Appendix 15.1). The latter combines business, organizational, and process innovations to provide high-quality, world-class cardiac surgery at less than one-tenth of the price prevailing in the United States, as well as telemedicine and insurance schemes for the rural poor.

Many inclusive innovations leverage ICT, providing traditional services to the poor in new ways and reaching remote populations at lower cost. For example, M-Pesa in Kenya has facilitated financial inclusion through mobile phone technology. New applications of mobile technology to health, education, and farming are emerging by the day (e.g., M-farm, Kenya; Text to Change, South Africa; and ReMedi Kiosks, India).

Pro-poor innovations have emerged across all economic and social sectors

There are several examples of inclusive innovation in agriculture (Jain irrigation and the Green Revolution), education (Aakash Tablet), finance (M-Pesa), and health (Narayana Hrundayalana heart surgery). Examples are available in many other sectors such as construction (Cemex Patrimonio Hoy) and retail sales (Drishtee retail supply network). It is possible to observe a wide range of innovations of different degrees of complexity within the same sector. For example, in the health sector, some innovations are simple, almost artisan in nature, such as the Jaipur foot in India. At the same time, GE's hand-held and battery-operated electrocardiogram (ECG) is a high-technology medical device offered at a very affordable price.

Geography of pro-poor innovations

Pro-poor innovations have sprung up around the globe, especially in developing countries. They may be sponsored by local agents, agents from developed countries, or global initiatives (Appendix 15.1). The most transformative pro-poor innovations have generally been sponsored by global foundations, global NGOs, or multilateral or bilateral institutions, and have emerged in larger developing countries, such as India and China. These countries have the advantage of larger markets in which to scale up and recoup the innovative effort and are therefore more attractive to MNCs. They also have greater innovation capabilities with a critical mass of researchers, large domestic firms, and NGOs.

Who are the agents participating in pro-poor innovations?

Compared to other innovations, pro-poor innovations involve a far more diverse group of agents. Multinational corporations, small and medium-sized enterprises, governments, public research institutions, foundations, NGOs, social enterprises, and development institutions all play important roles.

- *Multinationals in emerging markets* (e.g., Haier in China and Tata in India) are becoming interested in tapping the large market of low-income consumers, and MNCs from developed economies are following in their footsteps (e.g., Siemens with health care and ABB with mini-hydro solutions).
- *Foundations, universities, and NGOs from developed markets* (e.g., Path, D-Rev, and the Gates Foundation) as well as many from emerging markets are spurring inclusive innovation through prize competitions or by financing R&D. NGOs play a host of roles in the supply chain for pro-poor innovations, from R&D and product design (such as PATH) to delivery, diffusion, and education. Universities have also contributed through R&D.
- *International development institutions* are increasing their focus on pro-poor innovations through financing, coordination, and using their convening power to provide forums for the exchange of information and methodologies. A few large global public goods programs, such as the Green Revolution and the African River Blindness Eradication Program, have involved coordination at the global level. These initiatives entailed the participation of governments in emerging markets, multilateral development institutions, global research groups, and NGOs (Appendix 15.1).

Emerging business models

Business models underlying pro-poor innovations draw from models supporting other types of innovations. However, the challenging environments in which the products or services are delivered, the low and volatile income and educational characteristics of the target market, and the noncommercial motivations frequently underlying the innovation have required adaptations of these models or led to the emergence of new ones.

Simplicity of use and ease of maintenance

The low quality of basic infrastructure such as roads and electricity in the target market makes it costlier and more difficult to deliver the product or service. Poor infrastructure together with lower educational and income levels of potential users require cost-optimized products that are easy to use and maintain. For example, D-Rev, a US-based NGO focused on inclusive innovation, has developed a product that offers world-class treatment for newborns with jaundice symptoms and can function in the precarious conditions of rural health facilities in developing countries with scarce access to electricity and staff with relatively modest professional training.[4]

Recognizing gender and social norms

Inadequate understanding of the needs and cultural norms of the target population has often led to unsuccessful products. Social, cultural, and gender norms (e.g., constraints on women from making independent choices) may cause a product to be rejected or to have limited uptake. In Bangladesh, for example, a recent study showed that women's uptake of improved cooking stoves is low because they lack the authority to make purchases even though they have stronger preferences for better stoves – especially health-saving ones.

Frequent co-creation and collaboration with users

As discussed above, the failure to recognize users' needs, social norms, educational levels, and difficult physical environments can contribute to the failure of pro-poor innovations. For all these reasons, organizations and firms supporting pro-poor innovations are increasingly engaged in "co-creation," whereby potential users participate or collaborate in the design of the product/service and help to identify needs and constraints (whether social or infrastructure-based) and to define solutions that are relevant and acceptable to the potential user community.

Pricing and financing strategies

Understanding the financial and income constraints of users is crucial to ensure the sustainability of the product or service and reach the target market. Many low-income workers in developing countries are unemployed, self-employed, or employed in the informal sector, with incomes that are highly volatile. Mechanisms that have been used to tackle such constraints include pay-by-use, microleasing, microcredit, price segmentation, and hybrid financing models combining market- and donor-based approaches.

To overcome the problem of high upfront financial commitments, costs can be broken down into affordable units. "Pay-by-use" of services has been actively applied by mobile companies, and, in some cases, by water and energy providers. Other models have used microleasing. "Microcredit" is frequently offered together with the product or service. Some microcredit schemes try to match loan payments with the intended user's daily expenses on the product or daily income flow. In Nepal, for example, the solar-powered Tuki – a clean and affordable lantern compared to more dangerous and expensive kerosene lanterns – was offered together with a loan which had payment schedules roughly equivalent to the intended user's daily expense of kerosene for home lighting, facilitating its acceptance.

Price segmentation with cross-subsidized services is often used as a mechanism to achieve sustainability and reach the poor. Numerous examples exist in the health sector. LifeSpring Hospitals, a chain of small hospitals offering affordable care to low-income families in India, and the Narayana Hrudayalaya Hospital also use tier pricing and cross-subsidies from higher- to lower-income groups.

Others have used "hybrid financing models," a market-based approach combined with a donor-based one. The Community Cleaning Services in Nigeria, for example, began operations as a market-based project but shifted to a hybrid model where some services are fully recovered (e.g., toilet cleaning services) and social services such as educational and awareness work are partly covered by a donor.[5]

Partnerships

Partnerships are critical to the success of the product by bringing together the strengths and capabilities of different agents of the innovation ecosystem. Foundations and governments can contribute with funding, and governments can provide incentives to stimulate the supply and demand for inclusive innovations; NGOs, public research and technology centers, and universities can support R&D; enterprises can contribute with R&D, manufacturing, and distribution; and NGOs can also facilitate the diffusion of the product and provide training for potential users.

The aforementioned NGO D-Rev, for example, has the capacity to engage in R&D programs to support health products and income opportunities for the poor but does not have the financing or capacity to manufacture and deliver them on a large scale.

Scalability remains a challenge undermining potential impact

A review of more than 100 cases of inclusive innovation conducted as background research for this chapter revealed that most of the inventions did not get past the initial prototype stage. This is typical for innovations, but probably even more so for inclusive innovations, because the low income of the target market means that operating margins for commercial products are often very low. Many of those that made it to the market have not yet been scaled up.

Innovations supported by MNCs, global NGOs, and foundations typically have greater scalability because of their global reach. Partnerships with organizations have also facilitated scaling up by combining their different strengths and capabilities, for example by combining the R&D capacity of a public research center with the manufacturing capacity of an enterprise and the educational and distributional capacity of a local NGO.

Do pro-poor innovations differ from other types of innovations?

Pro-poor innovations share many common features with other types of innovations but also present some distinctive characteristics. Target markets are primarily located in developing countries, which along with other factors is changing the global geography of innovation. The ecosystem supporting inclusive innovation is complex, with a broader set of agents participating in the process. Limited access to infrastructure and the lower level of schooling of the primary users impose additional burdens on product or service design. Designs need to be simple to make products easy to use and maintain in difficult environments, and costs must be reduced to make them affordable. Co-creation with potential users can facilitate understanding of their needs, social norms, and other characteristics of the local environment. This process is likely to result in products and services with higher market acceptance and greater success. Pricing mechanisms have also been adapted to the realities and acute financial constraints of the target populations.

Public policies for inclusive innovation

Inclusive innovation is a new area of public policy that can yield high social and economic returns. However, too few governments are actively engaged in supporting it. Inclusive innovation can have an impact on productivity by reducing the costs of pro-poor programs and by increasing economic returns to the poor in their own activities in agriculture, manufacturing, and services through improved health and more efficient use of productive factors. In addition, many of the innovations developed for low-income markets can help reduce the cost of providing health care in high-income economies. This suggests that there is scope for promoting more inclusive innovation efforts in advanced countries as well.

Promoting inclusive innovation will require a shift in emphasis in current and new policy instruments. Public policies for inclusive innovation cannot be considered in isolation. They must take into account the entire innovation ecosystem, market failures affecting the development of inclusive innovations, and the roles played by different agents in the generation, transfer, and dissemination of innovations targeted at the poor and the excluded (see Table 15.2).

As can be seen from Table 15.2, the agents involved in inclusive innovation are a continuum from private firms, which are typically more narrowly focused on profit mining, to governments and international development organizations, which are more focused on improving public welfare. Foundations and NGOs are increasingly engaged in the financing and provision of public goods. Social entrepreneurs constitute another important and growing group of agents, which can be differentiated from private firms because of their willingness to accept lower profits in return for achieving greater social impact (Koh et al., 2012).

Each group of agents pursues different objectives, and its strengths and constraints for promoting inclusive innovations also differ.

Broad environment for innovation

Government policies designed to improve the broader environment for innovation are critically important. Most inclusive innovations are based on the recombination of existing technologies. This is particularly true of policies to facilitate access to international sources of innovation since

Table 15.2 Inclusive innovation: typical market failures and possible policy solutions

Failure	Possible solutions	Example
Prohibitively high cost of getting inclusive innovative product or service to the poor	• Using existing market or civil society networks to get the product or service to the poor	M-Pesa mobile banking service using cell phones, which partnered with other organizations to offer financial services
	• Innovating in the delivery of the good or service	Drishtee rural village store supply system
		Grameen and other microfinance systems
	• Leveraging ICT for delivery	Anhanguera educational services in Brazil
Low income of the poor does not permit them to buy the good or service	• Innovating financial services for the poor	Grameen Bank and many other microfinance programs
	• Partnering with existing financial interemediaries targetting the poor	Jain drip irrigation and SELCO solar lamps work with rural banks to finance purchases
	• Blending market with donor finance	Vestergraad Frandsen Life Straw uses carbon credit mechanism to finance donation of water filters
	• Innovating alternative way to finance	Cemex's Patrimonio Hoy has developed financing system for low-income purchases of construction materials
		Community Cleaning Services combining market with donor finance
Low education of the poor is a constraint on their ability to select or use the innovation	Bundle training with the supply of the innovation either • Directly or • Indirectly by collaborating with others who can provide the training	Drishtee supply and training for village kiosks in rural India Jain drip irrigation
Weak enterpreneurship capability of the innovators. Business models are not adequately designed for financial sustainability and growth	• Business incubators in science parks for university or research lab innovators • Build capacity of innovation intermediaries and networks to scale up their training services	Shenzhen Karva Diagnostic Bed Global Cycle Solutions from MITs D-Lab -Dasra, Innovation Alchemy, National Association of Social Enterprises in India

Source: Authors.

so much of the knowledge necessary for innovation is being developed outside the boundaries of any nation. Thus, a country should be open to all ways of tapping global knowledge, including an explicit focus on inclusive innovation. Growth policies can be more effective when there is a specific pro-poor focus, including a more explicit focus on inclusive innovation.

Effective policies to protect intellectual property rights (IPR) are also important. Developing the most appropriate IPR policies for inclusive innovation remains a challenge because of the trade-off between providing incentives for the innovative effort and aiming to maximize welfare by diffusing the innovation at the lowest possible cost. This is particularly important for innovations that are easy to copy, such as those by grassroots innovators, as well as many drugs and pharmaceutical products. One way to bridge this trade-off is for governments or other agents or foundations to buy the patent rights and make the technology available to others. The National Health Institute licensed a conjugate vaccine technology to PATH that obtained a critical part of the process needed to develop the meningitis vaccine for African countries and gave it to a developing country company (the Indian Serum Institute) that committed to produce vaccines at a cost of 50 cents per dose.

Innovation infrastructure

In promoting inclusive innovation, as well as innovation more generally, it is necessary to have supporting innovation infrastructure. This includes research institutes, metrology, standards, testing, and quality (MSTQ) institutions (Guasch et al., 2007; Racine, 2011), specialized consulting and management firms, a critical mass of technical human capital, information services, and venture capital firms.

Funding and complementary instruments

While inclusive innovation was traditionally not a priority when funding public research, some governments (particularly China and India) are starting to provide funding to spur applications-oriented research aimed at addressing the needs of the poor, particularly in the areas of health, education, and agriculture. More can be supported, however. In particular, more funding for public–private consortia could not only stimulate further relevant research but could also provide necessary links to the market in order to facilitate its future commercialization. In the case of public research institutes, financing incentives can be complemented by specific mandates to engage more actively in the support of inclusive innovation. This, for example, was the case of the government lab that helped Lupin develop a treatment for psoriasis. It was also the case of several of the Chinese examples of research labs and universities that developed medical devices (Appendix 15.1) and solar water heaters.

In addition, other financing (e.g., matching grants) could be made available to researchers and entrepreneurs to help commercialize existing research. Such financing could support the development of prototypes and pilot products up to a level ready for scaling up. Funding for research commercialization could be linked to other initiatives that facilitate networking and brokering services for intellectual property management (e.g., technology transfer offices) and/or mentoring for new entrepreneurs (incubators, venture accelerators, as well as virtual mentoring networks). An example of upstream funding of an inclusive innovation is DFID's support for the development of what became M-Pesa. An example of facilitating networking is the Indian government's support of the Honey Bee Network, which has catalogued hundreds of thousands of grassroots innovations.

Governments can also develop programs to stimulate the adoption, adaptation, and commercialization of available global technologies by entrepreneurs if not yet available in the local market. This can be supported through matching grants or other financing instruments that reduce the initial risk to the entrepreneur and help create a demonstration effect for other entrepreneurs.

Demand-side policies – whereby governments or foundations signal a need and a potential market – also offer promise for stimulating inclusive innovation. Public procurement is particularly important because it creates a concrete market. While it has helped to spur innovation in high-income countries (e.g., public procurement in the military has been a critical source of leading technological innovations with subsequent commercial applications), it has been underutilized as a policy instrument to stimulate innovations for the poor. This instrument is particularly attractive for larger countries, where governments can guarantee large-scale markets that will allow the developer to recoup the costs (e.g., the Aakash Tablet, which was procured by the government of India to offer e-learning to students at the very affordable price of US$35).[6]

Contests and grand challenges that invite solutions to solve specific inclusive innovation problems using crowdsourcing are another promising policy instrument. Governments, multilateral and bilateral institutions, and large foundations are using such grand challenges to stimulate the creation of public good innovations aimed at the very poor (e.g., Gates Foundation, Grand Challenges for Development launched by USAID in partnership with foundations, and the private X Prize Foundation, which is now also focusing on development challenges).

Effective demand-side policies, however, require a lot of work to identify not only the needs of the poor, but also how to best address them. This requires in-depth analysis of technical as well as social and cultural aspects that may stimulate or hinder the use of the product or service. Sustainability and opportunities for transferring technology to the local environment are other aspects to consider when developing demand-side policies for inclusive innovation. Hargreaves et al. (2011) compare two demand-side financing mechanisms that led to the development of lower-cost meningitis and pneumococcal vaccines with different sustainability and technology transfer implications. In the case of the meningitis vaccine, the Gates Foundation awarded a US$70 million grant to PATH to develop the vaccine at a cost of less than US$1 per dose. This price was based on wide consultations that determined the maximum price that would be affordable to ensure a large uptake around the developing world. PATH formed international partnerships to develop the vaccine, including a collaboration with the Serum Institute of India, resulting in technology transfer to a pharmaceutical company in an emerging market.[7] The latter committed to producing the vaccine at 50 cents a dose. By contrast, the development of the pneumococcal vaccine, which followed an advanced market commitment (AMC) mechanism, did not generate such strong incentives to reduce vaccine costs and foster technology transfer to an emerging market.[8]

Building bridges across institutions

Coordination among the different agents is very important because of the comparative advantages of the agents at different stages of the innovation cycle. Governments can help facilitate direct coordination efforts as well as foster coordination in the implementation of supply- and demand-side policies. For example, supply-side policies such as fiscal and financial incentives may be conditional on collaboration between firms, research institutes, and universities. Moreover, enhancing the productive activities of the poor in agriculture and other areas requires coordination among many agents in agricultural value chains, including private companies, NGOs, and specialized organizations. The same is true in health. The PATH example illustrates the importance of integrating the efforts of many agents to successfully develop a quality and low-cost meningitis vaccine for poor African countries.

How can other agents contribute?

Private companies and entrepreneurs

The cases presented in this chapter show that private companies and entrepreneurs can be profitable by innovating and delivering products and services that serve low-income markets. This was the key point made by Prahalad in his book *Fortune at the Bottom of the Pyramid*. Some developed-country MNCs, such as General Electric, Procter and Gamble, and Vodafone, as well as some developing-country MNCs, such as Haier and Tata, have discovered this and are actively pursuing these markets. However, many have not. This suggests that there is scope for wider dissemination of examples of private firms that have built profitable inclusive innovation business models.

Large domestic companies and MNCs have the organizational and financial resources to develop inclusive innovations as well as to integrate small producers into their supply and distribution chains. They can be far more involved in this area, not only for the sake of corporate social responsibility, but also profitable and sustainable business models. The NGO Enterprise Solutions for Poverty supports the involvement of the private sector in inclusive innovation. Its activity has two prongs. One is to collaborate with local and multinational leaders to build and expand competitive inclusive business models. The other is to encourage emerging entrepreneurs to build networks of enterprises that leverage economies of scale by connecting to the sourcing and distribution systems of large companies (Barry, 2007).

Universities

Universities can be important sources of inclusive innovation in four ways. First, they can train future entrepreneurs, scientists, and engineers with a focus on inclusive innovation. Second, they can engage in research that pursues the development of inclusive innovations. For example, the Embrace Baby Incubator came from Stanford University. Third, they can provide research, consulting services, and technical assistance to inclusive innovators. For example, Stanford University researchers collaborated in the development of a special knee joint for the Jaipur Knee. Fourth, they can establish network institutions, such as technology parks and business accelerators, that can help transform prototypes into business applications or innovative government programs.

Government laboratories

Government laboratories in developing as well as developed countries can do more to apply their scientific and technical knowledge to address the needs of the poor. There is a need to encourage staff to switch from more publications and patents to relevant goods and services. This requires in part an explicit orientation in their charters to have this focus. At a more micro level, it also requires that recognition for this effort be built into the incentive and promotion schemes of the researchers and staff. As for universities, it also requires linking them up with bridge institutions, businesses, and other agents that can help take innovations to the market.

NGOs

NGOs are becoming important players in addressing public needs that traditionally had been met by governments. Their actions should be seen as complementary to the role of government. They can help identify areas that need attention as well as approaches that work. Examples are D-Rev and PATH. They can also serve as important partners in the delivery of inclusive innovation goods and services.

Foundations

Foundations have become an important source of funding for addressing health, education, and other needs of poor people in developing countries. Through innovative financial instruments such as challenge grants, they are stepping in to provide public goods that were formerly the domain of governments. Their grant support has not only resulted in new products and services for the poor but also in the pioneering of new business models, as outlined in the Monitor Report *From Blueprint to Scale* (Koh et al., 2012).

International and bilateral development agencies

International and bilateral development institutions are well-placed to promote the inclusive innovation agenda. The roles they can play range from advocacy, dissemination, and advisory services to funding of specific initiatives. They can also help organize and fund major global public good efforts. Five major areas where international and bilateral development agencies can intervene are provided below.

First, international and bilateral development organizations can play a strong *advocacy role,* promoting more explicit focus on and investment in inclusive innovations by developed and developing country governments, NGOs, foundations, and private firms. They are well-placed to assume this role because of their broad experience analyzing and addressing developing-country needs across a wide spectrum of sectors. They also have well-developed global dissemination programs.

Second, these organizations can create an *information clearinghouse* on existing inclusive innovations (products, services, and new business and organizational arrangements). Many of the existing examples of inclusive innovation that have reached millions of people are not well-known, even though they may be applicable to millions more. The information clearinghouse could be an open and interactive virtual platform available to all interested parties. Once its basic information clearinghouse function is established, it could be expanded to provide a market exchange for inclusive innovations.

Third, besides facilitating knowledge exchange, they can contribute with further *policy analysis on effective models to generate, scale up, and disseminate inclusive innovations and provide policy advice and training* to governments and other interested parties on how to design and implement programs to spur inclusive innovations. They can also provide funding at the country level to implement these programs and to create the necessary capacities (skills and infrastructure) to support inclusive innovations.

Fourth, multilateral and bilateral organizations can *identify large-scale needs of the BoP which have not been successfully addressed and actively promote their resolution at the global level.* This can be implemented through global grand challenge funding mechanisms (as done by the Gates Foundation, the X Prize Foundation, and the government of the United States) or other advanced market commitments for the development of products and services that meet those needs. The experiences of developed countries with public procurement programs as well as of the foundations that are already doing this successfully would also prove useful.

Fifth, they can *identify large global public good BoP needs that require cross-national and multiagent participation and use their convening power as well as their own financing and funding raised from donors to create consortia to tackle these problems.* There is much scope for international coordination given increased global interdependence and new global challenges such as climate change and global disease pandemics. The Consultative Group on International Agricultural Research (CGIAR) and the Africa River Blindness Eradication Program are examples of what can be achieved through international collaboration. The implementation of these programs would not have been possible without the convening power of global institutions and the cross-country and multiagent coordination (including NGOs) that they facilitated.

Conclusions

The growing number of innovations aimed at the BoP is encouraging. Yet, millions remain deprived of access to basic goods and services. Developing countries with greater capabilities and larger markets seem to be benefiting more from the new trends, but many of these innovations have not reached lower-income countries. Even within the former, innovations by multinationals and large companies have been targeted at lower-income consumers, but rarely at those living on less than US$2 per day. The number of institutions involved in developing inclusive innovations is expanding. Unfortunately, to date, many of these initiatives have only touched a fraction of those who could benefit from them. Scaling up remains a big challenge.

The diversity of inclusive innovations presented in this chapter underscores the many steps of the innovation cycle. Research and development is not always necessary or sufficient. Much inclusive innovation is the result of recombining existing technologies, or the fundamental innovation lies in the new business or organizational model. This implies that more attention must be paid to entrepreneurship and the supporting business environment.

Where most innovation, particularly inclusive innovation, fails is in the steps between the prototype of the innovation and its further development, scale-up, and commercialization. Government policy needs to pay much more attention to how to support these subsequent steps. Lessons from successful and failed experiments alike highlight the importance of understanding the unique needs of the users. Co-creation and close consultation with users are emerging as good practices.

In fostering inclusive innovation, governments need to consider that their policies will take place within the broader innovation ecosystem and to leverage the opportunities afforded by its agents. Countries with weaker capacity will likely benefit the most from partnerships with NGOs, social entrepreneurs, and other agents who are increasingly active in this field. Although policy priorities will vary among countries, several key lessons for governments seeking to promote inclusive innovation are emerging.

First, facilitating access to international sources of innovation – through trade, joint ventures, technology transfer, linkages with the diaspora, and collaborations with knowledge centers abroad – is a priority for all. Much of the knowledge necessary for innovation is being developed outside the boundaries of any nation.

Second, it is necessary to further incentivize or support the private sector, as it is the main source of innovation in general and has the capacity to do a lot more inclusive innovation once it sees the business opportunities therein. Larger companies have the capacity to go from conception to production, scale-up, and commercialization around the world because they already have global systems in place.

Third, public procurement can be utilized more effectively to foster inclusive innovations, although, as was seen in the example of the Aakash Tablet, care must be taken to structure these programs carefully.

Fourth, new instruments, such as grand challenges and prizes, show great promise. They are crowdsourcing mechanisms for stimulating inclusive innovation. However, in order to move inclusive innovations beyond interesting prototypes, they have to be complemented by mechanisms to support scale-up, production, and mass commercialization.

Fifth, countries need to strengthen core innovation-supporting infrastructure and skills with a focus on priorities related to inclusive innovation (e.g., institutions providing technology extension services and R&D institutions focused on agriculture and basic health needs) and give them a strong mandate. Financing instruments can also be deployed more effectively to encourage more R&D aimed at improving the welfare of the very poor.

Sixth, many inclusive innovations require collaboration among different firms or between firms and public research laboratories. Many of the innovations produced by universities or grassroots innovators remain at the prototype stage. Therefore, policy makers need to encourage greater collaboration among agents with complementary comparative advantages.

Multilateral and bilateral development institutions and foundations can do more to promote the inclusive innovation agenda through advocacy, funding of initiatives at the country and global levels, and promoting knowledge exchange among countries and innovation agents. Multilateral agencies are well-placed to facilitate inclusive innovations that might greatly benefit from cooperation among multiple countries, such as the eradication of river blindness, and to help in the dissemination of successful experiences. A longer-term, more ambitious goal would be to raise funding to finance global public innovation activities to address the needs of the poor.

Appendix

The authors are grateful for the research assistance of Kathryn Hoffman and Pukar Malla (WBG consultants).

Appendix 15.1 Summary analysis of 6 inclusive innovation cases

Agent: Who? Where was it started?	What? Where was it started?	Why? How?	Lessons/special interest
Agriculture			
1. Green Revolution. International Consortium involving Ford and Rockefeller Foundations and CGIAR	Improved seed varieties and fertilizer led to increased agricultural productivity	Global concern about famines; advanced genetic research involving international network of R&D labs, government agricultural extension systems, seed and fertilizer providers, marketing and distribution	Example of major global public good R&D innovation effort that was started by international NGOs, expanded to CGIAR, and involved ministries of agriculture, provincial governments, and grassroots organizations
US, Mexico, India, Philippines, and other countries	**Mexico, South and East Asia**		
2. Private company Jain Irrigation Systems Ltd.	Microirrigation systems including drip irrigation	Commitment of company founder; works with farmer and large array of seed companies, fertilizer providers, agricultural finance, buyers, and distributors; provides training	Coordination and aggregation can help reap economies of scale and make very small farms profitable
India	From **India** expanded to the **Middle East, Europe, Central and South America, and US**		
Manufacturing			
3. Local	Household appliances	Originally a township and village enterprise making products for low-income rural market; acquired technological capability through joint ventures and R&D	Company became an MNC that competes with advanced products in developed markets
Multinational Haier China	**China and other developing countries**		Renewed focus on the needs of BoP in developing countries

(*Continued*)

Appendix 15.1   (Continued)

Agent: Who? Where was it started?	What? Where was it started?	Why? How?	Lessons/special interest
4. Foreign multinational GE China/India	GE medical equipment electrocardiogram in **India**, and ultrasound in **China**	Identified market need and feared domestic competition; used its strong R&D and engineering capacity to produce products for low-income markets	GE reverse innovation disrupting itself by selling products developed for lower income developing market in developed country markets
Services			
5. Government bank Grameen Bank and Grameen Phone **Bangladesh**	Microfinance and cell phones Started in **Bangladesh** and now present in several countries in **Asia and Africa**	US-trained economist develops system to provide credit to poor; experiments in villages; and receives government support to set microfinance institution	Started as NGO that turned into bank and expanded from Bangladesh to other countries
6. Domestic private company Aravind Eye care cataract surgery (Narayana Hrudayalaya heart surgery is a similar example) **India**	High-quality eye surgery at 1/10 the price of developed countries **India**	Desire to expand eye care/cataract surgery to low-income groups; innovated business model based on economies of scale in preoperative and surgery process, and sourcing inputs	Business process innovations which reduced costs. High capacity utilization of high-cost equipment

Source: This table draws on multiple sources of information including Immelt et al. (2009), Kumar and Puranam (2011), Prahalad and Mashelkar (2010), Utz (2010), and World Bank Vietnam Inclusive Innovation Project (2011).

Notes

1. Global Research Alliance. Cited at http://www.theglobalresearchalliance.org/What-we-do/Inclusive-Innovation.aspx (accessed July 31, 2014).
2. Some groups supporting inclusive innovation use higher thresholds than the international poverty line of US$2 per day to define their target income group.
3. Handicap and location are often causes of exclusion but are not always correlated with low income. They deserve explicit attention.
4. See D-Rev (2015): http://www.d-rev.org/about.html
5. The Kenya-based Community Cleaning Services (CCS), a social enterprise, seeks to improve youth employment and sanitation in Nairobi's low-income neighborhoods through a microfranchise model. It taps young entrepreneurs who were well-connected in their slum community to provide waste disposal services. These entrepreneurs formed microfranchise units, with CCS providing training, quality control, and access to cleaning supplies at bulk rates (see Thieme and DeKoszmovszky, 2012).
6. The case of the Aakash Tablet also highlights the importance of setting clear procurement evaluation criteria to avoid disruptions in the process (see Julka (2011) and Raina and Timmons (2011, 2012)).
7. This partnership required the purchase of patent rights from the US Food and Drug Administration to transfer a critical aspect of the technology to the Serum Institute of India.
8. A consortium of governments and the Gates Foundation committed US$1.5 billion to incentivize companies to produce 30 million doses each year for 10 years at a price of US$3.50 by topping up the price by US$3.50 per dose for 20 percent of the doses until the US$1.5 billion fund was exhausted. Pfizer and GlaxoSmithKline signed AMC contracts. However, the incentives to produce a very low-cost vaccine were weaker. It is unlikely that governments in the poorest countries, where the vaccines are most needed, will be able to afford the US$3.50 per dose.

References

Barry, Nancy. (2007). "Beyond Philanthropy: Enterprise Solutions for Poverty." Powerpoint presentation. Retrieved from wokai.typepad.com/files/nancy-barry-10.30.07.pdf

D-Rev. (2015). "About Us." Retrieved from http://www.d-rev.org/about.html

Guasch, J. L., Racine, J. L., Sanchez, I., and Diop, M. (2007). *Quality Systems and Standards for a Competitive Edge*. Washington, DC: World Bank.

Hargreaves, J. R., Greenwood, B., Clift, C., Goel, A., Roemer-Mahler, A., Smith, R., and Heymann, D. L. (2011). "Making New Vaccines Affordable: A Comparison of Financing Processes Used to Develop and Deploy New Meningococcal and Pneumococcal Conjugate Vaccines." *The Lancet*, 378 (9806): 1885–93.

Immelt, J. R., Govindarajan,V., and Trimble, C. (2009). "How GE is Disrupting Itself." *Harvard Business Review*. Retrieved from https://hbr.org/2009/10/how-ge-is-disrupting-itself/

Julka, H. (2011). "Tender for $35 Laptop Project Cancelled." *The Economic Times*. Retrieved from http://articles.economictimes.indiatimes.com/2011-01-18/news/28432799_1_laptop-project-bank-guarantee-sakshat

Koh, H., Ashish, K., and Katz, R. (2012). *From Blueprint to Scale: The Case for Philantrophy in Impact Investing*. Cambridge, MA: The Monitor Group in collaboration with the Acumen Fund.

National Bureau of Economic Research. (2009). "The Financing of R&D and Innovation." NBER Working Papers 15325. Cambridge, MA: Hall, Bronwyn, and Lerner.

Prahalad, C. K. (2005). *The Fortune at the Bottom of the Pyramid: Eradicating Poverty through Profits*. Upper Saddle River, NJ: Pearson Education/Wharton School Publishing.

Racine, J. L. (2011). *Harnessing Quality for Global Competitiveness in Eastern Europe and Central Asia*. Washington, DC: World Bank.

Raina, P. and Timmons, H. (2011). "Meet Aakash, India's $35 Laptop." *The New York Times*. Retrieved from http://india.blogs.nytimes.com/2011/10/05/meet-aakash-indias-35-laptop/

Raina, P. and Timmons, H. (2012). "The Tangled Tale of Aakash, the World's Cheapest Laptop." *The New York Times*. Retrieved from http://india.blogs.nytimes.com/2012/04/27/the-tangled-tale-of-aakash-the-worlds- cheapest-laptop/

Utz, A. (2010). "Chapter 11: Stimulating Pro-Poor Innovations." In World Bank (2010). *Innovation Policy: A Guide for Developing Countries* (335–370). Washington, DC: World Bank.

World Bank. (2014). *World Development Indicators*. Washington, DC: World Bank.

16
Conclusion

Ricardo Ernst and Jerry Haar

Innovation is and will continue to be the catalyst and a prime driver of the sustainable competitiveness of nations, industries, and individual companies. While the word "innovation" is most commonly associated with products or ways of doing things, its essence and contexts are much larger.

As presented in the introductory chapter, innovation is impacted by a broad and dynamic ecosystem – one which it helps shape as well. These include the ever-increasing access to finance, particularly the expansion of credit to the middle-class and lower-income groups not only via banking institutions but through retail establishments as well. The providers themselves have devised innovative financial products for consumers and institutions that have fueled spending and investment.

Another shaper of innovation is the proliferation of technology. One might even call this the "democratization" of technology, since its dissemination and diffusion enable not just industry leaders such as Cisco, Oracle, Google, and HP to benefit from technological advancements but smaller firms and start-ups, utilizing open source platforms and other low-cost mechanisms, to develop and launch products and services that span the universe of consumer and industrial customers. The incredible successes of small "born global" companies such as Skype, Kian, and Sproxil are testament to this most notable development. Technology as the "great equalizer" has hugely benefited emerging markets, witness Prezi from Hungary, Studio-C from Guatemala, and myVLE, an SasS online learning platform from Morocco.

The globalization of markets is another key feature of the ecosystem that impacts innovation. During the last two decades, trade, investment, finance, and commercial liberalization has swept over the globe at a rapid pace. Regional, subregional, and bilateral agreements such as NAFTA, the ASEAN Free Trade Agreement, G-3 (Mexico, Colombia, Venezuela), and the Gulf Cooperation Council have created an environment for innovation to flourish – mainly due to the features of these agreements such as intellectual property protection and trade in services. While there is still much to be done in these arenas, market liberalization is allowing commercial relations between industrialized and emerging markets, as well as between emerging markets themselves, to flourish and, in so doing, provide a hospitable environment for innovation.

Finally, the engine of economic growth in developing and emerging markets is spurring innovation to the benefit of all. Despite the ebb and flow of commodity prices and financial markets, and perpetual sluggishness in many industrial country markets in Western Europe and Japan, nonindustrial nations are reaping the benefits of prudent fiscal and monetary policies, deregulation, and improvements in infrastructure, health, education, and administrative and judicial systems. Innovation thrives in environments of growth and reform, and each region covered in this book – Asia, Africa, Central Europe, Latin America, and the Middle East/Gulf States – provides examples that make this case.

Moving beyond the ecosystem of innovation, there are three key drivers that have been the focus of this book: national innovation policy (clusters), facilitating institutions (R&D labs and industrial parks), and firm-level innovation. All emerging market regions have witnessed the infinite possibilities of these three drivers in enhancing their competitiveness through innovation. A large role of the state in national economies has always been a prime feature of emerging markets; therefore, it is no surprise that federal governments – and increasingly state/provincial and municipal governments – have played and will continue to play a major role in supporting innovation, be it directed by the public, private, or nonprofit sector. They do so through a plethora of fiscal, legal, and tax incentives as well as direct program assistance. Singapore, Korea, Malaysia, Brazil, Chile, Mexico, Nigeria, and Poland are good examples. Unquestionably, industrial and service "clusters" – most involving the triad of government, business, and academia – have been one of the most dynamic vehicles in catalyzing innovation in emerging markets. IT clusters in India and Brazil – Bangalore and Campinas, respectively – along with Russia's Skolkovo Innovation Center, the biotechnology industry cluster in the Prague metropolitan region, and the Cape Clothing and Textile Cluster in South Africa are prime examples.

As for facilitating institutions, the continuous upgrading of technological and scientific infrastructure in emerging markets and the expansion of laboratories (public and private) and growth of industrial parks have broadened and deepened the reach of innovation. Two excellent examples are Parque Austral in Buenos Aires and the Southern Taiwan Science Park. Finally, firm-level innovation taking place on the ground in emerging markets by way of both multinational firms and national companies (including start-ups and early-stage firms), along with industrial nation innovations transferred to (and in many cases modified for) developing nations, has been an immensely important driver as well.

One of the marvels of innovation in emerging markets is that all dimensions of innovation are represented – products, processes, services, and business models. The same creativity and ingenuity that fuel innovation in industrialized nations are alive and well in countries such as Brazil, Nigeria, Poland, China, and Lebanon. Again, the access and affordability of technology, combined with networks and affinity groups of tech-savvy young people with a passion to launch start-ups, are both responding to and creating demand among consumers of all demographic profiles. As for large firms, particularly multinationals, more and more are fostering intrapreneurship within their companies and designing products, processes, and services geared to emerging markets – for they know that is where the growth is and will continue to be. Establishing R&D outposts in the very markets they wish to serve is enabling them to more accurately gauge the market and enhance the likelihood of success through the introduction of appropriate goods and services.

As noted, innovation comes in many forms: reverse innovation, packaging and price, social media, distribution, advertising, sustainable innovation, digital innovation, disruptive innovation. In addition to reverse innovation, these other forms are features of emerging markets as well. Single-use packages of detergent, aimed at India's bottom of the pyramid class, as well as specially treated packaging for biscuits to protect the product from becoming soggy in high-humidity countries, are cases in point.

Contributors to this volume have addressed innovation in emerging markets along the lines of the key drivers that provide the template for this book – national policy, facilitating institutions, and firm-level innovation – as well as the sector and social dimensions of innovation. Let us highlight the findings.

National policy

Every emerging market region has experienced more robust national policies to support innovation. For Asia, the best example is Singapore, which has relied for the past few decades on government investment and programs to build a knowledge-based economy and stimulate innovation. Much of Singapore's push to upgrade its innovation credentials revolves around its National Research Foundation (NRF), which was set up in 2006 to develop policies to foster the Lion City's development technologies ranging from cybersecurity to the environment. India does not fall far behind the same wave: innovation has been strongly supported by the new government leadership.

Turning to Latin America, policy makers in that region time and again have been saddled with the dilemma of whether to promote innovation in the natural resource-based sectors that drive the economy at present or work towards building an innovation framework essentially based on R&D,

patents, and technology. In the field of policy initiatives, Chile´s National Development Agency (CORFO, Corporación de Fomento de la Producción, in Spanish) has launched focused programs to promote process innovations in the mining sector and to introduce new species of fish in the aquaculture sector. Similarly, in Argentina the development of dynamic clusters linked to natural resource-intensive sectors has received public funding from the country's technological fund (FONTAR, Fondo Tecnológico Argentino, in Spanish) to execute both individual and associative innovation projects.

In the Middle East and North Africa, more than half the economies have reformed their regulations during the last six years. In practical terms, however, not much has been done with the exception of the UAE – a country that has been ranked in the top ten for improving the most, although in its three core areas of success – registering property, getting credit, and protecting minority investors – only the last has a direct impact on start-ups. Overall, the UAE enacted a total of 30 reforms in 2013 and 2014 that protected minority investors. As for Africa, one of the most salutary results of political reform has been the change in macroeconomic policy across the continent, with governments eschewing populism in favor of sound money, balanced budgets, and debt reduction. Furthermore, African governments are increasingly allowing exchange rates to be guided by market forces rather than political expediency. Another consequence of political reform has been the changing of the guard in government ministries across the continent, with a new generation of globally experienced technocrats assuming leadership positions. Kenya is a case in point.

Assessing national policy in Central Europe, one common initiative focuses on fostering innovation, employment, and economic growth following the idea of comparative business advantages and specialization in innovation (commonly known as the concept of the "smart specialization strategies"). For example, the government of Hungary agreed in 2014 on the adoption of the National Smart Specialization Strategy (S3). Nonetheless, progress in innovation in Central Europe has been mixed. Poland, for example, remains a laggard in R&D spending, according to the World Bank – less than half of that spent by the neighboring Czech Republic. Regionally, however, some initiatives that have made an impact are the European Commission's proposal for cohesion policy in 2014–20 as a precondition for using the European Structural and Investment Funds (ESI), and policy developments for enterprise development, regional development, and clusters falling under the umbrella of the Small Business Act (SBA) for Europe.

In developing, planning, and implementing national policies for innovation, governments in emerging markets are increasing their attention to three key sectors – education, health, and financial services.

With regard to education, for the past several decades, policy makers in these regions have focused on expanding access to education and obtained some good results. According to the *Millennium Development Goals Report* of 2012, "many more of the world's children are enrolled in school at the primary level, especially since 2000." Girls have benefited the most. In Latin America, for example, according to the Inter-American Development Bank (IDB), school access for four- to five-year-olds increased from 36 percent in 1990 to more than 60 percent at the present day – including girls, making the region one of the most gender-equal ones of the world in terms of early access to school. In health care, there have been some supportive examples that have focused mainly on funding innovation and providing intellectual property frameworks. In Brazil, the government has supported health research with its Biotechnology Development Policy and a ten-year, $4 billion development program. Another major policy event was the creation of the Sistema Único de Saúde (SUS) or Unified Health System in 1996, which sought to embody the principle of a universal right to health established in the Federal Constitution of 1988. In the realm of financial services, governments in emerging markets noticed vast benefits offered to their citizens by new models of financial inclusion. So new policy initiatives mushroom across the developing world and financial innovations see the light of day every year. Governmental loans and guarantees for lending and insurance are playing crucial roles in stimulating entrepreneurship in emerging markets.

One important, powerful innovation that cuts across many dimensions of national policy, with significant socioeconomic and political implications, is social media. It is a tool that offers huge potential in reshaping national policy and promoting democracy that ordinary citizens may not have historically had. It also reshapes public diplomacy which transforms national policy across the globe. Latin America, Indonesia, and the Philippines are huge users of social media. Governments that seek to heavily connect with their populations – especially populist ones like Ecuador – use social media for political socialization purposes. According to the OECD, seven of the top ten national government Twitter accounts are those of emerging markets. E-government also (for social security, health information, voting, e-procurement) plays a vital role here. To illustrate, the Modi administration in India is using social media to augment its governing and policy making.

Facilitating institutions

For emerging market countries – in fact, for all countries – facilitating institutions play a unique role as a catalyst and nexus between national policy and private companies. R&D centers, laboratories, industrial parks, business associations, universities, think tanks, and other intermediary institutions acting alone or in collaboration with other facilitating entities form a vital part of the innovation ecosystem.

Often overlooked in discussions of facilitating institutions are the roles of social media and social enterprise. In the case of the former, collaborations between universities and privately held companies have generated innovations related to social media that have helped to address recent global challenges such as detecting illnesses (HealthMap from Boston Children's Hospital). Related to this is the role of social media in advancing inclusive innovation, whereby social as well as economic returns are realized. Some examples include: the Indian government's support of the Honey Bee Network, which has catalogued hundreds of thousands of grassroots innovations and facilitates networking, and the creation of international and bilateral development institutions that promote the inclusive innovation agenda since they can play a strong advocacy role, act as an information clearing house, and facilitate knowledge exchange.

As for social enterprise, there is an ever-expanding array of facilitating institutions that provide technical assistance, business mentorship, and networking opportunities tailored to drive both societal and business value. Global private institutions like Ashoka, Endeavor, the Skoll Foundation, and the Schwab Foundation for Social Entrepreneurship have turned the international spotlight toward players like Pandey, ushering them into once-exclusive domains. The recent acceleration in the number of social enterprises launching and attempting to scale in emerging markets is partially attributable to the fast-growing global network of organizations dedicated to supporting them, whether with equity investment capital, traditional philanthropic grants, or business know-how, mentoring, and technical training. To illustrate, the Schwab Foundation, among others, offers practical advice for policy makers that seek to create a more hospitable environment for social enterprise. These include developing regulatory frameworks that allow for both public and private provision of goods and services, such as schools or health clinics; this way nongovernmental actors (whether nonprofit or commercial) can fill gaps where government falls short. In terms of specific policies that recognize social enterprise as a unique legal form, most emerging market governments do not offer them. Each social enterprise is innovative at their firm level in one way or another since their goal is to solve social problems through new ways. An example is India's Husk Power Systems, an off-grid electricity company that converts rice husks into power that is reliable, eco-friendly, and affordable for families that can spend only $2 a month on power.

Facilitating institutions have garnered the attention of policy makers, academics, analysts, and the private and nonprofit sectors in emerging markets. In the case of Latin America, the analysis and findings from Innovalatino (2011) have produced recommendations which can facilitate the strengthening of the innovation capacity of the Latin American nations. Included are support of human capital as the catalyst of innovation, assistance to small and medium-sized enterprises (SMEs) through cluster policies, and aid that fosters social inclusion and sustainability. For the Middle East, the biggest challenges are access to capital and political risk insurance. Fortunately, some firms dedicate themselves to solely supporting other firms. One such firm having a real impact is Flat6Labs with over 80 mentors providing support. Another example is Oasis500 in Jordan, which provides training, mentoring, and funding for local start-ups. In Central Europe, comprehensive policy measures are being adopted for regional enterprise development and the creation of quasi-banking financial institutions (e.g. seed capital funds, business angel networks, regional and local loan funds), which are becoming increasingly common all over Central Europe. Finally, in Africa a wide range of institutions are playing the critical role of innovation facilitator. Foremost among these institutions is the growing number of innovation hubs and tech accelerators that cater to the continent's aspiring high-tech entrepreneurs. One of the most heralded innovation hubs is Kenya's iHub.

Firm-level innovations

Whether products, processes, services, or business models, emerging markets are hotbeds of innovation. Looking at *products*, Asia has rapidly moved up the value chain. Chinese developers are already beginning to leapfrog their foreign counterparts in consumer mobile Internet products, and are catching up fast in pretty much every other tech sector. Additionally, high-lumen LED lights from Taiwan and Vietnamese fishery and consumer products are illustrative. In Latin America, ethanol products from Brazil and Colombia immediately come to mind, along with Embraer jet aircraft and Natura cosmetic products. Regarding *processes*, China's Huawei in telecommunications, Mexico's Softtek in nearsourcing processes, and the Shoprite (South Africa) IT system for inventory management are noteworthy, while Jordan's Maktoob, often referred to as the "Yahoo! of the Middle East," and income-tax collection platforms from Tunisia and Madagascar have garnered the attention of authorities across the African continent.

In terms of *services*, Alibaba – China's eBay – has grown exponentially, while Peru's Astrid y Gastón international culinary empire has done much to promote indigenous agricultural products and established a chef school in a low-income neighborhood in Lima. In the education realm, Duolingo, a language teaching app from Guatemala, and Open English, a language learning system founded by the Venezuelan Andrés Moreno, have achieved great success. In the Middle East, RecycloBekia, founded in 2011, resells and recycles computers and electronics, and in Egypt has provided the first platform that enables users to report bad drivers instantaneously. In Central Europe, innovative services are provided to businesses in locations such as Infopark in Budapest and the Czech Technology Park in Prague. Finally, *business models* from emerging markets cover the gamut. Prepaid packages for mobile communications abound, such as those offered by Mexico's América Móvil, while M-Pesa from Kenya is the most notable mobile payments platform in the developing world.

This, then, is the world of innovation in emerging markets – dynamic, symbiotic, broadening, and expanding from North–South, South–North, and South–South. Emerging markets will continue to provide the most promising opportunities for companies both large and small; simultaneously, the rapid pace of innovation will continue unabated.

However, as Thomas Edison asserted: "Vision without execution is hallucination." The same may be said about innovation. The execution – implementing innovation along all its dimensions – is indispensable. Hopefully, other writers will pick up the mantle and address that vital issue as it relates to emerging economies.

Index